Organic Pollutants in Wastewater I

Methods of Analysis, Removal and Treatment

Edited by
Inamuddin
Abdullah M. Asiri
Ali Mohammad

Wastewater represents an alternative to freshwater if it can be treated successfully for re-use applications. Promising techniques involve photocatalysis, adsorption, nanocomposites, and membranes. The book focusses on the following topics:

Effluent detoxification and degradation kinetics of organic dyes using Fenton and photo-Fenton processes. Degradation of methylene blue using nanocomposites as a potential photocatalyst. Agricultural and agro-industries based wastes as low-cost biosorbents. Use of carbon quantum dots (CQDs) for photocatalytic degradation of organic pollutants. Detection, determination and removal of phenolic compounds from wastewater. Decomposition of organic dyes via photocatalysis. Oxide-semiconductor nanomaterials for photocatalytic wastewater purification. Photocatalytic efficiency of various ZnO composites for degradation of organic pollutants. TiO_2 based nanocomposites. Membrane filtration processes for the removal of organics from industrial wastewater.

Keywords: Wastewater Treatment, Organic Pollutants, Organic Dyes, Photocatalysis, Nanocomposite Photocatalysts, Photocatalytic Degradation, Adsorption, Membrane Filtration, Fenton Processes, Biosorbents, Phenolic Compounds, Carbon Quantum Dots, Methylene Blue Degradation, ZnO Composites, TiO_2 Based Nanocomposites

Organic Pollutants in Wastewater I
Methods of Analysis, Removal and Treatment

Edited by

Inamuddin[1,2], Abdullah M. Asiri[1,2] and Ali Mohammad[2]

[1] Chemistry Department, Faculty of Science, King Abdulaziz University, Jeddah 21589, Saudi Arabia

[2] Department of Applied Chemistry, Faculty of Engineering and Technology, Aligarh Muslim University, Aligarh-202 002, India

Published by **Materials Research Forum LLC**
Millersville, PA 17551, USA

Published as part of the book series
Materials Research Foundations
Volume 29 (2018)
ISSN 2471-8890 (Print)
ISSN 2471-8904 (Online)

Print ISBN 978-1-945291-62-3
ePDF ISBN 978-1-945291-63-0

Distributed worldwide by

Materials Research Forum LLC
105 Springdale Lane
Millersville, PA 17551
USA
http://www.mrforum.com

Manufactured in the United States of America
10 9 8 7 6 5 4 3 2 1

Table of Contents

Preface

The demand for freshwater resources has been continuously increasing, mainly in arid and semi-arid regions. Wastewater has been considered as an alternative water resource if it has been treated to a level that permits for re-use applications. The presence of organics in wastewater streams poses a great challenge and hence several treatment technologies have been developed for the remediation of organics from treated effluent streams. The presence of organics in wastewater has been associated with mining and oilfields, municipal and industrial effluents, and runoff from agricultural farms.

In this first volume of a 2-volume set, topics such as photocatalysis, adsorption, nanocomposites, and membranes are covered. For example, the first chapter presents the potential application of agricultural solid wastes as adsorbents for dye removal via physical and chemical activation methods. In the second chapter, effluent detoxification and degradation kinetics of organic dyes were investigated using Fenton and photo-Fenton processes through which several process operating conditions were optimized. The third chapter presents an interesting topic and discusses the degradation of methylene blue using nanocomposites as a potential photocatalyst. Several characterization methods were used supported by experimental studies. Furthermore, the fourth chapter is another application of the agricultural and agro-industries based wastes as low-cost biosorbents through which raw, processed and charred biomass were investigated. In the fifth chapter, great attention is given to the use of carbon quantum dots (CQDs) for photocatalytic degradation of organic pollutants. Different synthesis methods such as hydrothermal treatment, a microwave method, electrochemical and synthetic route are also highlighted. Another important topic is highlighted in the sixth chapter; which is the removal of phenolic compounds from wastewater. Chemical and physical methods of determining phenolic presence as well as the degradation of phenols via biological processes are discussed in this chapter. The seventh chapter highlights the evolution of photocatalysis process and its future trends as a promising technology in wastewater treatment. More attention is given to the fundamentals of titanium dioxide (TiO_2) based photocatalysts, including basic mechanism of heterogeneous photocatalysis, advantages, and challenges of TiO_2 based photocatalysts. The eighth chapter discusses the decomposition of organic dyes via photocatalysis process. It includes a critical review of oxide-semiconductor nanomaterials towards photocatalytic wastewater purification. The structural aspects, nanostructure formation process and the various parameters affecting catalytic activity, photocatalytic applications of oxides based catalysts for efficient photocatalytic degradation of organic dyes are reviewed. The ninth chapter summarizes recent progress in the fabrication and photocatalytic efficiency of various types of zinc oxide (ZnO) composites for degradation of organic pollutants. In the tenth chapter, TiO_2 based nanocomposites are thoroughly discussed. The eleventh chapter is dedicated to the membrane filtration process and its applicability in the removal of organics from industrial wastewater. It also focuses on different types of pollutants present in food and agro-alimentary industries and their membrane-based treatment methods.

We believe that this volume is a great source to a broad range of professionals such as industrial operators, professional engineers, academicians, undergraduate and graduate students, technicians, and many others.

Lastly, we extend our great appreciation to the contributing authors for their countless efforts towards the preparation of this book. We would like to thank all distributors, authors, and other who conceded consent to utilize their figures, tables, and schemes. However, every effort has been made to get the copyright approvals from the individual proprietors to join reference to the reproduced materials; we might need to offer our sincere articulations of disappointment to any copyright holder if unintentionally their benefit is being infringed.

Inamuddin[1,2], Abdullah M. Asiri[1,2], Ali Mohammad[2]

[1] Chemistry Department, Faculty of Science, King Abdulaziz University, Jeddah 21589, Saudi Arabia

[2] Department of Applied Chemistry, Faculty of Engineering and Technology, Aligarh Muslim University, Aligarh-202 002, India

Chapter 1

Green Applications of Magnetic Sorbents for Environmental Remediation

Tiyasha Kanjilal and Chiranjib Bhattacharjee*
Chemical Engineering Department, Jadavpur University, Kolkata-700032, West Bengal, India
cbhattacharyya@chemical.jdvu.ac.in, c.bhatta@gmail.com

Abstract

Adsorption of both inorganic and organic xenobiotics by utilizing a variety of biomaterials signifies promising prospects for the reduction of environmental pollution. For the purpose of improveing manipulation of the biosorbents, their potential magnetic derivatives have been investigated. These diamagnetic biomaterials of different origin (lignocellulosic material, prokaryotic and eukaryotic cells, food wastes, biopolymers etc.) can be magnetically modified in order to acquiring 'smart' biomaterial exhibiting a definite response to the external magnetic field. This chapter summarizes the synthesis procedures of magnetic modification and eco-friendly applications of these 'smart' biomaterials both in nano and micro-sized forms.

Keywords

Magnetic Separation, Magnetic Modification, Sorbents, Biomaterials, Carriers, Biocatalysts

Contents

1. Introduction

The treatment of contaminated soil and industrial effluents is a challenging task in respect of controlling water pollution, purification and soil remediation [1]. Several land areas, streams, and rivers get contaminated by untreated disposal of heavy metals, organic compounds, and other hazardous substances. Presently, increasingly metals are being used in process industries. Therefore, waste streams from these industries contain metal ions which cause devastating and permanent contamination of the environment because of their non-biodegradable and persistent nature [2]. The water pollution due to heavy metals (chromium, lead, mercury, zinc, cadmium etc. has gained more and more attention, due to increasing applications of heavy metals in the chemical industry for chrome plating, dyes, and battery etc. [3]. Presently, the removal of toxic materials from soil and effluents has attracted the attention of many researchers [4, 5]. Of late, scientists have investigated a range of innovative processes for the effective removal of pollutants from contaminated water and soil. Remediation techniques for environmental cleanups can be categorized as biological, chemical and physical processes [6]. For instance, several mechanisms such as biodegradation, electron-beam irradiation, extraction, adsorption, air sparging and incineration are being applied to remove or reduce the pollutants [7]. Chiefly, the adsorption of pollutants is widely utilized in this field due to simple applicability, cost-effectiveness and eco-friendliness of this approach [8]. Moreover, the removal by adsorption is mainly decided by the properties of various materials that have been used as adsorbents, namely activated carbon [9,10], zeolites [11,12], iron oxides [13,14] and silica [15]. Much interest lies in the utilization of biological materials such as bacteria, algae, and fungi for the removal and recovery of metals among these mechanisms [16-18]. The biomasses have revealed cost-effective and eco-friendly characteristics [19]. Bio sorbents are intrinsically broad sourced, low-cost, and rapidly useful for adsorption. In recent times, many reports are available regarding the correlative development on heavy metal remediation by immobilized microorganisms [20]. Magnetic separation has been a promising ways for an environmental remediation mechanism because of nonexistence of contaminants like flocculants production with an added advantage of large amount of waste removal within a short time span [1]. Magnetic particles have been used as an upcoming new sorbent to adsorb heavy metal ions where an external magnetic field is used to resolve the difficulty of separation [21]. The dual integrated techniques of bio-adsorption and magnetic separation has the advantage of being a green application, flexibility, eco-friendly features and economic viability. Hence the present chapter suggests the green application of magnetic sorbents for environmental cleanup.

2. Features of magnetic solid phase extraction (MSPE)

At present, the magnetic solid phase extraction (MSPE) with functionalized magnetic particles (FMPs) as adsorbents has created an innovative niche in the analytical community. During the 70s era, Robinson et al. (1973) first reported magnetic separation for biotechnological purpose [22]. Later, Wilkstrom et al. [23] reported magnetically susceptible additives like ferrofluids and external magnetic field for faster liquid-liquid phase extraction. In the 90s, Towler et al. [24] used for the first time manganese dioxide coated magnetite as an adsorbent for metal recovery from seawater, and in 1999, Safarikova et al. [25] first coined the term magnetic solid phase extraction for analytical purposes. For this purpose, an external magnetic field is employed to separate the target analyte from the aqueous solution by addition of FMPs into the sample solution and adsorption of the analyte onto the surface of magnetic beads. The FMPs can re-disperse in aqueous solution on the removal of the external magnetic field, which makes washing and desorption more convenient as depicted in Fig. 1. The FMPs can also facilitate mass transfer by increasing the interfacial area present between the analyte and sample solution [26]. Among the FMPs used, iron-oxides (Fe_3O_4 and Fe_2O_3) are the most common because of certain characteristics of simple and fast preparation for large-scale production; convenient modifications due to presence of abundant functional groups on the surface; easy mode of operation without centrifugation or filtration, due to their super-para-magnetic properties; appropriate recoverability and reusability as well as the excellent dispersibility in aqueous solution to reach rapid extraction equilibrium [26]. Unlike the conventional SPE; pretreatment of the sample is very convenient in MSPE, due to fast and easy phase separation by an external magnetic field [27]. The magnetic sorbents have been widely utilized as optimized materials for applications in medical, biological and engineering fields; in addition to their progressively use for trace metal analysis as well. Recently, several exciting techniques have been applied for analysis and on-line pre-concentration of heavy metal ions present in contaminant samples using micro-or nanomagnetic particles.

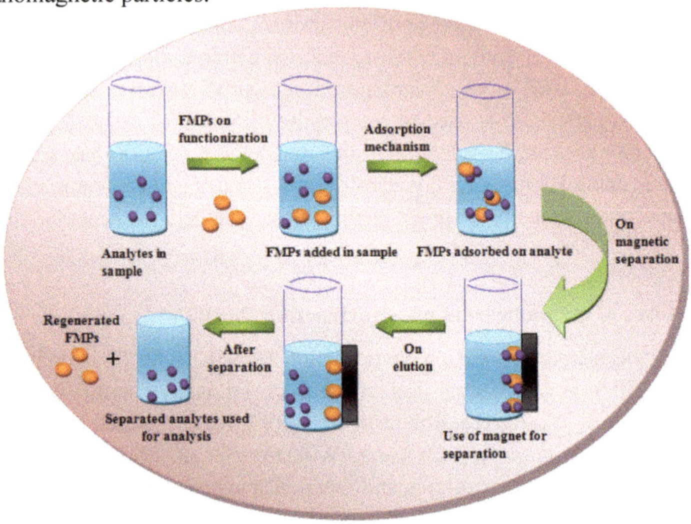

Figure 1. Schematic representation showing the technique applied for functionalized magnetic solid phase extraction (MSPE).

2.1 Magnetic sorbent materials

Different types of sorbent materials like ion exchangers, octadecyl (C_{18}), polymers, carbon nanotubes, activated carbon and magnetic particles have been effectively used in solid phase extraction. Usually, FMPs constitute of magnetic elements namely Fe, Ni, Co or their oxides and alloys with added super-para-magnetic or ferromagnetic characteristics. Upon magnetization, the FMPs act as small permanent magnets, so that due to magnetic interaction they can form aggregates or lattice-like structure. Moreover, the ferromagnetic particles are known to be a permanent magnet, so they can easily form lattice structure on the removal of the magnetic field, whereas the super-paramagnetic particles preserve no residual magnetism upon removal of the magnetic field. The FMPs are obtainable in a large range of sizes from micro-sized to nanometer-sized (1 to 100 nm) particles, which have drawn attention in technological research for their high dispersibility, large surface area and significant surface to volume ratio resulting inefficient adsorption capacity [28,29]. The nano-sized FMPs presently have a range of potential utilization in environmental, biological and food analysis due to their easiness for surface modifications, the simplicity of large-scale production and effective recycle [30]. A list of application of various magnetic sorbents for heavy metal removal has been provided in Table 1. The purely inorganic magnetic materials including Fe_3O_4 relatively intend to the formation of the lattice-like structure causing a change in their magnetic characters. Also, their possible low selectivity makes them un-adaptable for samples with composite matrices. Therefore, to overcome this limitation, an appropriate modification of the magnetic coating (core) with the attachment of specific active group is required. Such coatings onto the surface of FMPs are termed 'shell' as shown in Fig. 2. These are obtained by attachment of inorganic elements like silica and alumina or organic components like surfactant or any suitable polymer in order to increase their chemical stability, oxidation resistance and improving the uptake of specific ions [31-34]. These 'shells' which are attached to the FMPs are utilized for different analytical purposes [35-37] and environmental pollutant treatment [38-41]. For advanced environmental remediation through a fluidized bed which is magnetically stabilized, the FMPs can be fluidized and kept in a magnetic filter column while the waste effluent can be pumped through the column [42]. Complex FMPs like titano-magnetite have been often used as alternatives to fluidized magnetically stabilized beds possibly due to adequate retention of functionalized magnetic sorbents and free water flow for effective separation by its matrix [43]. Certain specialized microorganisms have the capability of forming inorganic FMPs like ferrous sulfate which has application for consuming heavy metals through metabolic processes, which makes removal of heavy metals simplified and eco-friendly [44]. Some other groups of naturally occurring magnetotactic bacteria have the ability to orient themselves in the direction of the magnetic field. These find an efficient application in organic contaminant removal from effluent by the enzymatic actions [45]. Surfactant covered magnetite particles have attracted the scientific community in the extraction of organic pollutants such as 2-hydroxyphenol [46]. Certain polymer coated floating magnetic sorbent materials having vermiculite iron oxide composite coating have been applied for fast and efficient separation of spilled oil pollutant in water [47].

Table 1: Applications of magnetic sorbents for toxic heavy metal removal.

Serial No.	Applications	Comments
1	For heavy metal uptake and magnetic separation, highly magnetic iron sulfide sorbent development	Certain microorganisms have the ability to produce magnetic iron sulfide. Likewise, it was for metal uptake
2	A clay substituted layered double hydroxide (LDH) type sorbent with Fe and Ni ions developed for As removal and magnetic separation	Under very high gradient field of about 5T; layered double hydroxide (LDH) synthesized from Fe and Ni demonstrated better uptake potential for As
3	Magnetic hydrogel synthesis for removal of heavy metal pollutants	Magnetic hydrogels with recycling features based on 2-acrylamide-2 methyl-1-propane sulfonic acid were employed for removal of heavy metal ions of Cd(II), Co(II), Fe(II), Ni(II), Cu(II), Cr(III) from liquid samples.
4	Bismuthioal-II-immobilized Functionalized magnetic particles (FMPs) for separation and preconcentration of heavy metals like Pb, Cu, Cr and their analysis using ICP-OES.	Bismuthioal-II-immobilized FMPs with recycling features demonstrated static sorption capacities of 8.6, 5.3 and 9.4 mg/g for Pb, Cu, and Cr, respectively. Specific sensitivity in detection and general advantage of utilizing magnetic assistance.
5	Magnetic Fe-Ni oxide synthesis from waste Ni liquid cause selective sorption of Cr (VI)	A magnetic Fe-Ni oxide with recycling features demonstrated maximum of 30 mg/g Cr (VI) uptake capacity.
6	Magnetic separation for magnetic ion and particle by magnetic chromatography with novel magnetic column.	Separation of magnetic materials or FMPs with varied susceptibilities Different fractionating column designed based on FMP size, susceptibility, material mobility and column height.
7	Synthesis and characterization of magnetic chelating resin made from chitosan for removal of Cu (II), Co (II) and Ni (II)	The magnetic carrier on the chitosan-based character.

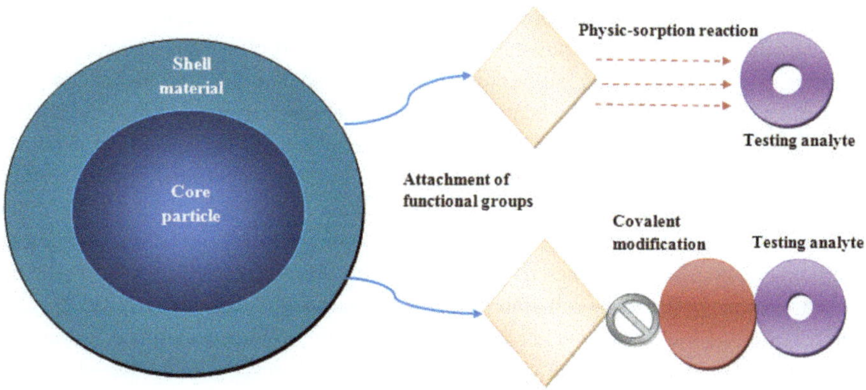

Figure 2. Reaction mechanism of a core-shell type structure of functionalized magnetic particles (FMPs).

2.2 Synthesis and characterization of magnetic sorbents

The synthesis of any magnetic particle generally involves three fundamental steps; namely, production of the magnetic material, coating formation of the magnet ore and the specific characterization of the magnetic core-shell structure. For the preparation of magnetite as a magnetic core, the chemical co-precipitation method of Fe^{2+} and Fe^{3+} ions in aqueous solution is commonly applied [31]. The synthesis process constitutes vigorous stirring of a solution mixture of $FeCl_3.6H_2O$ and $FeCl_2.4H_2O$ in ultrapure water at 70-85 °C in the presence of atmospheric nitrogen and rapid addition of liquid ammonia, resulting in a dark precipitate, confirming the formation of nanoparticles of magnetite. Alternatively, magnetite can also be prepared by solvothermal reaction process in the presence of ethylene glycol, oleic acid and hexadecylamine [48]. Conversely, the maghemite FMPs are derived from monodispersive nano-sized Fe_3O_4 particles upon successive dissolution in acid medium and heating at 90 to 100 °C for 30 min so that the magnetite particles (Fe_3O_4) get oxidized entirely [49]. Alternatively, cation diffusion is also a way to complete oxidation of maghemite FMPs by long-term exposure to air resulting in γ-magnetite FMP formation [50]. Recently, a new innovative reactor system; the submerged circulative impinging steam reactor unit (SCISR) was introduced for the synthesis of γ-maghemite FMPs, which has a special flow configuration, causing a perfect sequential circulation of mixing flow and non-mixing-flow and micro-mixing effect in the impingement zone, thereby providing the unique advantage of preparing γ-maghemite FMPs ultrafine powders as well as nanoparticles [51]. The magnetic core coating can be done using silica, alumina or any other oxides of metal and organic polymers. The cores of FMPs coated with silica are generally synthesized by the Stober method through a sol-gel reaction involving the use of tetraethoxysilane (TEOS) in acidic or basic solution in the presence of nitrogen. There is no requirement of any primer to promote the adhesion and deposition of silica since the oxides of iron a have strong bonding affinity toward silica. The silane coupling agents including amine, thiol, carboxylic and C18 are applied to introduce the functional groups onto the surface of the silica coated FMPs by silanation. Fig. 3 illustrates the schematic process for the preparation of silica coated FMPs modified with various functional groups. The generation of alumina coated FMPs are carried out by dissolving aluminum-isoproxide in a mixed solution of magnetite FMPs and ethanol with the help of ultrasonic, combined with the drop-wise mixing and vigorous shaking of water-ethanol into the suspension of the FMPs [37].

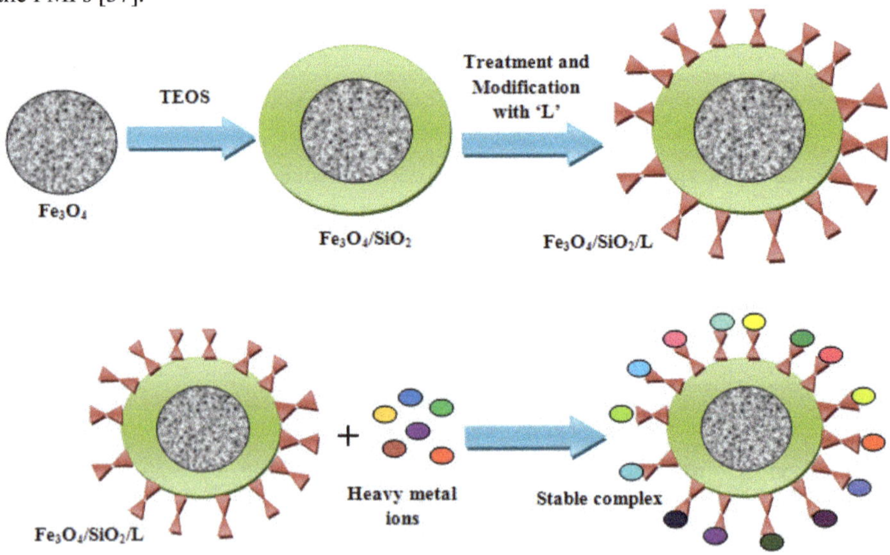

Figure 3. The mechanism for the synthesis of Schiff base functionalized iron-oxide- silica hybrid magnetic nanoparticles for pre-concentration of heavy metals like Hg (II), Pb (II), Cu (II), Cd (II) etc.

The fundamental techniques for the identification and characterization of FMPs include transmission electron microscopy (TEM), scanning electron microscopy (SEM) and X-ray diffraction (XRD). Transmission electron microscopy is used for determining differences in electron density, allowing analysis of shape and size of the FMPs while scanning electron microscopy is used for morphological and structural characterization of the FMPs. For determination of the crystalline structure specifically, a technique known as selected-area electron diffraction (SAED) is generally applied [52]. For determination of the composition and structural analysis of maghemite or magnetite in the solid matrix, X-ray diffraction is used. The porosity and mean diameter measurement along with the observation of the surface area of modified or unmodified FMPs are carried out using nitrogen adsorption-desorption porosimetry following the Brunauer-Emmett-Teller (BET) process [53]. The investigation of the adsorption mechanism of the target heavy metals and detecting the surface functional groups of solids is performed by Fourier transform infrared spectroscopy (FTIR) and X-ray photoelectron spectroscopy (XPS) [31]. Lately, the magnetic properties of the FMPs were studied using a Vibrating sample magnetometer (VSP) operating at large scale magnetic operations and maximum magnetic fields. These analytical techniques highlight important information regarding the FMPs. Also, certain innovative techniques like thermo-gravimetric analysis, Mossbauer spectrometry are applied for the estimation of thermal stability and magnetic behavioral properties of magnetic sorbents respectively [38]. Table 2 summarizes, in general, the magnetic solid phase extraction (MSPE) with inorganic and organic coating and detection techniques for removal of heavy metal ions from aqueous solutions.

Table 2: List of MSPE with inorganic and organic coating for removal of metal ions from water samples.

MSPE using inorganic coating material				
Target metal ions	**Coating material**	**Functional groups**	**Eluent**	**Technique**
Pd (II)	SiO_2	Metal-organic framework (Fe_3O_4-Py/MOF)	$NaOH + K_2SO_4$	Flame atomic absorption spectrometry (FAAS)
Cr (III), Al (III)	SiO_2	3-Mercapto-propionic acid (MPA)	HNO_3	Flame atomic absorption spectrometry and electrothermal atomic absorption spectrometry (FAAS, ETAAS)

Pb(II)	SiO_2	Zincon	HCl	ETAAS
Cu (II), Cd (II), Ni (II), Co (II), Pb (II)	SiO_2	Cetylpyridinium bromide	HNO_3 + Methanol	ETAAS
Cr (III), Cu (II), Pb (II)	SiO_2	Bismuthiol II	HNO_3	Inductively coupled plasma atomic emission spectrometry (ICP-AES)
Cr (III), Cu (II), Pb (II)	SiO_2	Dithizone	HNO_3	ICP-AES
Cd (II), Cr (III), Mn (II), Cu (II)	SiO_2	TiO2	HNO_3	Inductively coupled plasma mass spectrometry (ICP-MS)
$[UO2]^{2+}$	SiO_2	Quercetin	-	UV/Vis
Cd(II)	Al2O3	Sodium dodecyl sulphate	HNO_3	FAAS
Cr (III)	ZrO3	-	HNO_3	FAAS
MSPE using organic coating material				
Target metal ions	**Core particles**	**Shell material**	**Functional groups**	**Maximum adsorption capacity q_m (mg/g)**
Cd (II)	Iron oxide (Fe_3O_4)	Ammonium cellulose	-	42
Cr (VI), As (V)	Fe_3O_4	Chitosan	Chitosan anthranilic acid glutaraldehyde (Schiff's base)	58.48, 62.42
Cu (II), Cd (II), Ni (II)	Fe_3O_4	Chitosan	2-aminopyridine-glyoxal	124.23, 84.53, 67.35
Cr (VI)	Fe_3O_4	Polypyrrole	-	169.49
Cr (VI)	Fe_3O_4	Polyethylamine	Montmorillonite	8.8
Pb (II)	Fe_2O_3	Calcium alginate	-	50
Cu (II), Co (II), Ni (II)	Fe_3O_4	Polymethyl-methacrylate Divinylbenzene	Ethylene diamine	50.3, 51.7, 49.6
Cu (II), Co (II)	Fe_3O_4	Carbon	-	3.21, 1.23
Cd (II), Hg (II)	Fe_3O_4	Activated carbon	-	-
Co (II)	Fe_3O_4	Graphene oxide	-	12.98

3. Overview of magnetically responsive bio-sorbents

Biological compounds of various origins are essential sources of biologically active materials and other biomaterials derived substances and products. The magnetic nano-, as well as micro-particles (FMPs), have found several applications in different fields of medicine, biotechnology, biosciences and environmental technology. The varied responses of several sizes of FMPs enable different applications including targeting, localization and specific separation of these magnetically responsive FMPs. This is aided by an external magnetic field for uniform separation using a strong permanent magnet, generation of heat and rise in negative T2 contrast by magnetite FMPs. This is especially observed when there is a gradual increment in viscosity of the magneto rheological fluids upon exposure to an external magnetic field or in magnetic resonance imaging [54]. This is often applied for magnetic modification of diamagnetic materials including microbial cells and plant-derived substances. This is also used in case of biopolymers, magnetic labeling of bioactive substances and functional ligands such as enzymes, aptamers, and antibodies [54].

Several research groups are involved in applying the different processes leading to the synthesis of homogenous FMPs with uniform size, structure, composition, coating etc. [55]. Commercially several non-magnetic particulate compounds are available including adsorbents, chromatographic substances, catalysts, microbial cells etc. In most cases, the efficiency of such materials can be enhanced by their structural and functional modifications thereby causing the formation of functionalized magnetically responsive bio-sorbents. Such enhancement can significantly reduce the rigorous separation of magnetic complexes. Lately, it was observed that both functionalized magnetic nanoparticles as well as magneto-ferritin exhibit peroxidase-like activity [56].

There can be different possibilities for the separation of magnetic adsorbents and removal from treated aqueous solutions and suspensions, relating to the character of FMPs and volume of suspension.

In several cases, easily available and a cost-effective magnetic separator or a small piece of strong magnet is often utilized for separation of the micrometer-sized biological complexes.

The nanometer-sized magnetic compounds generally require more elaborate separation techniques like high gradient magnetic separator (HGMS). For large scale, magnetic separation processes using magnetic microparticles, the drum magnetic separators are utilized. Scientists are coming up with the strategy of new innovative and inexpensive magnetically responsive materials to improve the efficiency of magnetic separation processes for large-scale units. Therefore, magnetic modification of inexpensive biosorbents and carriers can genuinely direct the required substances for large-scale biotechnological and environmental applications.

3.1 Mechanism of magnetic biomaterial modifications

In addition to the currently applied mechanism on the selective development of distinct magnetic structures, alternative processes causing the formation of effective magnetic responsive biomaterials have been developed. These methods based on the principle of 'post-magnetization' of the already existing natural or experimental FMPs (diamagnetic or paramagnetic). The entire mechanism of magnetic biomaterial modification of nonmagnetic components was deduced by Mosbach at the end of the 70's and beginning of the 80's in the 20th century when magnetic modification of gel particles for column chromatography was done using a suitable magnetic fluid (Ferrofluid) [54]. Such magnetic modification processes led to the development of uniformly bio-specific magnetic functionalized derivatives with general application in ligand binding affinity chromatography. The mechanism of post-magnetization generally aims for the synthesis of heavy magnetic composite substances, where the original FMP (diamagnetic or paramagnetic) is basically accountable for biological, bio-adsorptive, bio-catalytic or bio-carrier function of the formed material, whereas the magnetic labels (which are often formed of magnetite or maghemite nano and microparticles deposited on the pores, gels or surface of treated materials) are accountable for strong magnetic behavior of the formed complex material [54]. The commonest approach for magnetic modification is based on the alkali precipitation and heating of ferrous and ferric salts in presence of the treated material [57]. On the other hands, erbium ions can be used to modify the non-magnetic materials which defend their remarkably high atomic magnetic dipole moment in several chemical structures [58]. The post-magnetization mechanism can be followed for several inorganic substances including clays, activated charcoal, synthetic polymer particles, biopolymer materials, waste plant products, microbial cell units, entire algal and cyanobacterial cell units and several others [54].

3.1.1 Magnetic fluid modification

As already discussed, for magnetic modification, various ionically and sterically stabilized magnetic fluids nanoparticles are utilized. Following the easy mechanism of addition of perchloric acid stabilized ferrofluids with methanol suspension of an inorganic waste powder material to be modified (eg. sawdust), which caused precipitation of FMPs from the ferrofluids on the particles surface [54].

In general, perchloric acid stabilized ferrofluids are used for the post magnetization modification of living baker's or brewer's yeast cells after washing with acetate or glycine-HCl buffer. Conversely, tetramethyl ammonium hydroxide stabilized ferrofluids are also applied for the modification of baker's yeast after washing with glycine-NaOH buffer thereby leading to the precipitation of nano-sized FMPs on the cell surface [57]. Several magnetic modifications can be conducted in varied physiological states of yeast cells, likewise; surface modifications can be obtained only in dormant yeast cells, whereas for actively growing cells, the modification causes the accumulation of magnetic modifier in the periplasmic space [59].

Fig. 4 (1,2) showing SEM and TEM images of ferrofluid stabilized *Saccharomyces cerevisae* cells attached to nano-sized functionalized magnetic particles (FMPs) and their aggregates on the surface of the cell as well as on the cell wall [59]. Effective and innovative post-magnetization mechanism is inclined on the application of water-based ferrofluid stabilized with perchloric acid, carefully mixed for proper modification and completely dried. This process has advantages due to its ease of application in various biological, inorganic and polymer materials to be converted to their magnetic components [60].

3.1.2 Microwave aided magnetic modification

Presently synthesis of FMPs from ferrous sulfate with the aid of microwave facility has been studied. Usually, such microwave irradiation can increase several chemical reaction phenomena in the production of organic and inorganic materials. Unlike conventional heating methods, chemical reactions assisted by microwave irradiate extremely rapidly and the products can be acquired within a short time span. The simple, one pot modification process includes treatment of iron (II) salt at high pH in the regular kitchen microwave oven (700-750 W, 2450 MHz) under specified time in presence of the treated substance. During the treatment process, several sub-micrometer FMPs in nano and micro sizes are deposited on the surface of the treated substances in the form of single units or aggregates [61]. For the modification of reasonably high pH and heat stable materials, direct microwave aided magnetization is followed. Nevertheless, for magnetic modifications of more sensitive and selective materials, an indirect microwave aided magnetization has been developed. The FMPs in nano and micro sizes (nano FMP ranges between 25 to 100 nm; which on synthesis form aggregates of size 20 μm) have been produced from ferrous sulfate at high pH in a microwave oven [62]. The substances to be magnetically modified were carefully added to the magnetite or maghemite particle suspension and completely dried at 70 °C which led to the deposition of functionalized magnetite or maghemite nanoparticle aggregates on the surface of the modified material as shown in Fig. 4 (10). The indirect post-magnetization mechanism applying microwave synthesized FMPs are used for the magnetic treatment of thermally sensitive substances like powdered enzymes and biopolymers [107]. These enzyme powders are suspended in aqueous media which prevent their solubilization (like highly concentrated glycol solution, ammonium sulfate, ethanol, and methanol) and cause cross-linkage with glutaraldehyde. The post magnetization process using nano and micro-sized magnetite or maghemite particles produced using microwave technique was successfully studied at a very low temperature in the freezer (-20 °C). The magnetically synthesized enzymes can generally take several weeks to dry and form the powder enzyme state completely [57]. Application of lower temperature drying techniques has been followed for such magnetically synthesized enzymes, which were immobilized on non-magnetic carriers and also for other sensitive biomaterials like starch grains.

3.1.3 Mechano-chemical magnetic modification

Of late, mechanochemical techniques were used to produce functionalized magnetite or maghemite particles and ferrites nanoparticles. The field of mechano-chemistry characterizes one of the several ways of chemical activation. In solid state mechano-chemistry, certain non-thermal chemical reactions occur due to the deformation and factorization of solids, which are mechanically processed by milling or grinding of the substances. During such processes, the extremely high mechanical energy causes chemical reactions and phase transformations [63]. The simple standard mechanism for mechano-chemical production of magnetic bio-composite materials includes grinding of hydrated ferrous and ferric chlorides in a mortar at room temperature for 10 minutes and addition of an excess of sodium chloride to the precursors to prevent agglomeration. This was followed by addition and thorough mixing of a specific amount of target non-magnetic powder material with the magnetic mixture for a definite time. Powdered alkaline hydroxide was added as the last mixing ingredient, followed by further grinding for about 10 minutes. For completing the mechano-chemical process, the magnetically modified substance was carefully washed with distilled water. Several varieties of inorganic and biological materials including starch, pollen and maize grains, potato, powdered lignocellulosic compounds (as shown in Fig. 4 (3,9)) have successfully modified magnetic mechanism [64,102].

4. Application of magnetically responsive bio-sorbents

With the development of scientific techniques, different magnetic derivatives of bio-sorbents have been widely studied and applied as adsorbents, biocatalysts, and carriers etc. Some bio-composites which have been extensively studied are listed below:

- Individual biopolymer sorbents of diverse origin such as polysaccharides (chitin, alginate, cellulose, agar, chitosan, agarose, plant gums etc.) proteins (keratin, casein, ovalbumin, gelatin, albumin) and poly-hydroxy alkanoates (e.g. poly-hydroxybutyrate).
- Complex biopolymer sorbents of various origins such as lignocellulosic materials of plant origin like spent barley grain, sawdust, straw, peanut husks, spent coffee and tea residues etc.
- Microbial and microalgae cellular biomass such as baker's yeast (*Saccharomyces cerevisiae*), fodder yeast (*Kluyveromyces marxianus*) and unicellular algae like *chlorella vulgaris.*
- Aquatic seagrass and macroalgae such as *Posidonia oceanic* (which forms Neptune balls) and various species of *Sargassum* [57].
- Inorganic biosorbents like egg shells, sheaths of *Leptothrix ochracea or* other different diatoms.

Table 3 lists the magnetically responsive bio-sorbent composites materials which have been applied for potential environmental sustainability. It is clear from Table 3 that several routes are available for the synthesis of different types of magnetic bio-sorbent composites. Both the nano and micro-sized bio-sorbent composites have been significantly utilized in model experiments. Presently, magnetic derivatives of chitosan open an attractive research area for scientists for inorganic as well as organic xenobiotics removal. Furthermore, magnetic derivatives from plant waste matter also pave a new pathway for green applications to magnetic biosorbent research.

Table 3: Magnetic responsive biocomposites for heavy metal removal.

Biopolymer	Composite type	Target metals	Matrix	Comments
Chitosan	Magnetically responsive chitosan resin	Au (II), Ag (I), Hg (II)	Aqueous solution	Reused with 0.1 M KI, glutaraldehyde cross-linking
	Amine nano-sized FMP chitosan composite	Pb (II), Cu (II), Cd (II)	Aqueous solution	Reused with weak acid and ultrasound
	Magnetic chitosan, quarternary amine resin	Cr (VI)	Alkaline aqueous solution	Have glutaraldehyde cross-linking
	Cross-linked chitosan-diacetyl monoxime (Schiff's base) resin	Co (II), Cu (II), Ni (II)	Aqueous solution	Have glyoxal cross-linking, reused with 0.01-0.1 M EDTA
	Saccharomyces cerevisae magnetically responsive chitosan particles	Cu (II)	Aqueous solution	The cells were immobilized on chitosan coated nano-sized FMP
Chitin	Magnetite particles immobilized on chitin	Cu (II)	Aqueous solution	Maximum adsorption potential 53.19 mg/g
Alginate	Calcium alginate encapsulated magnetic sorbent	AsO_4^{-3}	Aqueous solution	After biosorption, arsenate was reduced to arsenite
	Magnetically responsive alginate beads	Co (II)	Aqueous solution	Beads having Cyanex 272, reused using HNO_3
	Magnetically responsive alginate microcapsules	Ni (II)	Aqueous solution	Maximum adsorption potential 60.0 mg/g
Gum arabic	Biosorption of gum on surface of nanosized FMP	Cu (II)	Aqueous solution	Maximum adsorption potential 38.5 mg/g
Gellan gum	Magnetic nanosized particles with gel beads	Pb (II), Cr (III), Mn (II)	Aqueous solution	Reused with sodium citrate

4.1 Magnetically responsive biopolymer

The search for an effective sorbent that can be applied for environmental remediation has been obstructed by a preferred combination of simplicity to process, selectivity, and high affinity. Although, synthetic polymeric materials with these preferences have been developed from time to time; nevertheless the expense of ultrafiltration, toxicity of synthetic materials, generation of polymeric wastes and incompatibility of these materials with the environment have made them unsuitable for large-scale environmental application [65]. To resolve such difficulties, a plausible solution could be the mixtures of several types of natural biological substances. Thus, biopolymers which are a new generation of polymeric substances occurring that are still in the development stage have attracted attention for effective replacement of conventional synthetic materials due to a positive impact on sustainable development. One of the attractive biopolymers for the synthesis of non-magnetic as well as magnetic biocomposite adsorbents is chitosan. For proper biosorption of heavy metal ions, the amino and hydroxyl groups present in chitosan provide the main sorption sites. For the chelation process, nitrogen atoms of the amino chemically react with heavy metal cations under a neutral condition, whereas the protonated amino groups show a strong affinity with heavy metal anions by electrostatic forces under acidic conditions [66]. The technique for the uptake of certain metal ion may vary with respect to the pH and composition of the solution. The biosorption possibility of heavy metal ions increases with lower pH conditions (acidic) and on chitosan with decreasing rate of acetylation. For instance, chitosan has been reported to show higher sorption capacity than chitin for the heavy metal ions Pb(II), Hg(II), Cd(II), Cr(III) as well as for several dyes over a wide range of pH. Chitosan without post magnetization modification depicts reduced adsorptive capacity to alkaline and alkaline earth metal ions [67]. Generally, under acidic conditions, chemisorption plays an important role along with the main functional amino groups whereas, under alkaline conditions, physical-sorption dominates with chief functional hydroxyl groups. Fig. 4 (10) depicts optical microscopy image showing microwave stabilized magnetically responsive biopolymer chitosan microparticles [107]. Interactions of chitosan with organic dyes are through electrostatic attraction, hydrophobic forces and physic-adsorption but the molecular size and anionicity of dyes also affect the degree of biosorption [67]. Besides chitin and chitosan, other biopolymers like alginate and plant gums have also attracted attention in synthesis and application of magnetically responsive biosorbents for environmental remediation as listed in Table 3. Fig. 4 (4) depicts an image taken from an optical microscope showing magnetically responsive biopolymer derivative alginate microparticles [103]. In certain reported experiments, the functionalized magnetic biopolymers (FMBPs) have been used as a matrix to capture active sorbents for their modifications. For instance, the hydrophobic octadecyl nano-sized functionalized magnetite or maghemite particles were captured into hydrophilic barium alginate materials to produce a novel type of magnetic solid phase extraction (SPE) bio-sorbent utilized for the pre-concentration of polycyclic aromatic hydrocarbons and phthalate ester contaminants from marine waste samples. Here, the hydrophilicity of barium alginate capturing agent improvises the dispersibility of sorbents in marine samples while the magnetic properties of functionalized magnetite or maghemite particles facilitate the magnetic separation [68]. In a recently reported article, sodium alginate based magnetic carbonaceous material (Fe-SA-X, where X is heating temperature) was caged by carbothermal reduction of ferric alginate, which was intended for efficient removal of heavy metal ion like Cr (VI) from wastewater sample. The authors have studied the optimization, kinetics, and adsorption isotherm modeling parameters for Cr (VI) onto the samples and characterized using Fourier transform infrared spectroscopy (FTIR), X-ray photoelectron spectroscopy, nitrogen sorption, and Raman spectrometry. On observing the whole biosorption mechanism, the results depicted that the newly developed magnetic biosorbent is promising enough for the removal of toxic metal ions like Cr (VI) from water [69].

4.2 Magnetically responsive plant derivatives

The magnetically modified substances derived from plants are potential sorbents for the removal of organic and inorganic contaminants. Recent research on magnetic bio-sorbents is concentrated on developing and describing the sorption properties of naturally occurring eco-friendly, cost-effective and compatible green biomaterials like agriculture and wood industry by-products and bio-wastes like sawdust, barley straw, tea bark, peanut skins etc. There are plenty of agricultural by-products and bio-waste materials needing to be disposed and are therefore ready to be utilized, cheap, and easily available in good quantities. The use of such bio-wastes for removal of contaminants would be favorable for both the environment and industries. Research articles describing the application of such magnetically modified agro wastes (sawdust, spent tea and coffee residues) in the treatment of undesirable materials have attracted the scientific society, although there are only a few examples of FMPs which are derived from plants applicable for xenobiotics removal. These materials have been tested in laboratory experiments and large-scale functions are yet to be obtained. Various modification mechanisms have been followed like ferrofluid treatment for sawdust, peanut husks, spent grain, spent tea and coffee residues etc. and microwave aided treatment include straw or spent rooibos (*Aspalathus linearis*) tea biomass [70-74]. Fig. 4 (3) depicts an experimental image of magnetically responsive plant derivative spent grain before and after magnetically treated separation [102]. For instance, the water-based ferrofluid stabilized with perchloric acid has been utilized for laboratory scale magnetic modification of certain plant-derived substances which had the maximum adsorption capacity values reaching up to almost 100 mg/g. These biomaterials exhibit potential adsorption of not only heavy metal ions but also inorganic and complex water-soluble dyes, oils and various other undesirable pollutants from wastewater. Certain biomaterials (like wheat bran derived from wheat grain which was modified using a microwave magnetic procedure produced functionalized magnetite and maghemite particles or FMPs) have been applied for the sorption of uranium ions. These plant-derived magnetically responsive sorbent materials can also be used as cost-effective, biocompatible carriers for immobilization of different enzymes. For instance, spent grain which was magnetically modified by ferrofluids was used as cost-effective, magnetically responsive and compatible carrier for immobilization of *Candida rugosa* lipase. Another modification mechanism was also performed using native as well as poly (ethyleneimine) modified magnetic wheat grain. This material on post magnetization modification can be an effective low-cost magnetic carrier for immobilization of various enzymes which will find application in biotechnology and food and feed technology [75].

4.3 Magnetically responsive microalgae

The rising sources for biomass and metabolite production with substantial use in the fields of food, feed, marine, pharmaceutical, cosmetics and fuel industries are the prokaryotic blue-green algae and eukaryotic microalgae. The eukaryotic microalgae perform a pivotal part in several environmental applications by acting as potential nitrogen fixers, biocatalysts in biotransformation and degradation of organic contaminants, wastewater treatment material and also parts of biosensors for identification of polluting substances [76-78]. From a commercial point of view, the effectiveness of these eukaryotic organisms is based on their rich metabolic functions thereby producing valuable products. These are also easily cultivable, adaptable to various natural conditions and rapidly growing organisms. The expense of industrial process can be decreased by their application due to their ability to use natural sunlight, flue gas or wastewater as energy and nutrient sources. In addition, they can grow all around the year in enormous productivity independent of weather situations and in non-arable or wasteland unlike conventional marine plants or agricultural crops with much less requirement of land area [79]. In general, frequently occurring algal cells are able to interact with various nano and microparticles including gold, silica, platinum, palladium, silver and carbon nanotubes. These modified organisms mainly sustain their viability by themselves although the presence of such foreign particles on their cell surfaces, in intracellular organelles or protoplasm can offer certain supplementary functionalities. The modification with nano and micro-sized FMPs is chiefly important and such magnetically responsive microalgal cells have been considerably studied and applied in several applications [80]. The majority of microalgal cells exhibit diamagnetic characteristics. Fig. 4 (8) showing TEM image of a fine section of polyallylamine hydrochloride stabilized FMPs coated with microalgal cells of *Chlorella pyrenoidosa* [106]. The post magnetic modification of diamagnetic cells can be carried out by forming strong affinity of nano or micro-sized FMPs with the cell surface. The mechanism used here involves non-specific attachment of FMPs by ferrofluid treatment or magnetic microparticles binding onto cell surface, also polymer assisted FMP binding, covalent immobilization of FMPs onto microalgal cell surface or vice versa, selective interactions with immune-magnetic nano and micro-sized particles or entrapment onto biocompatible polymers [81-83]. The magnetic properties of the modifiers are basically due to the presence of nano or micro-sized functionalized magnetite or maghemite particles or their mixtures as well as ferrite particles. Generally, the phenotypic alteration or viable mutation of modified cells is not affected by bonding with the FMPs. In selective cases, the particles attached to the surface of the cell can be internalized by the treated cells and appears in the protoplasm. In most cases, the usual features of all magnetically modified or responsive cells are their definite interactions with external magnetic field.

The first discovered magnetotactic alga, from the coastal mangrove swamp near Fortaleza, Brazil, found in brackish mud and water samples were identified as *Anisonema platysomum.* The organism has cells enclosed with numerous, uniformly organized chains of bullet shaped magnetite crystal particles of sizes 80 to 180 nm in length and 40-50 nm in width. The coordinated structures of magnetosomes are so specifically arranged in this algae that this organism possibly bio-mineralizes and organizes endogenous magnetite crystal particles in a definite well-controlled fashion within the cell, where intracellular structural filaments play an important role in the production and functionalization of magnetosome chain as also shown by the magnetotactic bacteria [84].

Techniques for dewatering microalgae cultures, along with their advantages and disadvantages have been extensively reported in the literature. For instance, the harvesting of microalgae includes different solid-liquid separation techniques and steps such as centrifugation, gravity filtration, flocculation methods, sedimentation and flotation steps [85-87]. One of the critical bottlenecks in the large-scale production process is the biomass recovery since it claims almost 30-40% of biomass production expense. Generally, large microalgal biomass volumes are processed for concentrating the cells due to the fact of smaller size of microalgae (about 5-10 nm) so cell separation becomes difficult with low sedimentation velocity, also colloidal characteristics are present with repulsive negative surface charges and reduced biomass concentrations in culture broths (about 60.5 kg/m^3 dry biomass in large-scale systems) [84]. Therefore, it is very important to lower the volume of microalgal suspension after cultivation by a 2-step mechanism with concentration factors of 100-200 in the first and 2-10 in the second process to make the downstream processing cost effective. It is always essential to pre-concentrate the microalgal broth cultures using simple and low-cost techniques of flocculation before applying the rigorous cell separation techniques of centrifugation or filtration [84]. The different harvesting methods for microalgae are typically case dependent; likewise, the actual product price, selection of harvesting mechanisms can be varied per preferred moisture content in the product. Usually, a sludge formed due to gravity sedimentation has a higher amount of moisture content than the recovered microalgal biomass by centrifugation. This could significantly control the expense of product recovery when the related downstream processing step is an energy demanding (thermal-drying) process [88].

Nano and micro-sized FMPs have the capability of harvesting agents. The bio-separation processes employing such FMPs involve by the non-destructive feature of the magnetic field, the utility of simple techniques and instruments, biocompatibility of FMPs, effective manipulation and regeneration [84]. The magnetic harvesting process of microalgae using an external magnetic field after biosorption of a magnetic agent onto microalgal cells can also be a one-step synthesis process because flocculation and separation both occur simultaneously [87]. The preferred magnetic harvesting agents for various microalgal species are generally in the form of uncoated nano or micro-sized FMPs or as functionalized nanocomposites that constitute a magnetic core coated with silica. This silica coating can equally carry selective functional groups like polyethyleneimine (PEI) or cationic polyelectrolytes like chitosan (CS), polydiallyldimethyl ammonium chloride(PDDA) and cationic polyacrylamide (CPAM) [89].

The microalgal cells can be applied as a part of whole cell biosensors, also immobilized on transducers for different types of immobilization processes like physical-sorption, chemical cross-linking, matrix incorporation etc. Magnetic immobilization can overcome some important drawbacks of these processes including reduction of viability of algal cells. The magnetic immobilization technique constitutes of temporary attachment of FMPs on biosensor surface by a strong external magnetic field. The FMPs are held and manipulated on the electrode by minute magnets positioned under the electrode [90]. Presently, a whole cell amperometric herbicide biosensor on magnetically responsive microalgae and screen-printed electrodes has been studied. The microalgae cells coated with nano-sized FMPs were biocompatible and stabilized by polyallylamine hydrochloride with an average diameter of 15 nm. Suspension of algal cells was magnetized during vortexing after addition of FMP suspension. These modified cells were employed for the construction of amperometric biosensors having a screen-printed electrode, which was connected with supporting tetrafluoroethylene plate with slotted 2 mm cylindrical NdFeB magnet in the space under the working area of electrode thereby producing strong magnetic field surrounding the surface of the electrode. A drop of a suspension containing a mixture of an electron mediator ($K_3Fe\ (CN)_6$) and electrolyte (Na_2SO_4) was kept around the working area of the electrode and the magnetized algal cells were instantly covered above the magnetic field area. This particular biosensor was employed for analysis and quantification of triazine herbicides. The biosensor was capable of quantifying other xenobiotics including atrazine of 0.9 to 74 mM and propazine of 0.6 to 120 Mm with limits of detection values of 0.7 and 0.4 mM respectively [91].

Magnetically responsive diatoms have been considered as an effective drug carrier for *in-vivo* functions. This material exhibits low cytotoxic potential, due to the presence of low toxic silica and protein-coated FMPs. Also, this material depicted very slow biodegradability in body fluids [92].

Typical non-magnetic immobilized algal cells have been used for the removal of various pollutants and xenobiotics. However, magnetically modified cells of *C. vulgaris* were employed as the latest low-cost magnetic sorbent for the removal of a variety of water-soluble toxic dyes. The dried microalgal cells were magnetically modified using ferrofluids concentrated with perchloric acid. The experimental data from a published dye sorption process were fitted to Langmuir isotherm and the highest adsorption capacities were in between 24.2 and 257.99 mg/g of dried magnetically responsive algal biosorbents for Saturn blue LBRR and aniline blue respectively [84].

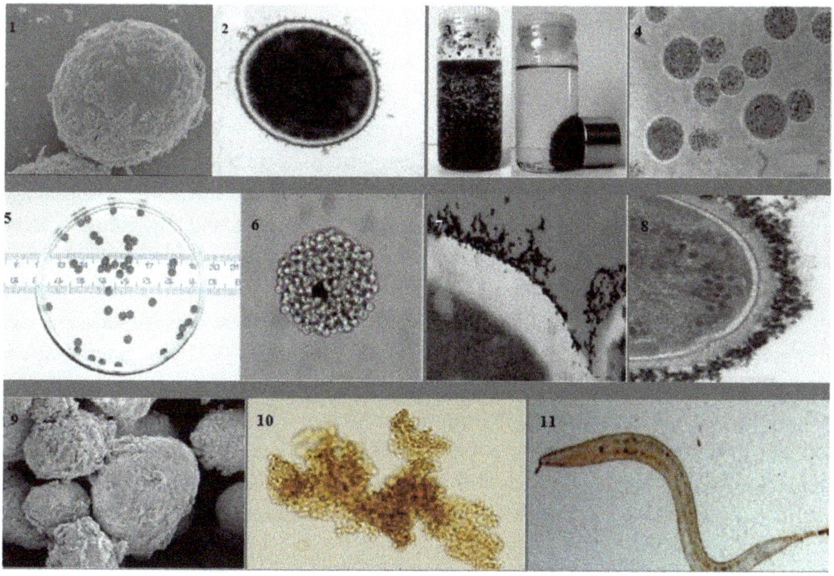

Fig. 4 (1) SEM image of ferrofluid stabilizedSaccharomyces cerevisiae cells showing attached functionalized magnetic particles (FMPs) and their aggregates on the surface of the cell [9] (2) TEM image showing ferrofluid modified Saccharomyces cerevisiae with attached magnetite nanoparticles on the cell wall [59], (3) Magnetically responsive plant derivatives including spent grain before and after magnetically treated separation [102], (4) Image taken from optical microscope showing magnetically responsive biopolymer derivative alginate microparticles [103], (5,6) Entrapment of Sachharomycescerevisaecells in magnetically responsive millimetre or micron-sized alginate bead particles and functionalized magnetic particles (FMPs). Scale bar showing the size of 50 μm [104], (7) TEM image showing magnetically responsive microbial cells including dried fodder yeast (Kluveromyces fragilis) [105], (8) TEM image showing a fine section of polyallylamine hydrochloride stabilized FMPs coated with microalgal cells of Chlorella pyrenoidosa [106], (9) SEM image showing mechanochemically modified native potato starch [64], (10) Image taken from optical microscope showing microwave stabilized magnetically responsive biopolymer chitosan microparticles [107], (11) Image taken from optical microscope showing a fine section of polyallylamine hydrochloride stabilized FMPs attached to Caenorhabditis elegans nematode [108].

4.4 Magnetically responsive activated carbons

One of the important and versatile materials applied as an effective adsorbent, catalyst and supporting catalyst with large surface area, high porosity, wide range of surface chemistry and surface reactivity is activated carbons (ACs) or activated charcoal. AC is used for several applications including treatment of contaminants from aqueous or gaseous phases, purification as well as recovery of materials. AC sorption is an immensely attractive domain to many industrial, economic and agricultural areas as varied as food, chemical, cosmetics, petroleum, agro products, pharmaceuticals, nuclear or automobile, treatment of industrial effluent, drinking water, domestic and urban waste water etc. The only disadvantage of activated carbons is their high production cost in comparison to other adsorbents. Nowadays, preparation of AC from green sources or waste materials has been a primary perspective of several types of research. The production cost can be lowered by opting easily available, naturally occurring raw materials [93]. One of the important possible approaches for improving the sorption efficiency of AC is the magnetic modification of such green activated carbons which can be simply separated from complex systems or positioned to a definite place using a strong external magnetic field.

The magnetic modification of activated carbons (ACs) is generally occurred due to the presence of nano or micro-sized FMPs within the pores of carbon particles or on the surface of carbon particles or inside the gels having co-entrapped carbon particles. Such nano or micro-sized FMPs modifying AC are chiefly magnetite or maghemite or different types of ferrites, although at times metallic iron and nickel have also been prepared and used as magnetic labels. The simplest and the most frequently applied magnetic modification is the alkaline precipitation of ferrous and ferric salts in presence of AC along with heat treatment of the liquid suspension forming particles of magnetite or maghemite or different types of ferrites. Alternatively, the ACs are impregnated with salts of iron or nickel by heat treatment at high temperature and the resulting modified FMPs formed; depends on the types of salt used, heating and atmosphere conditions. Alternatively, magnetic ACs can be synthesized using standard precipitation process by continuous stirring of carbon powder and magnetite or maghemite suspension mixture followed by heat drying at 40 °C. Interestingly ferrofluid stabilized triethanolamine oleate was employed for the rapid preparation of magnetically modified AC by step-by-step process of impregnation, washing and air drying at 90 °C [94]. Another reported mechanism showed treatment of AC and magnetite mixture in a high energy planetary ball mill. In another mechanism, the micro-sized FMPs were coated with a slim layer of epoxy resin which gradually enhanced the binding affinity of AC. Furthermore, the encapsulation of AC with FMPs inside a selective and specific biopolymer or synthetic polymer is another perspective of magnetically responsive activated carbons.

ACs has been frequently applied as potential sorbents for several organic and inorganic xenobiotics, radionuclides and noble materials [95]. In general, magnetically modified AC derivatives exhibit similar sorption mechanism for target substances as the native non-magnetic ACs. Of late, several studies reported the utilization of magnetic activated carbon for site remediation and xenobiotic removal from polluted water areas. Magnetic AC is an additional supplement to magnetically responsive biocomposites applied for such purposes. For instance, a novel magnetically separable composite photocatalyst, (titania based magnetic AC) was synthesized by deposition of anatase titania onto a magnetic AC surface and the photocatalytic activity of the samples was analyzed by degradation of reactive brilliant red X-3B under UV irradiation. This particular biocomposite could be regenerated and reused with a only slight reduction of its photocatalytic activity [96]. Modified AC with zero-valent Fe deposits demonstrated dehalogenation activity for chlorinated and brominated hydrocarbons. The contaminants were deposited and enriched at the surface of the magnetic AC and by reductive dechlorination, these contaminants were destroyed at the iron clusters. Regeneration studies of magnetically responsive ACs after adsorption processes have been performed to keeping the production expenses as low as possible. Recently, a magnetic $CuFe_2O_4$ AC composite was experimented by the chemical co-precipitation mechanism. The regeneration process was performed after adsorption of acid orange II by heating in an inert atmosphere. The results demonstrate that $CuFe_2O_4$ particles possibly catalyze thermal pyrolysis of the adsorbed dye material. The regeneration experiment predicted that full adsorption capacity of the composite sorbent was restored upon thermal treatment and could be reused for several cycles [95]. Alternatively, hydrogen peroxide has been widely used for such regeneration procedure and the magnetite or maghemite nano-sized particles were found effective for attaining high regeneration efficiency. Magnetic AC has been applied as a potential sorbent for dyes and heavy metal ions during the development of magnetic solid phase extraction (MSPE). About 460- fold enrichment of materials was obtained using magnetic AC as sorbent and water-soluble dyes and heavy metal ions as analytes. Moreover, magnetic AC can also become an exclusive easy to prepare the magnetic composite substance for smart innovative project materials in college or high school education. The simple laboratory experiment includes the preparation of magnetically responsive AC within 2 h using mostly easily available chemicals. Potential adsorption of dyes and metal ions by magnetic AC can be analyzed just by visual evaluation without the requirement of any spectrometer and the magnetic separation procedure can be easily performed using cheap, small NdFeB magnets [97].

4.5 Magnetically responsive microbial cells

The microbial cells in free or immobilized condition are applicable for the pre-concentration or treatment of metal ions, organic and inorganic dyes, xenobiotics or bioactive compounds. The adsorption of metal ions on microbial cells depends on the composition of their cell walls. Typical examples would be the cell walls of yeast constitute of numerous complex organic materials and their polymers including 28% of glucan, 31% of mannan, 13% of proteins, 8% of lipids and 2% of chitin and chitosan. The cell walls of brown algae constitute 10 to 40% of alginic acid, 5-20% of fucoidan and about 2-20% of cellulose; red algae constitutes agar, cellulose, pectin, carrageenan, and xylan while the cell wall of green algae constitutes pectin and cellulose. Such substances contain several functional groups participating in the heavy metal binding process [66]. Fig. 4 (5) and Fig. 4 (6) depicts the entrapment of *Saccharomyces cerevisiae* cells in magnetically responsive millimeter or micron-sized alginate bead particles and functionalized magnetic particles (FMPs) [104]. Research observation with fungal biomass and seaweeds showed a strong affinity for ion exchange metal binding to ionized amine, carboxyl and phosphate groups. For instance, the carboxyl groups of amino acids of peptidoglycan layer and the hydroxyl groups of polysaccharides present on the yeast cell walls form a strong attachment with the metal ion Au (III). The positively charged amino groups of yeasts are engaged in the adsorption of heavy metal like Cr (VI) while the imidazole groups of yeast histidine are engaged with As (V).The various physical and chemical treatment processes of microbial biomass reveal the metal attachment functional groups by disruption, modification and membrane permeability. Pre-concentration with HCl, NaOH, DMSO, ethanol or other organic solvents causes improved adsorption of certain heavy metal ions.

The magnetotactic bacteria are interesting microorganisms possessing the capability to produce intracellular biogenic magnetic functionalized nanoparticles, which allow a magnetic separation process. The other forms of prokaryotic and eukaryotic microbial cells are to be modified magnetically in laboratory conditions through complex formation with magnetite or maghemite. In general, the ferrofluids magnetic modification technique is frequently applied to induce strong attachment of magnetic nanoparticles on the surface of microbial cells through selective chemical interactions with immune-magnetic nano and micro-sized particles, biological-based precipitation of paramagnetic substances on the surface of the cell, covalent immobilization on magnetic carriers, cross-linking of the microbial cells or cell walls with bifunctional reagent in the presence of magnetic particles or by entrapment into biocomposite polymer [66]. Additionally, the magnetic modification can also be done by attachment of paramagnetic cations on acid groups present on the microbial cell surface. Of late, a novel mechanism for the synthesis of magnetically responsive microbial cells is based on suspending the cells in methanol and gradual mixing with perchloric acid based ferrofluid. During the process, a definite precipitation of FMPs on the outer surface of the treated microbial cells was observed [66]. The magnetically modified microbial cells find efficient application in removal of heavy metal ions such as Hg(II), Cu(II), Pb(II), and Cr(VI) along with organic xenobiotics. Important ones are low-cost biomaterials of baker's yeast like *Saccharomyces cerevisiae*, fodder yeast like *Kluyveromyces fragilis* and algae like *Chlorella Vulgaris* are of common interests [54]. Fig. 4 (7) shows a TEM image of magnetically responsive microbial cells of dried fodder yeast (*Kluveromyces fragilis*) [105]. It is reported that the biosorption capacities of magnetically modified yeast cells and algae cells for xenobiotics removal can be up to a level of 430 mg of the pollutant per gram of magnetic sorbent [98]. One of the promising sorbents is magnetic fodder yeast which can be used to concentrate and separate Sr^{2+} ions from effluents of spent fuel reprocessing in nuclear power plants [99]. The magnetic modification of living microbial cells can also form magnetically responsive whole-cell biocatalysts [100]. For this process, both direct and definite modifications of cell walls with ferrofluids and entrapment of cells into biocomposite polymers gels in presence of FMPs are carried out. Usually, the intracellular enzymatic reactions will not get reduced after the magnetic modification as seen in the case of hydrogen peroxide degradation and sucrose hydrolysis by the enzyme catalase and invertase which occurs inside magnetically responsive *Saccharomyces cerevisiae* cells [100]. The magnetically modified algal cells can be employed as part of interesting biosensor systems, both in microfluidics configuration and part of screen-printed electrodes. These are also used for measurement of genotoxic and cytotoxic potentials, quantifying the herbicides of atrazine and propazine [54]. Another frequently used technique for cell separation and magnetic modification is immunomagnetic separation. This mechanism is useful for the isolation of stem cells and cancer cells. Detection of disease-causing pathogenic microbes is essential in microbiological and parasitological fields, For example, *Salmonella* sp., *Listeria monocytogenes*, verocytotoxin producing *E. coli* and parasites including *Cryptosporidium* as well as *Giardia* sp. Fig. 4 (11) depict optical microscopy image showing a fine section of polyallylamine hydrochloride stabilized FMPs attached to *Caenorhabditis elegans* nematode [108]. These cells upon proper magnetic labeling with biocompatible nano-sized functionalized magnetic particles enable *in vitro* detection through Fe staining by the production of ferric ferrocyanide (Prussian blue) or *in vivo* detection through MRI visualization due to a specific reduction of T_2 relaxation time, causing the formation of the hypo-intense dark signal. The MRI can be utilized to analyze the engraftment of cells, also their migration time course and survival in the targeted tissues [101]. Table 4 summarizes the varied magnetically responsive microbial cells for removal of heavy metal ions.

Table 4: Magnetically responsive microbial cells for removal of heavy metal ions.

Microbial cells	**Composite type**	**Target metals**	**Matrix**	**Comments**
Pseudomonas putida	Magnetically responsive immobilized cells	Cu (II)	Waste effluent	Pretreatment of cells with HCl and regeneration with acid treatment
Pseudomonas putida	Magnetically responsive immobilized cells	Cu (II)	Industrial wastewater	Semi-continuous-adsorption; pretreatment of cells with HCl; and utilization with acid treatment
Stenotrophomonas sp.	Magnetotactic bacteria	Au (III)	Aqueous solution	Regeneration with thiourea
Saccharomyces cerevisiae	Cells immobilized on chitosan coated nano-sized FMP	Cu (II)	Aqueous solution	Maximum adsorption potential 144.9 mg g^{-1}
*Saccharomyces cerevisae*Subsp*uvarum*	Cells modified with water-based ferrofluids	Hg (II), Cu (II), Ni (II), Zn (II)	Simulated water solution	Desorption with 0.1 M HNO_3
*Saccharomyces cerevisae*Subsp*uvarum*	Cells modified with water-based ferrofluids	Water-soluble organic dyes	Aqueous solution	About five dyes tested
Saccharomyces cerevisiae	Cells modified with water-based ferrofluids	Water-soluble organic dyes	Aqueous solution	About six dyes tested
Kluveromyces fragilis	Cells modified with water-based ferrofluids	Sr (II)	Aqueous solution	Desorption with 0.1 M HNO_3

Kluveromyces fragilis	Cells modified with water-based ferrofluids	Water-soluble organic dyes	Aqueous solution	About seven dyes tested
Saccharomyces cerevisiae	Cells modified with water-based ferrofluids	Water-soluble organic dyes	Aqueous solution	Five dyes tested
Chlorella vulgaris	Algal cells modified with water-based ferrofluids	Water-soluble organic dyes	Aqueous solution	Six dyes tested

5. Conclusions and future outlook

The variety of naturally occurring, low cost, easily available biomaterials including plant-derived substances, microbial cells, activated carbons, microalgal cells etc. can be efficiently used as green sorbents for environmental remediation and sustainability. To modify and improve their sorption efficiency, conversion of these green sorbents was made into 'smart sorbents', which exhibited a response to a strong magnetic field; thereby facilitating their specific and definite magnetic separation from waste effluents. These 'smart' materials have a huge perspective to influence different areas of medicine, engineering technology, biosciences, environmental prospects, biotechnology etc. where varied applications both *in vitro* and *in vivo* have been discussed in this chapter. It is seen here, that these 'smart' magnetically responsive green sorbents cannot only be beneficial as sorbents for several organic and inorganic pollutants like heavy metal ions, EDCs (endocrine disruptors), water-soluble dyes, radionuclide, drug metabolites which are potential toxicants of the total ecosystem but also as biological carriers for cells and intracellular enzyme immobilization or as biosensors. Because of the presence of functionalized magnetic particles (FMPs) in nano or microforms, these biomaterials can be effectively employed as contrast agents in MRI or for targeting of potential drugs metabolites. Progress is expected from biotechnological as well as environmental point of view on the preparation of more commercially effective biocompatible magnetic composites. The utilization of green magnetic sorbents for large-scale applications is necessary as these are reasonably inexpensive and free from generating unwanted by-products and toxic waste. Efficacy, biocompatibility and safety studies of these 'smart' materials have to be tested from time to time especially their long-term toxicity potentials. The collaboration of scientists from different fields with similar mindset are utmost important to correlate the influence of green application of magnetic sorbents in the near future.

References

[1] V. Rocher, J.M. Siaugue, V.Cabuil, A. Bee, Removal of organic dyes by magnetic alginate beads, Water Res. 42 (2008) 1290-1298. https://doi.org/10.1016/j.watres.2007.09.024

[2] H. Li, Z. Li, T. Liu, X. Xiao, Z. Peng, L. Deng, A novel technology for biosorption and recovery hexavalent chromium in wastewater by bio-functional magnetic beads, Bioresource Technol. 99 (2008) 6271-6279. https://doi.org/10.1016/j.biortech.2007.12.002

[3] J. Zhu, S. Wei, M. Chen, H. Gu, S.B. Rapole, S. Pallavkar, T.C. Ho, J. Happer, Z. Guo, Magnetic nanocomposites for environmental remediation, Adv. Pow. Technol. 24 (2013) 459-467. https://doi.org/10.1016/j.apt.2012.10.012

[4] Z. Zhang, S. Xia, A.Yang, B. Xu, L. Chen, Z. Zhu, J. Zhao, J.R. Jaffrezic, D. Leonard, A novel biosorbent for dye removal: extracellular polymeric substances (EPS) of Proteus mirabilis TJ-1, J.Hazard. Mater. 15 (2009) 279-284. https://doi.org/10.1016/j.jhazmat.2008.06.096

[5] C.H. Niu, B. Volesky, D. Cleiman, Biosorption of arsenic(V) with acid-washed crab scheels, Water Res. 42 (2007) 2473-2478. https://doi.org/10.1016/j.watres.2007.03.013

[6] D.M. Hamby, Site remediation techniques supporting environmental restoration activities-a review, Sci. Total Environ. 191 (1996) 203-224. https://doi.org/10.1016/S0048-9697(96)05264-3

[7] R.J. Watts, Hazardous wastes: sources, pathways and receptors, John Wiley and Sons, Inc., 1997.

[8] T. Kanjilal, S. Babu, K. Biswas, C. Bhattacharjee, S. Datta, Application of mango seeds integument as bio-adsorbent in lead removal from industrial effluent, Des. Water Treat. 56 (2015) 984-996. https://doi.org/10.1080/19443994.2014.950999

[9] Q.L. Lu, G.A. Sorial, Adsorption of phenolics on activated carbon, impact of pore size and molecular oxygen, Chemosphere 55 (2004) 671-679. https://doi.org/10.1016/j.chemosphere.2003.11.044

[10] Y. Huang, S. Li, J. Chen, X. Zhang, Y. Chen, Adsorption of Pb (II) on mesoporous activated carbons fabricated from water hyacinth using H3PO4 activation:

adsorption capacity, kinetic and isotherm studies, Appl. Surface Sci. 293 (2014) 160-168. https://doi.org/10.1016/j.apsusc.2013.12.123

[11] R.V. Siriwardane, M.S. Shen, E.P. Fischer, Adsorption of CO_2 on zeolites at moderate temperatures, Energy Fuels 19 (2005) 1153-1159. https://doi.org/10.1021/ef040059h

[12] M.A. Hernandez, L.Corona, A.I. Gonzalez, F. Rojas, V.H. Lara, F. Silva, Quantitative study of the adsorption of aromatic hydrocarbons (benzene, toluene and p-xylene) on dealuminated elinoptilolites, Ind. Eng. Chem. Res. 44 (2005) 2908-2916. https://doi.org/10.1021/ie049276w

[13] B.H. Gu, J. Schmitt, Z.Chen, L.Y. Liang, J.F. McCarthy, Adsorption and desorption of different organic matter fractions on iron oxide, Geochim. Cosmochim. Acta 59 (1995) 219-229. https://doi.org/10.1016/0016-7037(94)00282-Q

[14] C.T. Yavuz, J.T. Mayo, W.W. Yu, A. Prakash, J.C. Falkner, S. Yean, L.L. Cong, H.J. Shipley, A. Kan, M. Tomson, D. Natelson, V. L. Colvin, Low field magnetic separation of monodisperse Fe_3O_4 nanocrystals, Science 314 (2006) 964-967. https://doi.org/10.1126/science.1131475

[15] L.A. Belyakova, N.N. Vlasova, L.P. Golovkova, A.M. Varvarin, D.Y. Lyashenko, A.A. Svezhentsova, N.G. Stukalina, A.A. Chuiko, Role of surface nature of functional silicas in adsorption of monocarboxylic and bile acids, J. Colloid Inter Sci. 258 (2003) 1-9. https://doi.org/10.1016/S0021-9797(02)00093-0

[16] R.S. Bai, T.E. Abraham, Studies on Chromium(VI) adsorption-desorption using immobilized fungal biomass, Bioresource Technol. 87 (2003) 17-26. https://doi.org/10.1016/S0960-8524(02)00222-5

[17] J.Z. Chen, X.C. Tao, J. Xu, T. Zhang, Z. L. Liu, Biosorption of lead, chromium and mercury by immobilized Microcystis aeruginosa in a column, Proc. Biochem. 40 (2005) 3675-3679. https://doi.org/10.1016/j.procbio.2005.03.066

[18] K.K. Nur, N.L. Jeppe, Y. Meral, D. Gonul, Characterization of a bacterial consortium for effective treatment of wastewaters with reactive dyes and Cr(VI), Chemosphere 67 (2007) 826-831. https://doi.org/10.1016/j.chemosphere.2006.08.041

[19] S.A. Sarabjeet, G. Dinesh, Microbial and plant derived biomass for removal of heavy metals from wastewater, Bioresource Technol. 98 (2007) 2243-2257. https://doi.org/10.1016/j.biortech.2005.12.006

[20] Y. Zhang, B. Charles, Factors affecting the removal of selected heavy metals using a polymer immobilized Sphagnum moss as a biosorbents, Environ. Technol. 26 (2005) 733-743. https://doi.org/10.1080/09593332608618515

[21] J. Hu, G.H. Chen, M.C.L. Irene, Removal and recovery of Cr(VI) from wastewater by maghemite nanoparticles, Water Res. 39 (2005) 4528-4536. https://doi.org/10.1016/j.watres.2005.05.051

[22] G. Giakisikli, A.N. Anthemidis, Magnetic materials as sorbents for metal /metalloid preconcentration and/or separation: A review, Analyte. Chim. Acta 789 (2013) 1-16. https://doi.org/10.1016/j.aca.2013.04.021

[23] P. Wikstrom, S. Flygare, A. Grondalen, P.O. Larsson, Magnetic aqueous two-phase separation: A new technique to increase rate of phase-separation, using dextran-ferrofluid or larger iron-oxide particles. Anal. Biochem. 167 (1987) 331-339. https://doi.org/10.1016/0003-2697(87)90173-4

[24] P.H. Towler, J.D. Smith, D.R. Dixon, Magnetic recovery of radium, lead, and polonium from sea water samples after preconcentration on a magnetic adsorbent of manganese dioxide coated magnetite, Anal. Chim. Acta 328 (1996) 53-59. https://doi.org/10.1016/0003-2670(96)00080-3

[25] M. Safarikova, I. Safarik, Magnetic solid phase extraction, J. Magn. Magn. Mater. 194 (1999) 108-112. https://doi.org/10.1016/S0304-8853(98)00566-6

[26] X.S. Li, G.T. Zhu, Y.B. Luo, B.F. Yuan, Y.Q. Feng, Synthesis and applications of functionalized magnetic materials in sample preparation, Trends Anal. Chem. 45 (2013) 233-247. https://doi.org/10.1016/j.trac.2012.10.015

[27] K. Aguilar-Arteaga, J.A. Rodriguez, E. Barrado, Magnetic solids in analytical chemistry: A review, Anal. Chim. Acta 674 (2010) 157-165. https://doi.org/10.1016/j.aca.2010.06.043

[28] T.A. Rangreez, Inamuddin, A.M. Asiri, B.G. Alhogbi, M. Naushad, Synthesis and ion-exchange properties of graphene th(IV) phosphate composite cation exchanger: Its applications in the selective separation of lead metal ions, Int. J. Environ. Res. Public Health. 14 (2017) 828. https://doi.org/10.3390/ijerph14070828

[29] A. Kumar, M. Naushad, A. Rana, Inamuddin, Preeti, G. Sharma, A.A. Ghfar, F.J. Stadler, M.R. Khan, ZnSe-WO_3 nano-hetero-assembly stacked on Gum ghatti for photo-degradative removal of Bisphenol A: Symbiose of adsorption and

photocatalysis, Int. J. Biol. Macromol. 104 (2017) 1172–1184. https://doi.org/10.1016/j.ijbiomac.2017.06.116

[30] Y.H. Zhal, S. Duan, Q. He, X.H. Yang, Q. Han, Solid phase extraction and preconcentration of trace mercury(II) from aqueous solution using magnetic nanoparticles doped with 1,5- diphenylcarbazide, Microchim. Acta 169 (2010) 353-360. https://doi.org/10.1007/s00604-010-0363-8

[31] H. Mei-Jiang, Z. Peng-Yan, Y. Zhao, X. Hu, H. Zhen-Lian, Zincon immobilized silica coated magnetic Fe_3O_4 nanoparticles for solid phase extraction and determination of trace lead in natural and drinking waters by graphite furnace atomic absorption spectrometry, Talanta 94 (2012) 251-256. https://doi.org/10.1016/j.talanta.2012.03.035

[32] A.E. Karatapanis, Y. Fiamegos, C.D. Stalikas, Silica modified magnetic nanoparticles functionalized with cetyl pyridinium bromide for the pre concentration of metals after complexation with 8-hydroxyquinoline, Talanta 84 (2011) 834-839. https://doi.org/10.1016/j.talanta.2011.02.013

[33] M. Faraji, Y. Yamini, A. Saleh, M. Rezace, M. Ghambarian, R. Hassani, A nanoparticle based solid phase extraction procedure followed by flow injection inductively coupled plasma-optical emission spectrometry to determine some heavy metal ions in water samples, Anal. Chim. Acta 659 (2010) 172-177. https://doi.org/10.1016/j.aca.2009.11.053

[34] Y.F. Huang, Y. Li, Y. Jiang, X.P. Yan, Magnetic immobilization of amine-functionalized magnetic nanoparticles in a knotted reactor for on-line solid phase extraction coupled with ICP-MS for speciation of trace chromium, J. Anal. At. Spectrom. 25 (2010) 1467-1464. https://doi.org/10.1039/c004272b

[35] G. Cheng, M. He, H. Peng, B. Hu, Dithizone modified magnetic nanoparticles for fast and selective solid phase extraction of trace elements in environmental and biological samples prior to their determination by ICP-OES, Talanta 88 (2012) 507-515. https://doi.org/10.1016/j.talanta.2011.11.025

[36] L. Chen, B. Li, Application of magnetic molecularly imprinted polymers in analytical chemistry, Anal. Methods 4 (2012) 2613-2621. https://doi.org/10.1039/c2ay25354b

[37] Y. Wang, X. Luo, J. Tang, X. Hu, Q. Hu, C.Yang, Extraction and preconcentration of trace levels of cobalt using functionalized magnetic nanoparticles in a sequential injection lab on value system with detection by electro thermal atomic

absorption spectrometry, Anal. Chim. Acta 713 (2012) 92-96. https://doi.org/10.1016/j.aca.2011.11.022

[38] G.C. Silva, F.S. Almeida, A.M. Ferreira, V.S.T. Ciminelli, Preparation and application of a magnetic composite (Mn_3O_4/Fe_3O_4) for removal of As(III) from aqueous solutions, Mater. Res. 15 (2012) 403-408. https://doi.org/10.1590/S1516-14392012005000041

[39] Y. Ren, N. Li, J. Feng, T. Luan, Q. Wen, Z. Li, M. Zhang, Adsorption of Pb(II) and Cu(II) from aqueous solution on magnetic porous ferrospinel $MnFe_2O_4$, J. Colloid. Interface Sci. 367 (2012) 415-421. https://doi.org/10.1016/j.jcis.2011.10.022

[40] I. Larraza, M. Lopez-Gonzalez, T. Corrales, G. Marcelo, Hybrid-materials: Magnetite-polyethylenimine-montmorillonite, as magnetic adsorbents for Cr(VI) water treatment, J. Colloid Interface Sci. 385(2012) 24-33. https://doi.org/10.1016/j.jcis.2012.06.050

[41] A. Idris, N.S.M. Ismail, N. Hassan, E. Misran, A.F. Ngomsik, Synthesis of magnetic alginate beads based on maghemite nanoparticles for Pb(II) removal in aqueous solution, J. Ind. Eng. Chem. 18 (2012) 1582-1589. https://doi.org/10.1016/j.jiec.2012.02.018

[42] R.D. Ambashta, M. Sillanpaa, Water purification using magnetic assistance: A review, J. Hazard. Mater. 180 (2010) 38-49. https://doi.org/10.1016/j.jhazmat.2010.04.105

[43] Z. Yang, A.G. Langdon, The use of magnetic bed conditioning and pH control to enhance filtration by natural titanomagnetite, Water Res. 38 (2004) 3304-3312. https://doi.org/10.1016/j.watres.2004.04.017

[44] J.H.P. Watson, D.C. Ellwood, Biomagnetic separation and extraction process for heavy metals from solution, Miner. Eng. 7 (1994) 1017-1028. https://doi.org/10.1016/0892-6875(94)90030-2

[45] A.S. Bahaj, P.A.B. James, F.D. Moeschler, Efficiency enhancements through the use of magnetic field gradient in orientation magnetic separation for the removal of pollutants by magnetotactic bacteria, Sep. Sci. Technol. 37 (2002) 3661-3671. https://doi.org/10.1081/SS-120014825

[46] Z.G. Peng, K. Hidayat, M.S. Uddin, Extraction of 2-hydroxyphenol by surfactant coated nano-sized magnetic particles, Korean. J. Chem. Eng. 20 (2003) 896-901. https://doi.org/10.1007/BF02697295

[47] L.C.R. Machado, F.W.J. Lima, R. Paniago, J.D. Ardisson, J. Sapag, R.M. Lago, Polymer coated vermiculite-iron composites: Novel floatable magnetic adsorbents for water spilled contaminants. Appl. Clay. Sci. 31 (2006) 207-215. https://doi.org/10.1016/j.clay.2005.07.004

[48] Y. Hou, J. Yu, S. Gao, Solvothermal reduction synthesis and characterization of super para-magnetic nanoparticles, J. Mater. Chem. 13 (2003) 1983-1987. https://doi.org/10.1039/b305526d

[49] J.C. Lyon, D.A. Fleming, M.B. Stone, P. Schiffer, M.E. Williams, Synthesis of iron oxide ore/Au shell nanoparticles by iterative hydroxylamine seeding, Nano Letters 4 (2004) 719-723. https://doi.org/10.1021/nl035253f

[50] M. Monier, D.M. Ayad, Y. Wei, A.A. Sarhan, Preparation and characterization of magnetic chelating resin based on chitosan for adsorption of Cu(II), Co(II) and Ni(II) ions, Reactive Funct. Polym. 70 (2010) 257-266. https://doi.org/10.1016/j.reactfunctpolym.2010.01.002

[51] X. Luo, L. Zhang, High effective adsorption of organic dyes on magnetic cellulose beads entrapping activated carbon, J. Hazard. Mater. 171 (2009) 340-347. https://doi.org/10.1016/j.jhazmat.2009.06.009

[52] P.L. Lee, Y.C. Sun, Y.C. Ling, Magnetic nano-adsorbent integrated with lab on valve system for trace analysis of multiple heavy metals, J. Anal. At. Spectrom. 24 (2009) 320-327. https://doi.org/10.1039/b814164a

[53] H. Huang, Y. Ji, Z. Qiao, C. Zhao, J. He, H. Zhang, Preparation, characterization and application of magnetic Fe-SBA-15 mesoporous silica molecular sieves, J. Antom. Methods Manag. Chem. 2010 1-7.

[54] I. Safarik, K. Pospiskova, K. Horska, M.Safarikova, Potential of magnetically responsive (nano) biocomposites, Soft Matter. 8 (2012) 5407-5413. https://doi.org/10.1039/c2sm06861c

[55] S. Laurent, D. Forge, M. Port, A. Roch, C. Robie, L.V. Elst, R.N. Muller, Magnetic iron oxide nanoparticles: Synthesis, stabilization, vectorization, physicochemical characterizations and biological applications, Chem. Rev. 108 (2008) 2064-2110. https://doi.org/10.1021/cr068445e

[56] H. Wei, E. Wang, Fe_3O_4magnetic nanoparticles as peroxidase mimetics and their applications in H_2O_2 and glucose detection, Anal. Chem. 80 (2008) 2250-2254. https://doi.org/10.1021/ac702203f

[57] I. Safarik, K. Pospiskova, E. Baldikova, M. Safarikova, Magnetically responsive biological materials and their applications, Adv. Mater. Lett. 7 (2016) 254-261. https://doi.org/10.5185/amlett.2016.6176

[58] M. Safarikova, I. Safarik, The applications of magnetic techniques in biosciences, Mag. Elect. Sep. 10 (2000) 223-252. https://doi.org/10.1155/2001/57434

[59] M. Safarikova, Z. Maderova, I. Safarik, Ferrofluid modified Saccharomyces cerevisae cells for biocatalysis, Food Res. Int. 42 (2009) 521-524. https://doi.org/10.1016/j.foodres.2009.01.001

[60] I. Safarik, K. Horska, K. Pospiskova, M.Safarikova, One step preparation of magnetically responsive materials from non-magnetic powders, Pow. Technol. 229 (2012) 285-289. https://doi.org/10.1016/j.powtec.2012.06.006

[61] I. Safarik, K. Horska, K. Pospiskova, Z. Maderova, M.Safarikova, Microwave assisted synthesis of magnetically responsive composite materials, IEEE Trans. Magn. 49 (2013) 213-218. https://doi.org/10.1109/TMAG.2012.2221686

[62] I. Safarik, M. Safarikova, One step magnetic modification of non-magnetic solid materials, Int. J. Mat. Res. 105 (2014) 104-107. https://doi.org/10.3139/146.111009

[63] M.K. Beyer, M. Clausen-Schaumann, Mechanochemistry: The mechanical activation of covalent bonds, Chem. Rev. 105 (2005) 2921-2948. https://doi.org/10.1021/cr030697h

[64] I. Safarik, K. Horska, K. Pospiskova, J. Filip, M.Safarikova, Mechanochemical synthesis of magnetically responsive materials from non-magnetic precursors, Mat. Lett. 126 (2014) 202-206. https://doi.org/10.1016/j.matlet.2014.04.045

[65] I. Wagner-Dobler, H. von Canstein, Y. Li, K.N. Timmis, W.D. Deckwer, Removal of mercury from chemical wastewater by microorganism in technical scale, Environ. Sci. Technol. 34 (2000) 4628-4634. https://doi.org/10.1021/es0000652

[66] I. Safarik, K. Horska, M.Safarikova, Magnetically responsive biocomposites for inorganic and organic xenobiotics removal, in: P. Kotrba, M. Machova, T. Macek (Eds.), Microbial biosorption of metals, Springer Publications, Netherlands, 2011, pp. 301-320. https://doi.org/10.1007/978-94-007-0443-5_13

[67] C.B. Li, S. Hein, K.Wang, Biosorption of chitin and chitosan, Mater. Sci. Technol. 24 (2008) 1088-1099. https://doi.org/10.1179/174328408X341771

[68] S.X. Zhang, H.Y. Niu, Y.Q. Cai, Y.L. Shi, Barium alginate caged Fe_3O_4-C-18 magnetic nanoparticles for the preconcentration of polycyclic aromatic

hydrocarbons and phthalate esters from environmental water samples, Anal. Chim Acta 665 (2010) 167-175. https://doi.org/10.1016/j.aca.2010.03.026

[69] Z. Lei, S. Zhai, J. Lv, Y. Fan, Q. An, Z. Xiao, Sodium alginate based magnetic carbonaceous biosorbents for highly efficient Cr(VI) removal from water, RSC Adv. 5(2015) 77932-77941. https://doi.org/10.1039/C5RA13226F

[70] I. Safarik, M. Safarikova, F. Weyda, E. Mosiniewiez-Szablewska, A. Slawska-Waniewska, Ferrofluid-modified plant based materials as adsorbents for batch separation of selected biologically active compounds and xenobiotics, J. Magn. Magn. Mater. 293 (2005) 371-376. https://doi.org/10.1016/j.jmmm.2005.02.033

[71] I. Safarik, M. Safarikova, Magnetic fluid modified peanut husks as an adsorbent for organic dyes removal, Physics Procedia 9 (2010) 274-278. https://doi.org/10.1016/j.phpro.2010.11.061

[72] I. Safarik, K. Horska, B. Svobodova, M.Safarikova, Magnetically modified spent coffee grounds for dyes removal, Eur. Food Res. Technol. 234 (2012) 345-350. https://doi.org/10.1007/s00217-011-1641-3

[73] E. Baldikova, M. Safarikova, I. Safarik, Organic dyes removal using magnetically modified rye straw, J. Magn. Magn. Mater, 380 (2015) 181-185. https://doi.org/10.1016/j.jmmm.2014.09.003

[74] I. Safarik, Z. Maderova, K. Horska, E. Baldikova, K. Pospiskova, M.Safarikova, Spent Rooibos (Aspalathus linearis) tea biomass as an adsorbent for organic dye removal, Bioremed. J. 19 (2015) 183-187. https://doi.org/10.1080/10889868.2014.979279

[75] K. Pospiskova, I. Safarik, Magnetically modified spent grain as a low cost, biocompatible and smart carrier for enzyme immobilization, J. Sci. Food Agric. 93 (2013) 1598-1602. https://doi.org/10.1002/jsfa.5930

[76] N. Abdel-Raouf, A.A. Al-Homaidan, I.B. Ibraheem, Microalgae and wastewater treatment, Saudi, J. Biol. Sci. 19 (2012) 257-275. https://doi.org/10.1016/j.sjbs.2012.04.005

[77] G. Markou, E. Nerantzis, Microalgae for high value compounds and biofuels production: a review with focus on cultivation under stress condition, Biotechnol. Adv. 31 (2013) 1532-1542. https://doi.org/10.1016/j.biotechadv.2013.07.011

[78] T. Haigh-Florez, C. de la Hera, E. Costas, G. Orellana, Microalgae dual head biosensors for selective detection of herbicides with fiber active luminescent O_2

transduction, Biosens. Bioelectron. 54 (2014) 484-491. https://doi.org/10.1016/j.bios.2013.10.062

[79] M. R Granados, F.G. Acien, C. Gomez, J.M. Fernandez-Sevilla, E. Mollina-Grima, Evaluation of flocculants for the recovery of freshwater microalgae, Bioresour Technol. 118 (2012) 102-110. https://doi.org/10.1016/j.biortech.2012.05.018

[80] K.T. Semple, R.B. Cain, S. Schmidt, Biodegradation of aromatic compounds by microalgae, FEMS Microbiol. Lett. 170 (1999) 291-300. https://doi.org/10.1111/j.1574-6968.1999.tb13386.x

[81] P.Y. Toh, B.W. Ng, L.A. Ahmad, D.C. Chieh, J. Lim, The role of particle to shell interactions in dictating nanoparticle aided magnetophoretic separation of microalgal cells, Nanoscale 6 (2014) 12838-12848. https://doi.org/10.1039/C4NR03121K

[82] R. Venu, B. Lim, X.H. Hu, I. Jeong, T.S. Ramulu, C.G. Kim, On chip manipulation and trapping of microorganisms using a pattered magnetic pathway, Microfluidics Nanofluidics 14 (2013) 277-285. https://doi.org/10.1007/s10404-012-1046-z

[83] E. Eroglu, N.J. D'Alonzo, S.M. Smith, C.L. Raston, Vortex fluidic entrapment of functional microalgal cells in a magnetic polymer matrix, Nanoscale 5 (2013) 2627-2631. https://doi.org/10.1039/c3nr33813d

[84] I. Safarik, G. Prochazkova, K. Pospiskova, T. Branyik, Magnetically modified microalgae and their applications, Crit. Rev. Biotechnol. 36 (2016) 931-941.

[85] C.Y. Chen, K.L. Yeh, R. Aisyah, D.J. Lee, J.S. Chang, Cultivation, photobioreactor design and harvesting of microalgae for biodiesel production: A critical review, Bioresour. Technol. 102 (2011) 71-81. https://doi.org/10.1016/j.biortech.2010.06.159

[86] L. Christenson, R. Sims, Production and harvesting of microalgae for wastewater treatment, biofuels and bioproducts, Biotechnol. Adv. 29 (2011) 686-702. https://doi.org/10.1016/j.biotechadv.2011.05.015

[87] D. Vandamme, I. Foubert, I. Fraeye, B. Meesschaert, K. Muylaert, Flocculation of Chlorella vulgaris induced by high pH: Role of magnesium and calcium and practical implications, Bioresour. Technol. 105 (2012) 114-119. https://doi.org/10.1016/j.biortech.2011.11.105

[88] T.M. Mata, A.A. Martins, N.S. Caetano, Microalgae for biodiesel production and other applications: A review, Renew. Sust. Energy Rev. 14 (2010) 217-232. https://doi.org/10.1016/j.rser.2009.07.020

[89] S.K. Wang, F. Wang, Y.R. Hu, C.Z. Liu, Magnetic flocculant for high efficiency harvesting of microalgal cells, ACS Appl. Mater. Interfaces 6 (2014) 109-115. https://doi.org/10.1021/am404764n

[90] E.A. Naumenko, M.R. Dzamukova, R.F. Fakhrulin, Magnetically functionalized cells: Fabrication, characterization and biomedical applications, in: E. Katz (Eds.), Implantable bioelectronics, Wiley-VCH, Weinham, 2014, pp. 7-26.

[91] A.I. Zamaleeva, I.R. Sharipova, R.V. Shamagsumova, A.N. Ivanov, G.A. Evtugyn, D.G. Ishmuchametova, R.F. Fakhrullin, A whole cell amperometric herbicide biosensor based on magnetically functionalized microalgae and screen printed electrodes, Anal. Methods 3 (2011) 509-513. https://doi.org/10.1039/c0ay00627k

[92] T. Todd, Z.P. Zhen, W. Tang, H. Chen, G. Wang, Y.J. Chuang, K. Deaton, Z. Pan, J. Xie, Iron oxide nanoparticle encapsulated diatoms for magnetic delivery of small molecules to tumors, Nanoscale 6 (2014) 2073-2076. https://doi.org/10.1039/c3nr05623f

[93] J.M. Dias, M.C.M. Alvim-Ferraz, M. F. Almeida, J. Rivera-Utrilla, M. Sanchez-Polo, Waste materials for activated carbon preparation and its use in aqueous phase treatment: A review, J. Environ. Manage. 85 (2007) 833-846. https://doi.org/10.1016/j.jenvman.2007.07.031

[94] K. Kekalo, V. Agabekov, G. Zhavnerko, T. Shutava, V. Kutavichus, V. Kabanov, N. Goroshko, Magnetic nanocomposites for sorbents and glue layers, J. Magn. Magn. Mater. 311(2007) 63-67. https://doi.org/10.1016/j.jmmm.2006.10.1159

[95] I. Safarik, K. Horska, K. Pospiskova, M.Safarikova, Magnetically responsive activated carbons for bio and environmental applications, Int. Rev. Chem. Eng. 4 (2012) 346-351.

[96] Y.H. Ao, J.J. Xu, D.G. Fu, C.W. Yuan, Photocatalytic degradation of X-3B by titania coated magnetic activated carbon under UV and visible irradiation, J. Alloys. Comp. 471 (2009) 33-38. https://doi.org/10.1016/j.jallcom.2008.04.001

[97] L.C.A. Oliveira, R.V.R.A. Rios, J.D. Fabris, R.M. Lago, K. Sapag, Magnetic particle technology. A simple preparation of magnetic composites for the adsorption of water contaminats, J. Chem. Edu. 81 (2004) 248-250. https://doi.org/10.1021/ed081p248

[98] I. Safarik, M. Safarikova, Magnetically modified microbial cells: A new type of magnetic adsorbents, China Particuol. 5 (2007) 19-25. https://doi.org/10.1016/j.cpart.2006.12.003

[99] Y. Q. Ji, Y.T. Hu, Q. Tian, X.Z. Shao, J.Y. Li, M.Safarikova,I. Safarik, Biosorption of strontium ions by magnetically modified yeast cells, Sep. Sci. Technol. 45 (2010) 1499-1504. https://doi.org/10.1080/01496391003705664

[100] I. Safarik, Z. Sabatkova, M. Safarikova, Hydrogen peroxide removal with magnetically responsive Saccharomyces cerevisae cells, J. Agric. Food Chem. 56 (2008) 7925-7928. https://doi.org/10.1021/jf801354a

[101] E. Sykova, P. Jendelova, Magnetic resonance tracking of implanted adult and embryonic stem cells in injured brain and spinal cord, Ann. N. Y. Acad. Sci. 1049 (2005) 146-160. https://doi.org/10.1196/annals.1334.014

[102] I. Safarik, K. Horska, M. Safarikova, Magnetically modified spent grain for dye removal, J. Cereal Sci. 53 (2011) 78-80. https://doi.org/10.1016/j.jcs.2010.09.010

[103] M. Safarikova, I. Roy, M.N. Gupta,I. Safarik, Magnetic alginate microparticles for purification of α-amylases, J. Biotechnol. 105 (2003) 255-260. https://doi.org/10.1016/j.jbiotec.2003.07.002

[104] I. Safarik, Z. Sabatkova,M. Safarikova, Invert sugar formation with Saccharomyces cerevisaecells encapsulated in magnetically responsive alginate microparticles, J. Magn. Magn. Mater. 521 (2009) 1478-1481. https://doi.org/10.1016/j.jmmm.2009.02.056

[105] I. Safarik, L.F.T Rego, M. Borovska, E. Mosiniewicz-Szablewska, F. Weyda, M. Safarikova, New magnetically responsive yeast basedbiosorbent for the efficient removal of water soluble dyes, Enzyme Microbiol. Technol. 40 (2007) 1551-1556. https://doi.org/10.1016/j.enzmictec.2006.10.034

[106] R.F. Fakhrullin, L.V. Shlykova, A.L. Zamaleeva, D.K. Nurgaliev, Y.N. Osin, J. Garcia-Alonso, V.N. Paunov, Interfacing living unicellular algae cells with biocompatible polyelectrolyte stabilized magnetic nanoparticles, Macromol. Biosci. 10 (2010) 1257-1264. https://doi.org/10.1002/mabi.201000161

[107] I. Safarik, K. Pospiskova, Z. Maderova, E. Baldikova, K. Horska, M. Safarikova, Microwave synthesized magnetic chitosan microparticles for yeast cells immobilization, Yeast 32 (2015) 239-243.

[108] R.T Minullina, Y.N. Osin, D.G. Ishmuchametova, R.F. Fakhrullin, A direct technique for magnetic functionalization of living human cells, Langmuir 27 (2011) 7708-7713. https://doi.org/10.1021/la2006869

Chapter 2

Photocatalytic Decomposition of Organic Dyes

Zubera Naseem[1], Haq Nawaz Bhatti[1], Munawar Iqbal[2,*]

[1]Environmental and Material Chemistry Laboratory, Department of Chemistry, University of Agriculture, Faisalabad, Pakistan

[2]Department of Chemistry, The University of Lahore, Lahore, Pakistan

bosalvee@yahoo.com*

Abstract

Effluent detoxification and degradation kinetic of Acid Yellow 216 were investigated using Fenton and photo-Fenton processes. pH, contact time, Fe^{+2} concentration, H_2O_2 concentration, dye initial concentration, and temperature were optimized. The decolorization (87%) was considerably higher under experimental conditions of 20 min irradiation, pH 3, 50 mg/L initial dye concentration, 1.5 mM Fe^{+2}, 5 mM H_2O_2 and 40°C temperature. However, in the photo-Fenton process dye degradation was achieved up to 98%. Among, first-order, second-order and Behnajady–Modirshahla–Ghanbery (BMG) kinetic models, BMG kinetic model fitted well to experimental data. Under optimized conditions, up to 83 and 94% degradation of textile effluents were achieved using Fenton and Photo-Fenton process along with 56 and 76% COD reduction, respectively.

Keywords

Fenton oxidation, Acid Yellow, Effluents, Dyes, Photo-Fenton, Textile effluents, Kinetics model, Detoxification

Contents

1. Introduction

The effluents discharged from textiles and other related industries constitute hazardous organic and inorganic components, especially dyes, which are stable and cause environmental and aesthetic problems [1-7]. Besides, dyes and their intermediates are toxic and carcinogenic hereby posing serious threats to living organisms. The genotoxicity and mutagenicity cause genetic alteration that appeared in future generations [8-12]. Therefore, the remediation of dyes is important. As a result, different biological, physical and chemical processes have been applied for the treatment of waste and wastewaters containing dyes [13-18].

Among the treatment methods, an advanced oxidation process (AOP) has emerged as an efficient technology for the degradation and mineralization of organic pollutants [19, 20]. The AOP is a chemical oxidation method based on *in situ* generation of reactive radicals (i.e., hydroxyl radical, $^{\bullet}OH$) and refers to a set of different methods like Fenton and Fenton-like reactions (Fe^{2+}/H_2O_2, Fe^{3+}/H_2O_2), photo-Fenton and photo-Fenton-like reactions ($UV/H_2O_2/Fe^{2+}$, $UV/H_2O_2/Fe^{3+}$), UV/H_2O_2, UV/TiO_2, and ionizing radiation which have been successfully employed for the removal of pollutants [20]. Among all investigated AOPs, the Fenton reaction is considered to be an important source of $^{\bullet}OH$ radicals since the catalytic breakdown of hydrogen peroxide yielded $^{\bullet}OH$ radicals in the presence of Fe ions as a catalyst [9] (Eqs. 1-2). The strong oxidizing species like $^{\bullet}OH$

radicals are produced *in situ*, which break down the complex organic molecule into harmless substances such as CO_2, H_2O, and inorganic ions through chain reactions [11].

$$Fe^{2+} + H_2O_2 \rightarrow Fe^{3+} + {}^{\bullet}OH + {}^{-}OH \qquad \text{(Eq. 1)}$$

$$Fe^{3+} + H_2O + UV \rightarrow Fe^{2+} + {}^{\bullet}OH + H^{+} \qquad \text{(Eq. 2)}$$

Fenton and photo-Fenton processes degrade AY 216 dye in aqueous solution and textile effluent were investigated. The effects of the solution pH, concentration levels of H_2O_2 and dye concentration, UV exposure duration and Fe^{2+} dose were examined comprehensively in a series of decolorization of AY 216 by batch experiments. The experimental data were also subjected to the first-order, second-order and Behnajady–Modirshahla–Ghanbery kinetic models.

2. Material and methods

2.1 Chemical and reagents

The chemical and reagents were purchased from Sigma-Aldrich and dye the sample was procured from the local textile market, Faisalabad, Pakistan. Dye 1*M* solution (1L) was prepared and working solutions of different concentrations (mg/L) were prepared by dilution method. An aqueous solution of NaOH (0.1 M) and H_2SO_4 (0.1 M) were used for pH adjustment (HI-8014 HANNA). Fenton's reagent (Fe^{2+} + H_2O_2) (100 mL) was prepared by adding $FeSO_4.7H_2O$ (2 mM) and 35% H_2O_2 (5 mM). The quenching of the oxidation of reaction mixture was done by Na_2SO_3.

2.2 Treatment process

A water bath shaker (PA 9/250 U) with controlled temperature system was used in all the experiments for dye solution. The UV irradiation (UV- 4S/L lamp, 365 nm) was done at different shaking and intensities (261, 243, 212, 176 and 96 Counts/Sec) of UV radiation with the wavelength of 365 nm. For analysis, samples were drawn at regular intervals and dye residual concentration was monitored at λ_{max} = 440 nm (CE Cecil 7200, UK) and percentage degradation was estimated using the relation shown in Eq. 3. Where C_i and C_t are initial and final concentrations of dye.

$$Degradation\ (\%) = \left[\frac{C_i - C_t}{C_i}\right] x\ 100 \qquad \text{(Eq. 3)}$$

2.3 Analytical procedures

The kinetics of the decolorization process of dye was determined by 1st order, 2nd order and BMG models. The cytotoxicity, COD, and TOC were evaluated as reported previously [11, 20]

2.4 Statistical analysis

All experiments were performed in triplicate and data were averaged and reported as mean ±SD. The correlation coefficient R^2 values of kinetic models were determined by using statistical functions of Microsoft Excel, 2007 (Version Office XP, Microsoft Corporation, USA).

3. Results and discussion

3.1 Effect of pH

The key factor to examine the efficiency of Fenton's reagent is based on the pH of the solution. The generation of $^{\bullet}OH$ radicals and the concentration of Fe^{2+} are mainly controlled by the pH of the solution. The pH range 3-8 for decolorization of AY 216 was premeditated with the experimental conditions of 50 mg/L dye initial concentration, 2 mM Fe^{2+}, 5 mM H_2O_2, 30°C and duration 1 h at 100 rpm. The degradation pattern obtained is shown in Fig. 1(A) and UV-Vis spectrum can be seen in Fig. 2(A). Maximum dye removal of 84% was achieved at pH 3.0 and on further increasing the pH to 8.0, the percentage removal efficiency decreased to 69%. This decreasing trend might be due to the precipitation of Fe as hydroxide above pH 4. At lower pH (pH < 3), the dye removal was also less which can be attributed to the low activity of $^{\bullet}OH$ radical, probably due to the formation of hydronium ions (H_3O^+) by solvating the proton and hydrogen peroxide. The hydronium ion enhances the stability of hydrogen peroxide and presumably reduced the reactivity with ferrous ion significantly [21]. Therefore, the $^{\bullet}OH$ radical reactivity decreased at low pH values and resultantly, the removal efficiency decreased. These findings are in consonance with previous studies i.e., the methyl violet dye degradation by Fenton's reagent conducted at different pH values of 2-4. It was reported that the pH strongly affects the oxidation process of methyl violet and pH 3 was found favorable for dye degradation. The lower degradation of dyes at pH less than 3 is probably due to the formation of Fe complexes i.e., $[Fe(H_2O)_6]^{2+}$,$[Fe(H_2O)_6]^{3+}$ and $[Fe(H_2O)_5OH]^{2+}$. The reaction slowed down in the presence of hydrogen peroxide concentration greater/less than optimized level because of hydroperoxyl radical (Eq. 4) instead of $^{\bullet}OH$ radical [22].

$$Fe^{3+} + H_2O_2 \rightarrow Fe^{2+} + HO_2^{\bullet} + H^+ \qquad \text{(Eq. 4)}$$

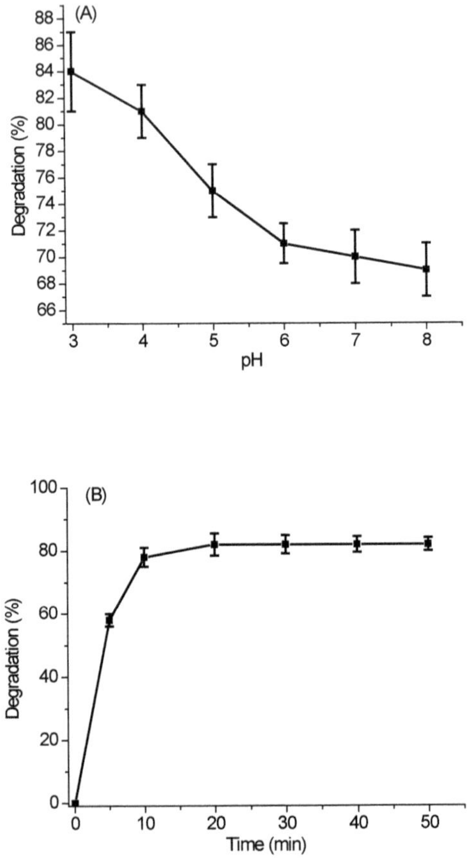

Figure 1 (A) Effect of pH on the decolorization of AY 216 (dye concentration—50 mg/L, time—60 min, Fe^{+2} concentration—2 mM, H_2O_2—5 mM, temperature—30°C, shaking speed—150 rpm) and (B) Effect of contact time (min.) on the decolorization of AY 216 (dye concentration—50 mg/L, pH—3, Fe^{+2} concentration—2 mM, H_2O_2—5 mM), temperature—30°C, shaking speed—150 rpm).

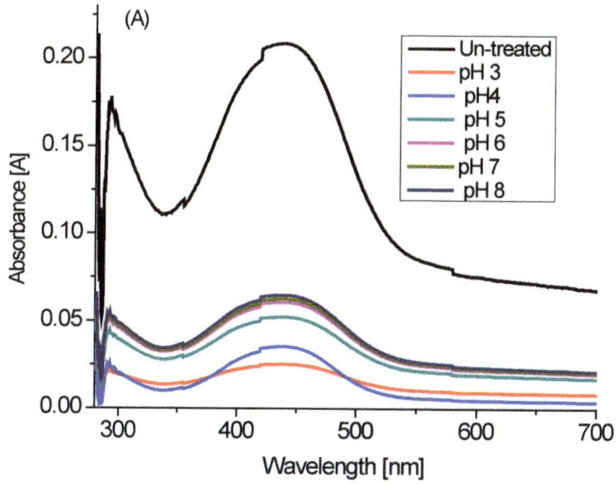
(A)
Un-treated
pH 3
pH4
pH 5
pH 6
pH 7
pH 8
Absorbance [A]
0.00
0.05
0.10
0.15
0.20
300
400
500
600
700
Wavelength [nm]

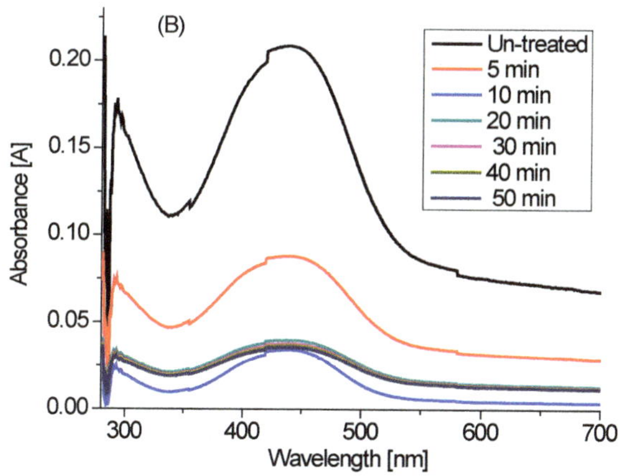
(B)
Un-treated
5 min
10 min
20 min
30 min
40 min
50 min
Absorbance [A]
0.00
0.05
0.10
0.15
0.20
300
400
500
600
700
Wavelength [nm]

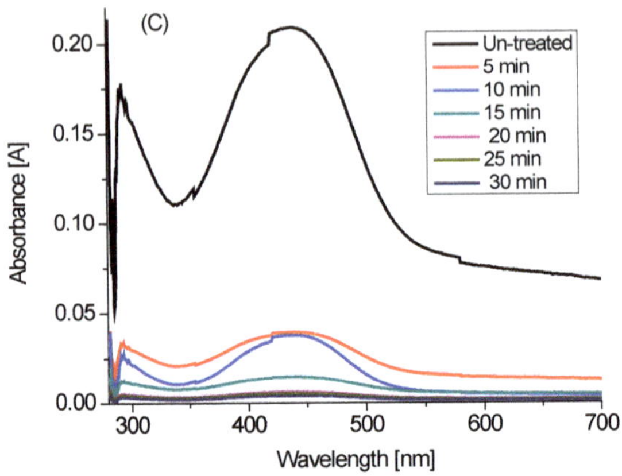

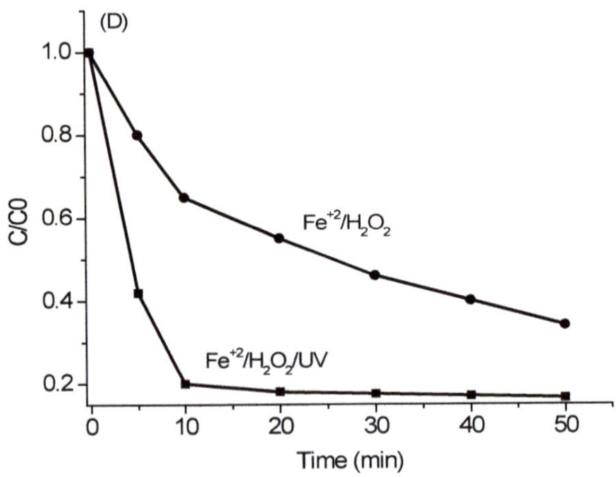

Figure 2 (A) UV–Vis absorption spectra of AY 216 untreated and treated at different initial pH values, (B) UV–Vis absorption spectra of AY 216 untreated and treated at different time intervals by Fe^{+2}/H_2O_2 (C) UV–Vis absorption spectra of AY 216 untreated and treated at different time intervals by $Fe^{+2}/H_2O_2/UV$ (D) Decrease of the AY 216 as a function of time by Fenton and photo-Fenton processes.

3.2 Effect of contact time

The equilibrium time is one of the important parameters for a cost-effective wastewater treatment. In decolorization efficiency, the utmost dye removal was achieved within 20 min of reaction. The reaction was preceded at optimal pH 3, while keeping Fe^{2+} concentration, H_2O_2 dose, initial dye concentration, temperature and shaking speed constant. The selected time ranges were 0, 5, 10, 20, 30, 40, 50 and 60 min (Figs. 1B) and Figs. (2B and 2C). Maximum percentage removal of 82% of AY 216 was achieved within 20 min followed by no change up to 60 min. It was assumed that the direct reduction of $^{\bullet}OH$ radicals by metal ions was responsible for reducing the activity. The decolorization of RY 86 within 20 min was 96% when the Fe^{2+} concentration was 5.0×10^{-4} mol/L. The initial concentration of H_2O_2 also affects the decolorization of RY 86. When the sample was irradiated by UV radiation, the decolorization rate of RY 86 increased with increasing the concentration of H_2O_2. It was noticed that RY 86 did not achieve 90% decolorization within 20 min when the concentration of H_2O_2 was less than $5.0{\times}10^{-2}$ mol/L [23].

3.3 Effect of Fe^{+2} dose

The removal of dye enhanced as the concentration of ferrous ion increased to a certain extent. The optimal dosage of ferrous ion is characteristic of Fenton's reagent for the dyes degradation. Decolorization of AY 216 varied on varying the concentration of Fe^{+2} ions. In the present study, the concentration range of 0.25, 0.5, 1, 1.5, 2.0, 2.5, 3.0, 3.5 and 4.0 mM Fe^{+2} ions was examined at pH 3, contact time 20 min and 5 mM H_2O_2 at 30°C temperature with shaking speed 150 rpm. As evident from Fig. 3(A), a 85% dye removal was possible with 1.5 mM Fe^{+2} ions. Starting from 72% degradation at a lower concentration of Fe^{+2} ions, it reached a maximum at 1.5 mM of Fe^{+2} ions and then, decreased to 71% at 4 mM concentration of Fe^{+2}. The UV-Vis absorption spectrum of dye degradation is shown in Fig. 4(A). It has been reported that Fe^{2+} is efficient to a certain limit and beyond optimum concentration, the efficiency may decrease significantly. Previously, a similar effect of Fe^{2+} was reported. Based on TOC reduction, it was observed that 0.27 mM concentration of Fe^{+2} was better and further increase in Fe^{+2} concentration decreased the removal efficiency. An excessive dosage of Fe^{+2} leads to other competitive and undesirable reactions (produced less active species) which ultimately lower the quantity of available radicals required to oxidize organics (Eqs. 5a and 5b) [24, 25].

$$Fe^{2+} + HO^{\bullet} \rightarrow Fe^{3+} + OH^{-} \qquad \text{(Eq. 5a)}$$

$$Fe^{2+} + HO_2^{\bullet} \rightarrow Fe^{3+} + HO_2^{-} \quad \text{(Eq. 5b)}$$

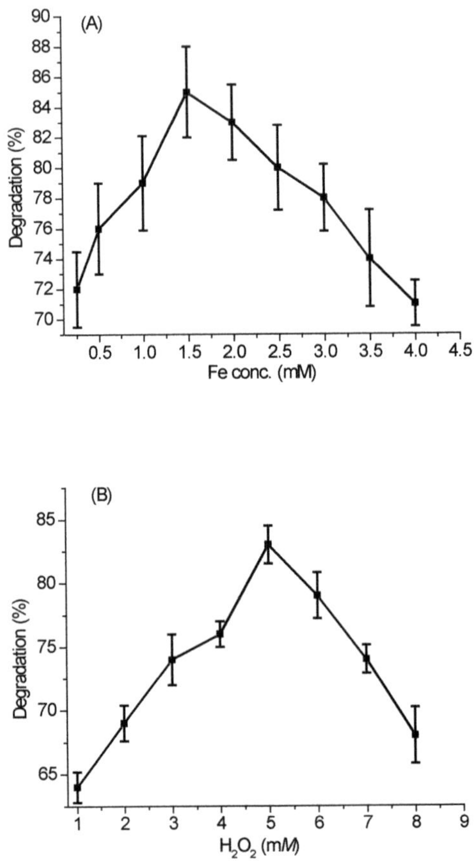

Figure 3 (A) Effect of Fe^{2+} on the decolorization of AY 216 (dye concentration—50 mg/L, time—20 min, pH—3, H_2O_2—5 mM, temperature—30°C, shaking speed—150 rpm) and (B) Effect of H_2O_2 on the decolorization of AY 216 (dye concentration—50 mg/L, time—20 min, pH—3, Fe^{+2} concentration—1.5 mM, temperature—30°C, shaking speed—150 rpm).

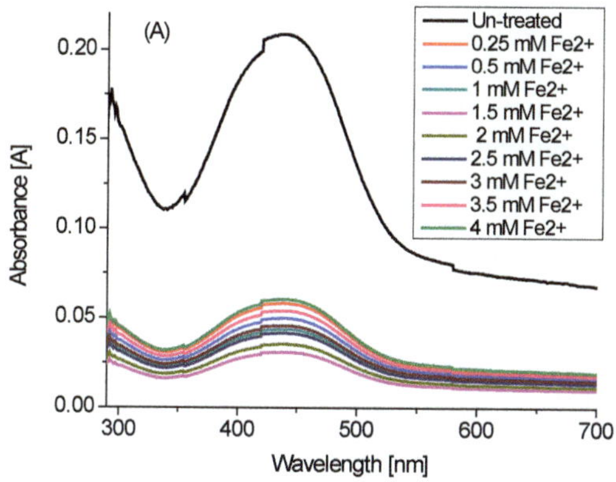

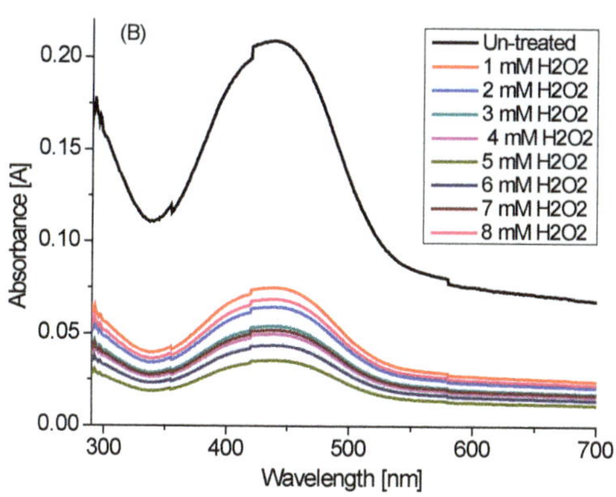

Figure 4 (A) UV–Vis absorption spectra of AY 216 untreated and treated at different doses of Fe^{+2}*, (B) UV–Vis absorption spectra of AY 216 untreated and treated in the presence of different* H_2O_2 *concentrations.*

3.4 Effect of H_2O_2 concentration

The decolorization AY 216 was also checked by varying the H_2O_2 dose. The H_2O_2 doses of 1, 2, 3, 4, 5, 6, 7 and 8 mM were investigated, whereas other parameters were kept constant (i.e., pH 3, contact time 20 min, 1.5 mM of Fe^{+2}, 30°C temperature and shaking speed 150 rpm) and the results thus, obtained are shown in Fig. 3(B). It was observed that the maximum dye removal (83%) was achieved using 5 mM H_2O_2 concentration. At lower concentration, the dye removal was 64%, which increased to 83% at 5 mM of H_2O_2 and then, again decreased to 68% at 8 mM concentration of H_2O_2. Previously, methyl orange also showed similar behavior regarding H_2O_2 concentration effect in the range of 4.5-48 mmol/L.

3.5 Effect of initial dye concentration

The percentage removal of AY 216 was also investigated at different dye initial concentrations. The initial dye concentration was investigated in the range of 10, 20, 30, 40, 50, 60, 70 and 100 mg/L, whereas pH 3, 1.5 mM Fe^{+2} concentration, 5 mM H_2O_2 concentration, 30°C and 150 rpm were kept constant. Degradation percentage and UV-Vis spectrum are shown in Fig. 5(A) and Fig. 6(A). It was observed that dye removal was 60% at a lower concentration that reached to 85% for 50 mg/L of dye initial concentration and finally, reduced to the 60% at 100 mg/L of dye initial concentration. It is reported that $^{\bullet}OH$ radicals generation, which are responsible for dye degradation remained invariable at lower concentration and at higher concentration, the dye molecules increased but the $^{\bullet}OH$ radical concentration remained constant and therefore, the removal efficiency decreased. In UV irradiation process at higher dye concentration photons, penetration into solution decreased facilitating the formation of $^{\bullet}OH$ radicals [26].

3.6 Effect of temperature

Temperature is a significant parameter to understand the nature of the reaction. The effect of temperature on the dye removal was examined at 30, 40, 50, 60 and 70°C, whereas pH 3, 1.5 mM concentration of Fe^{+2}, 5 mM H_2O_2 concentration, 50 mg/L initial dye concentration and 150 rpm remained constant. The rate of reaction with Fenton's reagent increased by increasing the temperature and reached to 87% at 40°C and then, decreased gradually to 64% at 70°C (Fig. 5(B) (degradation) and Fig. 6(B)) (UV-Vis spectra). It is known that the reaction between H_2O_2 and Fe^{2+} is affected by temperature and therefore, the kinetics of dye degradation was also influenced by the temperature variation. In the Fenton process, the optimum temperature was around 50°C for maximum removal of methyl violet dye [27].

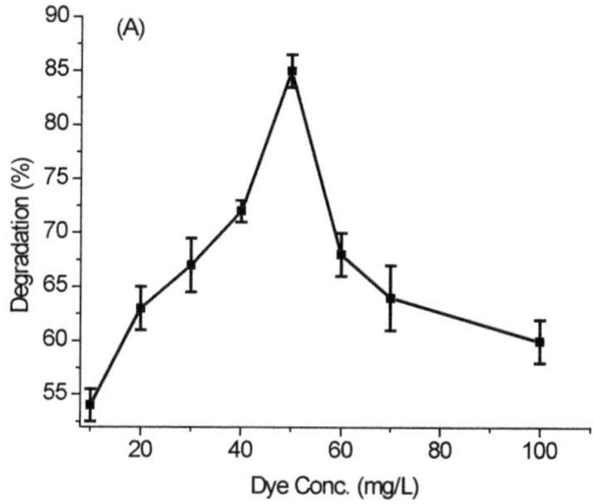

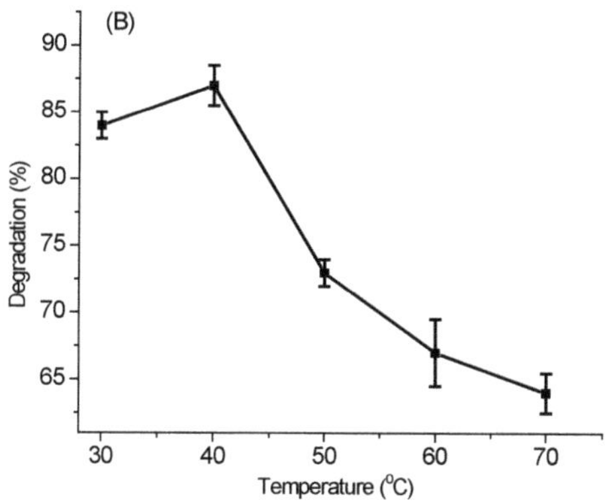

Figure 5 (A) Effect of initial dye concentration on the decolorization of AY 216 (reaction time—20 min, pH—3, Fe^{+2} concentration—1.5 mM, H_2O_2—5 mM, temperature—30°C, shaking speed—150 rpm) and (B) Effect of temperature (°C) on the decolorization of AY 216 (dye concentration—50 mg/L, time—20 min, pH—3, Fe^{+2} concentration—1.5 mM, H_2O_2—5 mM, shaking speed—150 rpm).

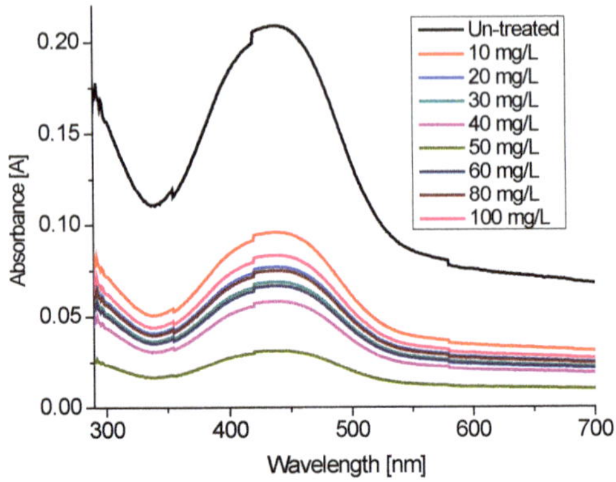

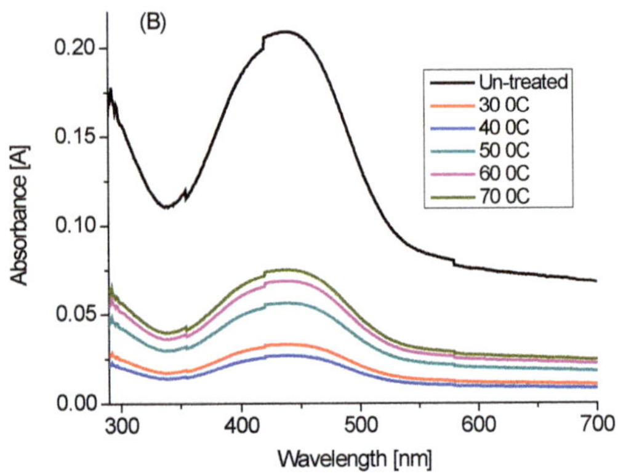

Figure 6 (A) UV–Vis absorption spectra of untreated and treated as a function of different initial dye concentration (B) UV–Vis absorption spectra of AY 216 untreated and treated at different temperatures.

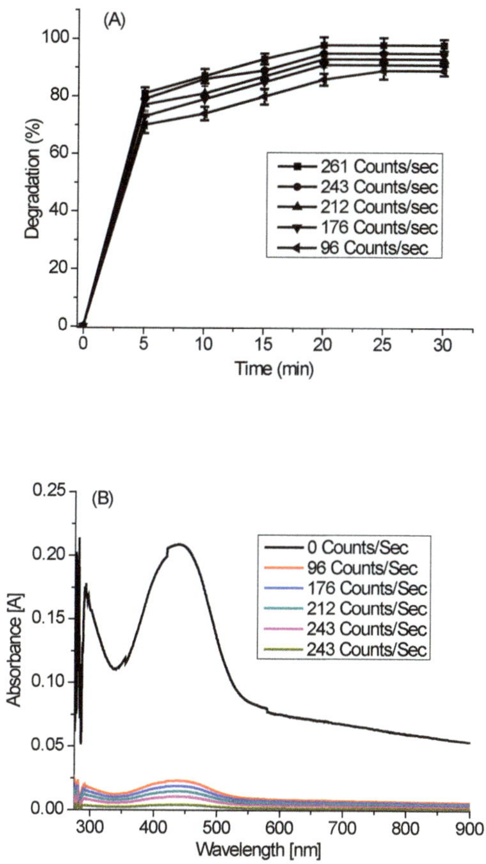

Figure 7 (A) Effect of UV intensity on the decolorization of AY 216 (dye concentration—50 mg/L, time—20 min, pH—3, Fe^{+2} concentration—1.5 mM, H_2O_2—5 mM, temperature—40°C, shaking speed—100 rpm), (B) UV–Vis absorption spectra of AY 216 untreated and treated at different UV intensities.

3.7 Effect of UV radiation

The dye degradation rate enhanced by UV radiation in combination with Fenton reagent and UV intensity also found to be effective in degrading dye since the dye degradation was different at different UV intensities i.e., 261, 243, 212, 176 and 96 counts/sec. The percent dye removal reached to 98% under the influence of UV radiation (Fig. 7(AB)) in combination with Fe^{+2}/H_2O_2. The degradation of AY 216 was considerably enhanced by increasing UV intensity. This was due to the fact that the $^{\bullet}OH$ radicals generation is enhanced under irradiation. It is also evident that Photo-Fenton efficiency was considerably higher than Fenton treatment. The increase in $^{\bullet}OH$ radical formation leads to enhancement in dye removal efficiency. The hydrolysis rate of H_2O_2 increased with UV irradiation and in Fenton reagent, the generation of hydroxyl radical was less under UV irradiation [28]. The Fenton oxidation process was studied by varying the intensity of UV radiation from 8.6 to 45.3 W/m^2. By increasing the UV intensity from 8.6 to 45.3 W/m^2, within 2 min the dye removal rate increased from 57.3 to 96%, which indicates that the dye degradation can be enhanced significantly by using UV radiation in a combination of Fenton process (Fig. 7(A)).

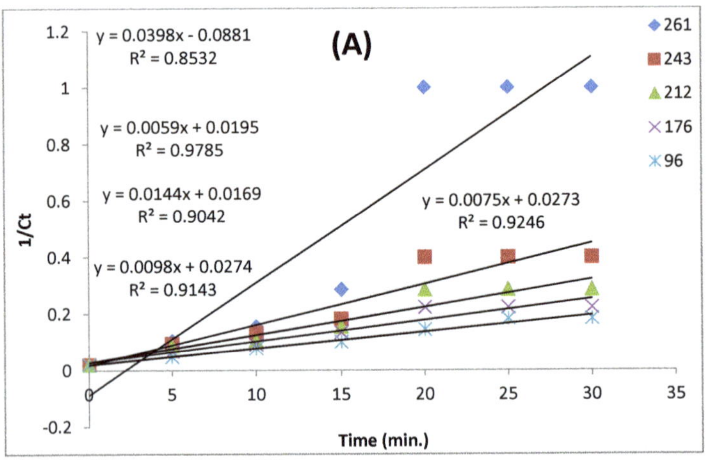

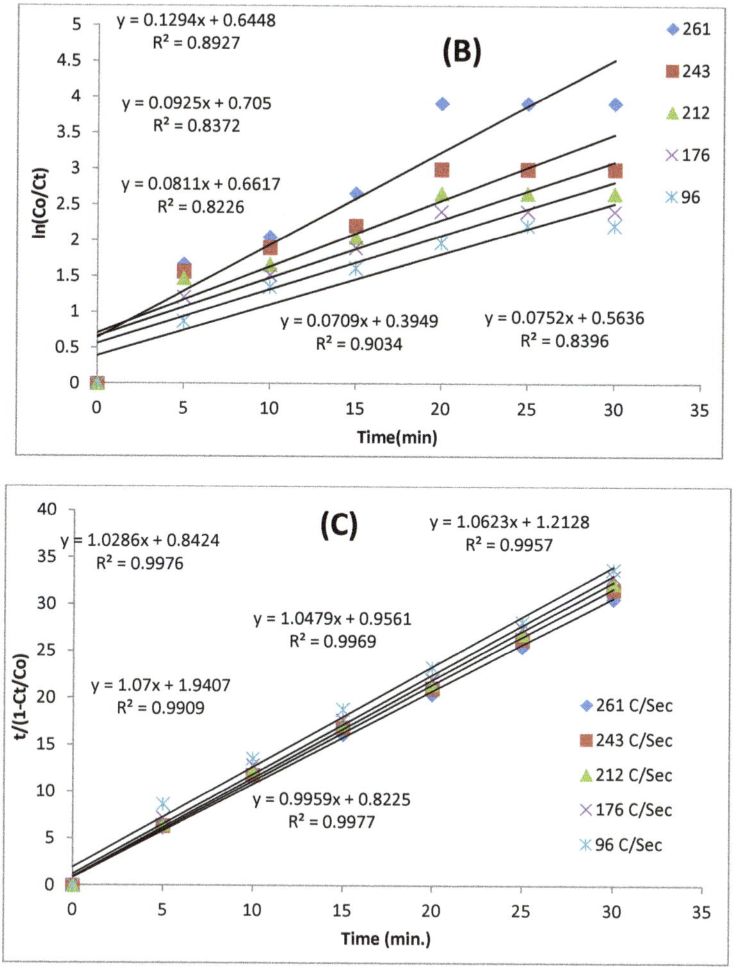

Figure 8 Kinetics models fit for AY 216 degradation (A) First-order, (B) second-order and (C) Behnajady–Modirshahla–Ghanbery.

4. Kinetic modeling

The data obtained from AY 216 oxidation degradation were tested using three kinetic models i.e., first order, second order and BMG models [29]. The linear forms of first-order, second-order and BMG models are given Eqs. 6-8. Where C_0 and C_t are representing initial at "0" and time t, respectively, m and b are the constants, which represent the relation between oxidation capacity and reaction kinetics. A Higher value of 1/m is indicative of faster removal rate of dye. The maximum removal of a fraction of dye (1/b) is equal to the maximum oxidation capacity of Fenton process at the end of reaction when the time approaches to infinity [30].

$$ln\frac{C_0}{C_t} = K_1 . t \qquad \text{(Eq. 6)}$$

$$ln\frac{1}{C_t} - \frac{1}{C_0} = K_2 . t \qquad \text{(Eq. 7)}$$

$$\frac{t}{1(C_t - C_0)} = m + b . t \qquad \text{(Eq. 8)}$$

The kinetic model's parameters for the decolorization of AY 216 at different reaction conditions were calculated by applying for the first order, second order and BMG models. The calculated results for AY 216 are represented in Table 1 and Figs. 8 (ABC). The experimental data of Fenton process fitted well to the BMG kinetic model with a higher correlation coefficient (R^2) of 0.9986 as compared to first and second order models. Previously, acid red 66 and direct blue 71 have also shown similar kinetic behavior [29]. In another study, the kinetics of decolorization of dyes was also studied and it had been revealed that the experimental data of dyes degradation were best explained by BMG kinetic model [31]. The kinetic model's parameters for the decolorization of AY 216 at different UV intensities i.e., 96, 176, 212, 243, and 261counts/sec were also calculated by employing these three models. The obtained results are shown in Table 2. The R^2 values for BMG model again are higher than those of the first-order and the second-order kinetic models. Therefore, BMG kinetic model is fitted well to the decolorization of AY 216 at different intensities of UV radiation.

Table 1 Comparison of first-order, second-order and Behnajady–Modirshahla–Ghanbery Kinetic models for AY 216 decolorization.

First order Kinetic		Second order Kinetic		BMG Kinetic model		
K_1 (1/min)	R^2	K_2 (1/M. min)	R^2	1/m	1/b	R^2
0.026	0.553	0.002	0.641	1.0123	0.837	0.999

Table 2 Comparison of first-order, second-order Behnajady–Modirshahla–Ghanbery (BMG) Kinetic models for Acid Yellow 216 decolorization at different intensities of UV radiation intensities.

Intensity (counts/sec.)	First order Kinetic		Second order Kinetic		BMG Kinetic model		
	K_1 (1/min)	R^2	K_2 (1/M. min)	R^2	1/m	1/b	R^2
96	0.071	0.903	0.006	0.979	0.515	0.935	0.991
176	0.075	0.839	0.008	0. 925	0.825	0.941	0.996
212	0.081	0.823	0.009	0.914	1.046	0.954	0.997
243	0.093	0.837	0.014	0.904	1.187	0.972	0.996
261	0.129	0.893	0.039	0.853	1.216	1.004	0.998

5. Degradation, water quality and cytotoxicity of effluents

The treatment conditions optimized for the degradation of AY 216 were employed for the treatment of real samples textile effluents. The treatment conditions i.e., pH 3, 1.5 mM Fe^{+2}, 5 m*M* H_2O_2, 40 °C temperature, reaction time 30 min and UV radiation 261 counts/sec were applied for the treatment of textile effluents and treatment efficiency was evaluated on the basis of degradation, water quality parameters and reduction in cytotoxicity. Among water quality parameters, TOC and COD were selected and cytotoxicity was evaluated using bioassays such as haemolytic and brine shrimp lethality tests. The COD determines the oxygen required for the chemical oxidation of organic matter and non-biodegradable matter present in wastewater. COD is also an important pollution indicator which reflects the chemical quality of effluent [11], whereas TOC is suitable for determining organic matter content and provides more accurate evaluation of the total organic compounds present in a water/wastewater, which is estimated by measuring the CO_2 generated when the organic compounds are oxidized [32]. Bioassays have been used to evaluate the toxicity of raw and treated industrial effluents i.e., brine shrimp and heamolytic tests [10, 11, 19, 20, 33]. The textile wastewater was treated by both Fenton and photo-Fenton processes which exhibited excellent efficiency for the degradation of dyes in textile effluents (Fig. 9AB). The degradation to the levels of 83% and 94% was achieved under optimized conditions by Fenton and photo-Fenton. Besides, the water quality improvement, the reduction in COD was observed up to 56.3 and 76% for Fe^{+2}/H_2O_2 and $Fe^{+2}/H_2O_2/UV$, respectively. Similarly, the TOC was reduced up to 52.2 and 67.8% by Fenton and photo-Fenton processes, respectively. In view of considerable degradation as well as COD and TOC reduction, the cytotoxicity was also reduced significantly as a result of Fenton and Photo-Fenton treatments. The cytotoxicity reductions according to heamolytic test were 42 and 57% for Fe^{+2}/H_2O_2 and $Fe^{+2}/H_2O_2/UV$, respectively, whereas up to 51 and 66% cytotoxicity reductions were observed in case of brine shrimp assay. Previously, the efficiencies of AOPs such as TiO_2/UV, O_3, electro-Fenton, wet-air oxidation, UV/electro-Fenton, Photo-Fenton, sunlight irradiation, TiO_2 based photocatalysis, H_2O_2/UV, O_3/UV, and TiO_2/H_2O_2/UV [34] were also evaluated for toxicity reduction. Present investigation revealed that both treatment processes are capable to degrade the mixed dyes in textile wastewaters and subsequently, the improvements in water quality and toxicity reduction. These findings are in line with previous studies that advanced oxidation processes are capable to detoxify the toxic compounds in simulated aqueous solution and in industrial wastewater as well i.e., the effect of gamma radiation treatment on detoxification of textile effluents was evaluated on the basis of *Allium cepa*, heamolytic, brine shrimp, and Ames tests. The microbial load was reduced to zero, and root length and root count were increased by

31.1 and 38.3%, respectively, at 5 kGy treatment. The reductions in cytotoxicity were recorded as 39. 6, 49.7, and 79.6% in the case of human and sheep red blood cells and brine shrimp, respectively. Before treatment, the samples were mildly mutagenic and after treatment mutagenicity was not detected [35]. The cytotoxicity and mutagenicity of textile effluents, treated by gamma radiation in combination with hydrogen peroxide were investigated. The *Allium cepa*, heamolytic, brine shrimp bioassays were used for cytotoxicity evaluation, whereas mutagenicity was tested using Ames tests. Before treatment, textile effluents showed a significant cytotoxicity and mutagenicity signs which were and reduced significantly after treatment. *A. cepa* showed a reduction of 50% in cytotoxicity, whereas the reductions of 56–59% in case of heamolytic test and up 93% in cytotoxicity were recorded by brine shrimp assays. The mutagenicity of gamma radiation in the presence of hydrogen peroxide-treated effluents reduced up to 59% and 54% in case of TA98 and TA100, respectively [11]. Gamma radiation/H_2O_2 treatment of nonylphenol polyethoxylates was performed and treatment effect was evaluated on the basis of degradation, water quality and toxicity reduction. In response of 90% degradation of nonylphenol ethoxylates, up to 68.7, 77, and 94% reductions in cytotoxicity were observed in case of *A. cepa*, heamolytic and shrimp assays, respectively. The mutagenicity reduced up to 62, 74, and 79% (TA98) and 68, 78, and 82% (TA100), respectively of NPEO-6, NPEO-9, and NPEO-30 irradiated to the absorbed dose of 15 kGy/4.6% H_2O_2. Photo-catalytic detoxification of industrial wastewater also showed a significant reduction in toxicity along with water quality improvement [19]. Authors subjected wastewater to ultraviolet radiation in the presence of hydrogen peroxide and titanium dioxide and cytotoxicity, mutagenicity, microbial load reduction were measured. *allium cepa*, heamolytic and Ames tests showed considerable reductions in cytotoxicity and mutagenicity because of photo-catalytic and UV/H_2O_2/TiO_2 treatments of soap and detergent wastewaters. It was revealed that the advanced oxidation treatment could be used for removal of cytotoxicity and mutagenicity and treated wastewater could possibly be used for irrigation purposes [33]. Therefore, in view of water quality improvement and toxicity reduction, the present investigation, as well as previous reports, support the application of Fe^{+2}/H_2O_2 and Fe^{+2}/H_2O_2/UV for the treatment of industrial wastewater.

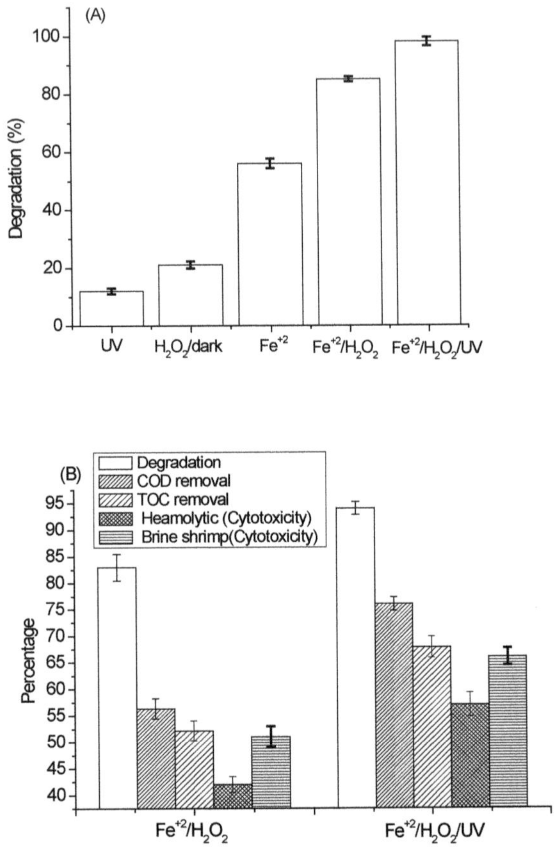

Figure 9 (A) Comparison of different treatment combinations for AY 216 degradation, (B) Degradation and effect of Fenton and Photo-Fenton treatment on TOC, COD and cytotoxicity reductions (pH 3, 1.5 mM Fe^{+2} concentration, 5 mM H_2O_2, 40°C temperature and UV exposure time reaction time 30 min).

6. Conclusions

The Fenton and Photo-Fenton processes were investigated for decolorization of AY 216 and textile effluents. Results revealed that Fenton process was efficient for the decolorization of AY 216 dye in wastewater. About 85% decolorization of dye was achieved at pH 3, with dye initial concentration (50 mg/L), in the presence of 1.5 m*M* Fe^{+2} and 5 m*M* H_2O_2, at 40 °C with 20 min of reaction time. The decolourization enhanced to 98% if reaction carried out under UV irradiation instead of sun light. Under optimized conditions, up to 83 and 94% degradation along with significant reductions in TOC, COD and cytotoxicity were also achieved in response of Fe^{+2}/H_2O_2/UV treatment of textile effluent. The Behnajady–Modirshahla–Ghanbery kinetic model fitted well to experimental data for AY 216 degradation. The results indicated that Photo-Fenton process can be used for efficient degradation of dyes in textile effluents.

References

[1] A. Banazadeh, H. Salimi, M. Khaleghi, S. Shafiei-Haghighi, Highly efficient degradation of hazardous dyes in aqueous phase by supported palladium nanocatalyst—A green approach, J. Environ. Chem. Eng. 4 (2015) 2178-2186. https://doi.org/10.1016/j.jece.2015.09.007

[2] N.K.R. Bogireddy, H.A.K. Kumar, B.K. Mandal, Biofabricated silver nanoparticles as green catalyst in the degradation of different textile dyes, J. Environ. Chem. Eng. 4 (2016) 56-64. https://doi.org/10.1016/j.jece.2015.11.004

[3] R. Dhanabal, A. Chithambararaj, S. Velmathi, A.C. Bose, Visible light driven degradation of methylene blue dye using Ag_3PO_4, J. Environ. Chem. Eng. 3 (2015) 1872-1881. https://doi.org/10.1016/j.jece.2015.06.001

[4] M. Ge, Y. Chen, M. Liu, M. Li, Synthesis of magnetically separable $Ag_3PO_4/ZnFe_2O_4$ composite photocatalysts for dye degradation under visible LED light irradiation, J. Environ. Chem. Eng. 3 (2015) 2809-2815. https://doi.org/10.1016/j.jece.2015.10.011

[5] S. Mandal, S. Natarajan, Adsorption and catalytic degradation of organic dyes in water using $ZnO/ZnxFe_3{-}xO_4$ mixed oxides, J. Environ. Chem. Eng. 3 (2015) 1185-1193. https://doi.org/10.1016/j.jece.2015.04.021

[6] S.P. Patil, V.S. Shrivastava, G.H. Sonawane, S.H. Sonawane, Synthesis of novel Bi_2O_3–montmorillonite nanocomposite with enhanced photocatalytic performance in dye degradation, J. Environ. Chem. Eng. 3 (2015) 2597-2603. https://doi.org/10.1016/j.jece.2015.09.005

[7] J. Wang, H. Zhu, C. Hurren, J. Zhao, E. Pakdel, Z. Li, X. Wang, Degradation of organic dyes by P25-reduced graphene oxide: Influence of inorganic salts and surfactants, J. Environ. Chem. Eng. 3 (2015) 1437-1443. https://doi.org/10.1016/j.jece.2015.05.008

[8] W. Abbas, T.H. Bokhari, I.A. Bhatti, M. Iqbal, Degradation study of disperse red F3BS by gamma radiation/H_2O_2, Asian J. Chem. 27 (2015) 282-286. https://doi.org/10.14233/ajchem.2015.17773

[9] N. Bilal, S. Ali, M. Iqbal, Application of advanced oxidations processes for the treatments of textile effluents, Asian J. Chem. 26 (2014) 1882.

[10] M. Iqbal, Vicia faba bioassay for environmental toxicity monitoring: a review, Chemosphere 144 (2016) 785-802. https://doi.org/10.1016/j.chemosphere.2015.09.048

[11] M. Iqbal, J. Nisar, Cytotoxicity and mutagenicity evaluation of gamma radiation and hydrogen peroxide treated textile effluents using bioassays, J. Environ. Chem. Eng. 3 (2015) 1912-1917. https://doi.org/10.1016/j.jece.2015.06.011

[12] H. Shindy, Basics in colors, dyes and pigments chemistry: A review, Chem. Int. 2 (2016) 29-36.

[13] A. Kumar, M. Naushad, A. Rana, Inamuddin, Preeti, G. Sharma, A.A. Ghfar, F.J. Stadler, M.R. Khan, ZnSe-WO3 nano-hetero-assembly stacked on Gum ghatti for photo-degradative removal of Bisphenol A: Symbiose of adsorption and photocatalysis, Int. J. Biol. Macromol. 104 (2017) 1172–1184. https://doi.org/10.1016/j.ijbiomac.2017.06.116

[14] A. Babarinde, G.O. Onyiaocha, Equilibrium sorption of divalent metal ions onto groundnut (*Arachis hypogaea*) shell: kinetics, isotherm and thermodynamics, Chem. Int. 2 (2016) 37-46.

[15] M. Iqbal, R.A. Khera, Adsorption of copper and lead in single and binary metal system onto *Fumaria indica* biomass, Chem. Int. 1 (2015) 157b-163b.

[16] M.A. Jamal, M. Muneer, M. Iqbal, Photo-degradation of monoazo dye blue 13 using advanced oxidation process, Chem. Int. 1 (2015) 12-16.

[17] K. Qureshi, M.Z. Ahmad, I.A. Bhatti, M. Iqbal, A. Khan, Cytotoxicity reduction of wastewater treated by advanced oxidation process, Chem. Int. 1 (2015) 53-59.

[18] G. Sharma, S. Bhogal, M. Naushad, Inamuddin, A. Kumar, F.J. Stadler, Microwave assisted fabrication of La/Cu/Zr/carbon dots trimetallic nanocomposites with their adsorptional vs photocatalytic efficiency for

remediation of persistent organic pollutants, J. Photochem. Photobiol. A Chem. 347 (2017) 235–243. https://doi.org/10.1016/j.jphotochem.2017.07.001

[19] M. Iqbal, I.A. Bhatti, Re-utilization option of industrial wastewater treated by advanced oxidation process, Pak. J. Agric. Sci. 51 (2014) 1141-1147.

[20] M. Iqbal, I.A. Bhatti, Gamma radiation/H_2O_2 treatment of a nonylphenol ethoxylates: Degradation, cytotoxicity, and mutagenicity evaluation, J. Hazard. Mater. 299 (2015) 351-360. https://doi.org/10.1016/j.jhazmat.2015.06.045

[21] F.P. Van der Zee, G. Lettinga, J.A. Field, Azo dye decolourisation by anaerobic granular sludge, Chemosphere 44 (2001) 1169-1176. https://doi.org/10.1016/S0045-6535(00)00270-8

[22] M.-W. Chang, J.-M. Chern, Decolorization of peach red azo dye, HF6 by Fenton reaction: Initial rate analysis, J. Taiwan Instit. Chem. Eng. 41 (2010) 221-228. https://doi.org/10.1016/j.jtice.2009.08.009

[23] K. Kümmerer, The presence of pharmaceuticals in the environment due to human use–present knowledge and future challenges, J. Environ. Manage. 90 (2009) 2354-2366. https://doi.org/10.1016/j.jenvman.2009.01.023

[24] M.S. Lucas, J.A. Peres, Decolorization of the azo dye reactive black 5 by fenton and photo-fenton oxidation, Dye. Pigment. 71 (2006) 236-244. https://doi.org/10.1016/j.dyepig.2005.07.007

[25] N.N. Mahamuni, Y.G. Adewuyi, Advanced oxidation processes (AOPs) involving ultrasound for waste water treatment: a review with emphasis on cost estimation, Ultrason. Sonochem. 17 (2010) 990-1003. https://doi.org/10.1016/j.ultsonch.2009.09.005

[26] M. Muruganandham, M. Swaminathan, Advanced oxidative decolourisation of reactive yellow 14 azo dye by UV/TiO_2, UV/H_2O_2, UV/H_2O_2/Fe^{2+} processes—a comparative study, Separat. Purificat. Technol. 48 (2006) 297-303. https://doi.org/10.1016/j.seppur.2005.07.036

[27] M. Pera-Titus, V. García-Molina, M.A. Baños, J. Giménez, S. Esplugas, Degradation of chlorophenols by means of advanced oxidation processes: a general review, Appl. Catal. B: Environ. 47 (2004) 219-256. https://doi.org/10.1016/j.apcatb.2003.09.010

[28] M. Rauf, S.S. Ashraf, Radiation induced degradation of dyes—an overview, J. Hazard. Mater. 166 (2009) 6-16. https://doi.org/10.1016/j.jhazmat.2008.11.043

[29] M. Behnajady, N. Modirshahla, F. Ghanbary, A kinetic model for the decolorization of CI acid yellow 23 by fenton process, J. Hazard. Mater. 148 (2007) 98-102. https://doi.org/10.1016/j.jhazmat.2007.02.003

[30] M. Behnajady, N. Modirshahla, N. Daneshvar, M. Rabbani, Photocatalytic degradation of an azo dye in a tubular continuous-flow photoreactor with immobilized TiO_2 on glass plates, Chem. Eng. J. 127 (2007) 167-176. https://doi.org/10.1016/j.cej.2006.09.013

[31] J.-H. Sun, S.-P. Sun, M.-H. Fan, H.-Q. Guo, L.-P. Qiao, R.-X. Sun, A kinetic study on the degradation of p-nitroaniline by Fenton oxidation process, J. Hazard. Mater. 148 (2007) 172-177. https://doi.org/10.1016/j.jhazmat.2007.02.022

[32] C.F. Bustillo-Lecompte, M. Mehrvar, E. Quiñones-Bolaños, Cost-effectiveness analysis of TOC removal from slaughterhouse wastewater using combined anaerobic–aerobic and UV/H_2O_2 processes, J. Environ. Manage. 134 (2014) 145-152. https://doi.org/10.1016/j.jenvman.2013.12.035

[33] M. Iqbal, I.A. Bhatti, M. Zia-ur-Rehman, H.N. Bhatti, M. Shahid, Efficiency of advanced oxidation processes for detoxification of industrial effluents, Asian J. Chem. 26 (2014) 4291-4296. https://doi.org/10.14233/ajchem.2014.16959

[34] L. Rizzo, Bioassays as a tool for evaluating advanced oxidation processes in water and wastewater treatment, Water Res. 45 (2011) 4311-4340. https://doi.org/10.1016/j.watres.2011.05.035

[35] M. Iqbal, M. Abbas, M. Arshad, T. Hussain, A.U. Khan, N. Masood, M.A. Tahir, S.M. Hussain, T.H. Bokhari, R.A. Khera, Gamma I radiation treatment for reducing cytotoxicity and mutagenicity in industrial wastewater, Polish J. Environ. Stud. 24 (2015) 2745-2750. https://doi.org/10.15244/pjoes/59233

Chapter 3

Effective Degradation of Methylene Blue using ZnO:Fe:Ni Nanocomposites

Pooja Dhiman[1]*, Gagan Kumar[1], Khalid Mujasam Batoo[2], Amit Kumar[3], Gaurav Sharma[3], M. Singh[4]

[1]Department of Physics, IEC University, Solan, India

[2]King Abdullah Institute for Nanotechnology, King Saud University, P.O. Box 2455, Riyadh, 11451, Saudi Arabia

[3]School of Chemistry,Shoolini University, Solan, India

[4]Department of Physics,Himachal Pradesh University, Shimla, India

Email: dhimanpooja85@gmail.com*

Abstract

This work supports the novelty of ZnO based semiconductor composites as a potential photocatalyst for the degradation of methylene blue (MB) dye. ZnO:Fe:Ni nanocomposites were prepared by solution combustion route. The prepared product was characterized for structural, morphological, optical and photocatalytic activities. The crystallinity and the structure of the samples were determined by X-Ray diffractometer and the relevant structural parameters were obtained using Rietveld fitting of the X-ray spectra. The X-ray photoelectron spectroscopy confirms the elemental presence and the valence state of the elements present in the sample. UV–visible absorption spectra have been utilized to calculate the band gap of the prepared samples. The prepared nanocomposite showed good photodegradation of MB dye. In general, many photocatalytic degradation studies have been reported using the man-made light source, however in the present chapter work on ZnO:Fe:Ni composite used in degradation of methylene blue in presence of direct sunlight is discussed.

Keywords

Rietveld Fitting, X-ray Photoelectron Spectroscopy, HRTEM, Photocatalysis

Contents

1. Introduction

Our environment is getting degraded by polluting wastewater emerging from the industries. The organic dyes as the major component of the wastewater of the industries like textile, food, coloring, and leather etc. are continuously causing their harmful effects on human health [1-3]. The dyes have been broadly classified according to their usage and their chemical structures. Dyes consist of the group of atoms or chromophores which are responsible for their characteristic colors. The wastewater coming from the industries generally consists of residual stuff of dyes and which accounts for their colored nature and toxicity [4]. Dyes are also known as natural dyes, food dyes, synthetic dyes and other dyes etc. Among all, azo dyes are widely used in the pharmaceutical, textile, cosmetic and other coloring industries. It is still an important and less resolved issue on how the impact of the pollutants can be reduced. Since India is a developing country the issue becomes more important as we have limited number of techniques available for this purpose. Thus, the development of a simple and economic method for the efficient degradation of dyes has gained greater significance. However, nowadays, semiconductor-based photocatalysis is being widely used for the degradation of these dyes [5]. Photocatalysis is better and more acceptable approach for this problem as this method can mineralize the dyes into minerals (H_2O and CO_2) without pollution or health hazards. A lot of semiconductors which includes TiO_2, Fe_2O_3, Cu Fe_2O_4, ZnO, CdS, ZnS, or their composites have been tried as a photocatalyst. These materials are relatively cheap, easily available, and effective and are environment-friendly as well [6-14]. According to the recent reports, ZnO has emerged an effective photocatalyst for the degradation or de-coloring of these organic dyes. ZnO has been utilized for the degradation of dyes like Methyl orange, methylene blue, malachite green, rhodamine B, congo red, and benzopurpurine 4B etc. In the present chapter,we are limiting ourselves to methylene blue

(MB) which are a heterocyclic aromatic chemical compound having IUPAC name 3, 7-bis(dimethylamino)-phenothiazin-5-ium chloride with the molecular formula $C_{16}H_{18}N_3SCl$. In the recent studies, ZnO nanoparticles have been employed for the degradation of methylene blue with very promising results [15-20]. These reports differ in the synthesis route opted with different photocatalysis results. Some composites of ZnO have also been reported by researchers which are also potential photocatalyst for degradation of MB [21-23]. We have synthesized ZnO: Fe: Ni nanocomposites by an easy, fast and economic synthesis route of solution combustion for the effective degradation of MB.

2. Materials and methods

For the preparation of ZnO:Fe:Ninanocomposites, zinc, iron, and Ni nitrates were taken as raw materials. Glycine was used as fuel for the combustion process of aqueous solutions of metal nitrates. Oxidizer to fuel (O/F) ratio was kept unity using total oxidizing and reducing valencies of the oxidizer and the fuel [24-25]. The mixture was heated at 373K with constant stirring on a magnetic hot plates. Finally obtained powder was kept in a muffle furnace at a temperature of 773 K for 4 h. The prepared samples were named as ZNF1, ZNF3, and ZNF5 respectively with 1% Ni and Fe in the range of 1-5%. The resulting powder was ground to fine powder and characterized with various instrumental techniques.

Crystal structure changes and the relevant structural parameters were obtained using a XPERTPRO diffractometer using Cu Kα radiation. Transmission electron microscope has been used to analyze the particle size and the particle shape was determined using high-resolution transmission electron microscope (HRTEM) (JEOL, USA). The elemental composition and the valence state information were collected using X-ray photoelectron spectrometer.

The study of the optical properties was performed on SYSTRONICS double beam spectrophotometer 2202 by using suspension of 2 mg of prepared sample in distilled water.

The UV–Vis studies were performed using double beam spectrophotometer. The photocatalytic efficiency of the samples was examined for degradation of methylene blue (MB) (Merck, 98.9%) in presence of natural sunlight in a slurry type batch reactor. All experiments were repeated three times and average values were reported.

2.1 X-Ray diffraction

The Rietveld fitted X-ray diffraction spectra for the prepared samples have been presented in Fig.1. It can be seen from the spectra that the diffraction peaks exhibit a pure phase wurtzite structure having space group $P_{63}mc$ (JCPDS#36-1451). No traces of impurity or secondary phase formation have been observed within the detection limit of the X-ray diffractometer. The possibility of cluster formation is rare because of the very small content of transition metals. Any cluster cannot be present in the sample because the annealing temperature of 500°C is quite high which automatically converts the content of the cluster into vapors. The impurity phases can be like $ZnFe_2O_4$, NiO, Fe_2O_3, Fe_3O_4 etc. The position of the peak (0 0 2) which is the characteristic peak of hexagonal wurtzite structure changes with an increase in Fe content. The intensity of characteristic peaks decreases with increase in Fe percentage. In general, only two possibilities are there (a) transition metal can substitute Zn sites and (b) they can replace the interstitials. Since energy barrier required for the substitution of interstitials is smaller than that of forming interstitial atoms, therefore the probability of substituting atoms is higher than that of forming interstitial atoms [26]. With the higher iron content, the change in lattice parameters is attributed to the difference in the ionic radii of iron, nickel and zinc ions.

For the wurtzite structure of ZnO, different parameters can be calculated using the following equations:

$$\frac{1}{{d_{hkl}}^2} = \frac{4}{3}\frac{h^2 + hk + k^2}{a^2} + \frac{l^2}{c^2} \tag{1}$$

and Zn-O bond length is expressed as:

$$L = \sqrt{a^2/3 + (\tfrac{1}{2} - u)^2 c^2} \tag{2}$$

and unit cell volume:

$$V = 0.866 \times a^2 \times c \tag{3}$$

The theoretically calculated and refined parameters are listed in Tables 1 and 2, respectively. The fitted parameters and the manually calculated parameters are almost same.

Table 1 Different structural parameters derived manually from XRD pattern using FWHM method.

Sample	Lattice parameter a=b (A^0)	Lattice parameter c (A^0)	c/a ratio	u parameter	Crystallite volume V($Å^3$)	Bond length (Zn-O)
ZNF1	3.2491	5.2062	1.603	0.3785	47.595	1.9796
ZNF3	3.2476	5.2029	1.602	0.3797	47.521	1.9767
ZNF5	3.2598	5.2209	1.601	0.3886	48.044	1.9698

Table 2: Structural parameters derived from Rietveld Fitting of XRD pattern.

Sample	Lattice parameter a=b (A^0)	Lattice parameter c (A^0)	c/a ratio	u parameter	Crystallite volume V($Å^3$)	Bond length (Zn-O)
ZNF1	3.2498	5.2053	1.6020	0.3799	47.610	1.9776
ZNF3	3.2497	5.2053	1.6019	0.3799	47.608	1.9776
ZNF5	3.2477	5.2019	1.6017	0.3799	47.518	1.9763

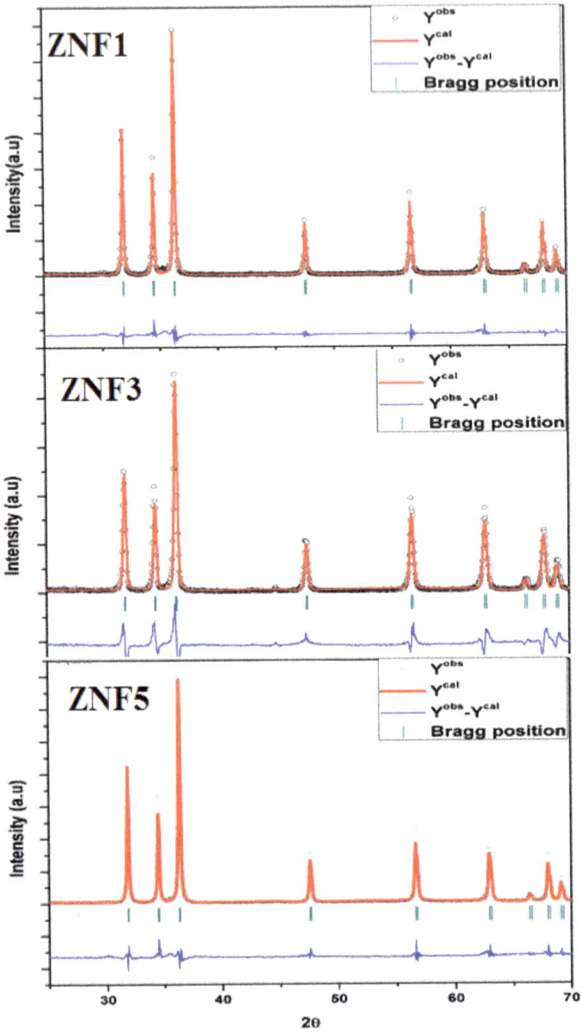

Figure 1 X-Ray Diffraction spectra for prepared samples.

2.2 X-Ray photoelectron spectroscopy

X-ray photoelectron spectroscopy has been used for the quantitative analysis and the valence state determination of the transition metal ions. XPS spectra for ZNF1 and ZNF3 nanocomposite are presented in Fig. 2. The obtained spectrum reveals the presence of peaks which are assigned to Fe, Ni, C, Zn, and O peaks. However, this is an integral spectrum obtained but for more qualitative and detailed information, core level XPS scan has been done. After de-convoluting the peaks according to the shape symmetry of the peaks using XPS peak fit software, peak positions of the different elements are presented in Table 3. Fig. 3 presents the core level de-convoluted scan for O 1s peak containing three deconvoluted peaks in which the low binding energy side peak remains at the same position while the other two get slightly shifted indicating the variation of the oxygen environment in both samples. It can be clearly seen from the spectra of ZNF3 sample that the area under the medium binding energy curve slightly decreases for the ZNF3 in comparison to ZNF1 sample. This implies that the oxygen deficiency slightly decreases in the ZNF3 but the loosely bound oxygen or adsorbed oxygen increases in comparison to ZNF1. Fig. 4(a-b) presents the fitted narrow scan, Gaussian fitted XPS spectra for Fe of ZNF1 and ZNF3 respectively. From the earlier reports, peak position of metallic Fe occurs at 719.9 eV (Fe $2p_{1/2}$) and 706.5 eV (Fe $2p_{3/2}$) and that of Fe^{2+} occur at 709.30 eV for Fe $2p_{3/2}$, and 722.30 eV for Fe $2p_{1/2}$ respectively, meanwhile the peaks of Fe $2p_{3/2}$ and Fe $2p_{1/2}$for Fe^{3+} in the Fe_2O_3 are located at 710.70 and 724.30 eV, respectively. The peaks of FeO occur at 722.3 eV and at 709.3 eV [27-28].

Fig. 5(a-b) presents the Ni 2p HR-XPS spectra of ZNF1 and ZNF3 respectively. To investigate the disturbing effects of Ni element, Gaussian fitting was performed on the Ni 2p spectrum. From the literature, the positions of Ni $2p_{3/2}$ peaks for metallic Ni occur at 852.7 eV, for NiO at 853.8 eV, and for Ni_2O_3 at 856.7 eV respectively [29-30]. The position of the satellite peak of 2p $_{3/2}$occurs at 861.3 eV [31-32]. The energy difference for ZNF1 between Ni peaks is 18.42 eV which is greater than the mentioned value in the literature (17.49 eV). These results confirm that Ni ions are only present in the 2+ state and all other types of impurity phases related to Ni are absent in the samples. Therefore, it can be concluded from XPS analysis that the Fe, Zn, and Ni are in valance states of 3+, 2+, and 2+ respectively. XPS analysis also gives information regarding defects like oxygen vacancies and zinc interstitials present in the samples.

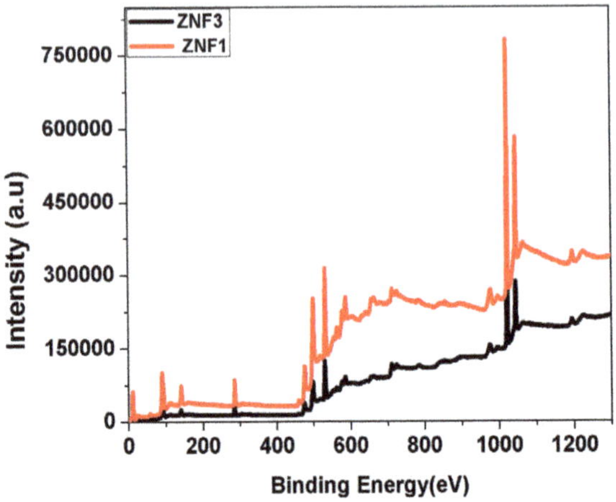

Figure 2 XPS survey spectra for ZNF1 and ZNF3 samples.

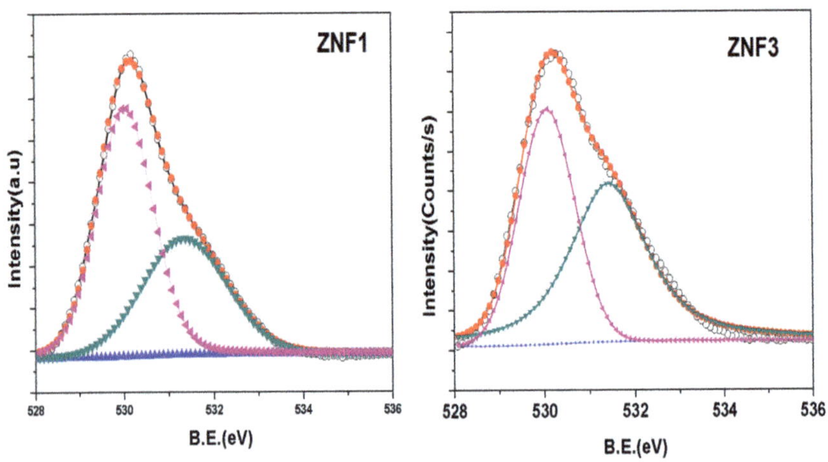

Figure 3 De-convoluted O 1s spectrum for ZNF1 and ZNF3 nano-composites.

Table 3 Various peak positions of de-convoluted Fe, O, and Ni XPS peaks.

O1s peak				
Sample	**Peak 1 position (eV)**	**Peak 2 position (eV)**	**Peak 3 position (eV)**	
ZNF1	530.15	531.64	532.62	
ZNF3	530.15	531.46	532.33	
Fe 2p peak				
Sample	**Peak 1 position(eV)**	**Peak 2 position (eV)**	**Satellites peaks (eV)**	
ZNF1	710.35	724.14	712.40	718.34
ZNF3	710.33	724.19	712.26	718.32
Ni 2p peak				
ZNF1	853.71	855.22	861.11	
ZNF3	854.51	855.42	861.36	

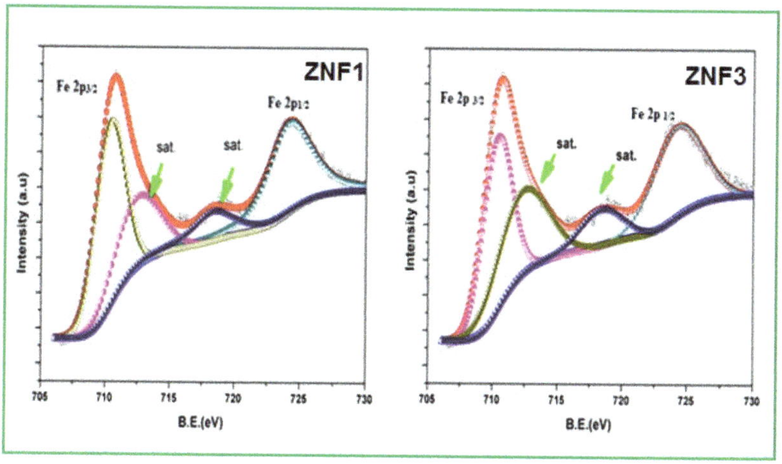

Figure 4 De-convoluted Fe 2p spectrum for ZNF1 and ZNF3 nanocomposites.

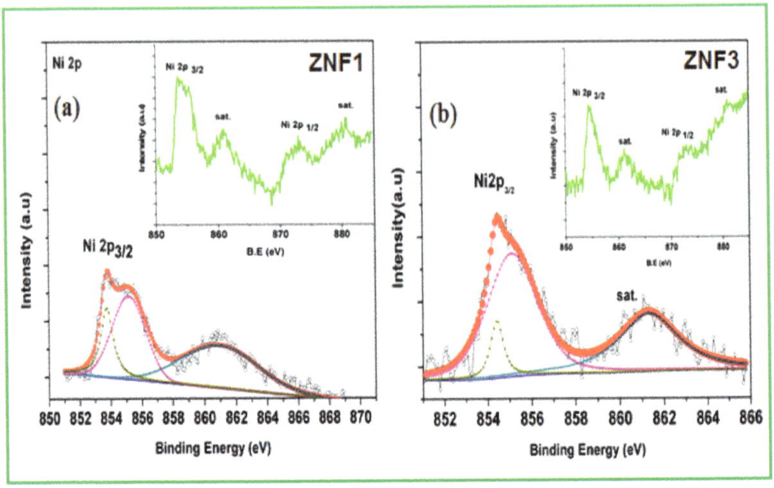

Figure 5 De-convoluted Ni 2p spectrum for ZNF1 and ZNF3 nanocomposites.

2.3 Transmission electron microscopy (TEM)

Fig. 6(a-c) represents the low-resolution TEM images for all prepared nanocomposite samples. The samples have been found to consist of roughly spherical particles with a wide range of particle size distribution. Particle ranges from 20 to 45 nm for all samples. A slight variation in the particle diameter can be observed from ZNF1 to ZNF5. However, a slight degradation in the crystallinity from ZNF1 to ZNF5 can also be seen from the high-resolution TEM presented in Fig. 7(a-c). The d spacings can be clearly seen from the inset image in the Fig. 7(a-c).

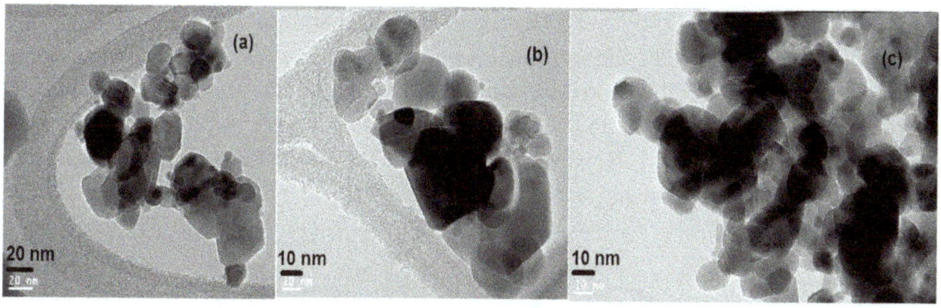

Figure 6(a-c) TEM images of prepared nanocomposites.

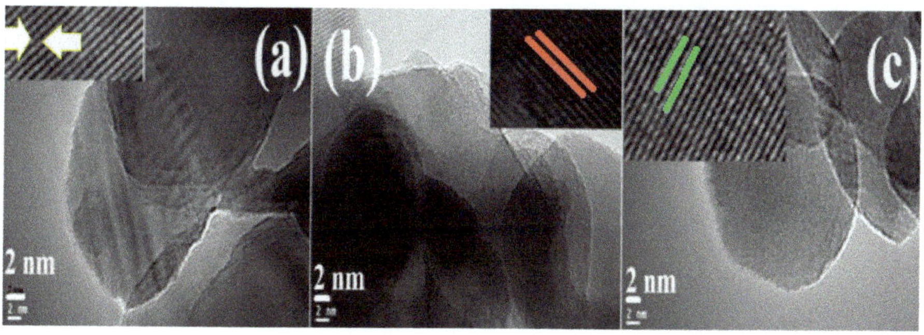

Figure 7(a-c) HRTEM images of prepared nanocomposites.

2.4 UV –Visible spectroscopy

The UV-Vis spectra of the samples ZNF1, ZNF3 and ZNF5 are presented in Fig. 8. It is observed from the absorption spectra that the absorption band edge slightly shifted to the higher wavelength side starting from ZNF1 to ZNF5 associating with the decrease in band gap [33]. This observed red shift in the absorption spectra is assigned to the formation of shallow levels inside the band gap due to the incorporation of transition metal in the structure. The increased absorption in the wavelength range from 400–600 nm, is attributed to the d-d transition [34]. The band gap value obtained by using Tauc's plot was found to decrease with increase in iron content. The band gap value varies from

3.14 eV (ZNF1) to 3.08 (ZNF5) eV. The decrease in the band gap is attributed to the sp-d exchange interactions. Many authors have reported increase or decrease in band gap on doping Fe or Ni in ZnO. In the recent work, the band gap value for $Fe_{0.01}$ $Ni_{0.01}Zn_{0.98}$ O nanoparticles (~23nm) was found to be 3.17eV, which is very near to our calculated value for the band gap [35]. However the effect of increased Fe content in ZnO:Fe:Ni nanocomposites on the band gap is not reported anywhere.

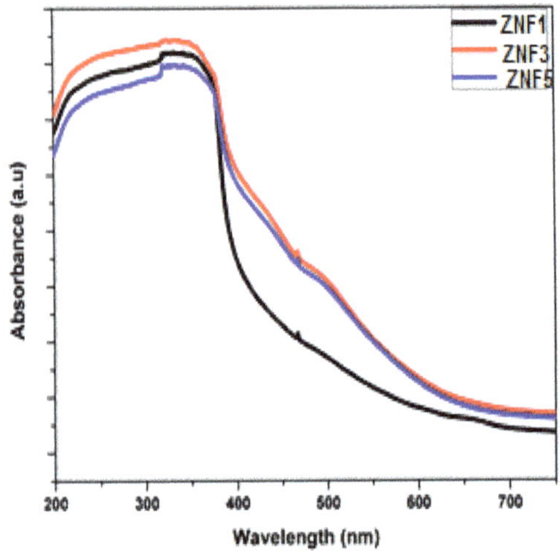

Figure 8 UV-vis. Spectra for ZNF1, ZNF3, and ZNF5 samples at room temperature.

2.5 Photocatalytic activity

The photocatalytic activity of all prepared samples was tested under natural sunlight illumination using methylene blue as a target pollutant [36-37]. The optical absorption spectroscopy is a great tool for monitoring the degradation of dye. For photocatalytic efficiency and many other practical applications, it is required that any photocatalyst should absorb UV as well as visible light as visible radiation accounts for 45% of energy in the solar radiation while UV light less than 10% [38]. It is therefore of great significance to construct a material that can absorb both UV and visible light to increase the range of a photocatalyst. This can be done by creating surface defects in a material which ultimately results into increase in the surface area and thereby increase in photo-degradation efficiency. There have been several reports of increased photocatalytic activity of different morphologies of nanoparticles [39-40]. The defects present in any system can cause either improvement or worsen of the photocatalytic activity, depending on the type, nature,and location of native defects. In the mechanism of absorption of radiation, generation of electron-hole pairs takes place. These electron-hole pairs on reaction with water and dissolved oxygen create hydroxyl free radicals and superoxide ion radical which attack the conjugation in the dyes and thus degrades them. The photocatalytic efficiency of the samples was analyzed for degradation of methylene blue as a pollutant in presence of natural sunlight in a slurry type batch reactor. Photocatalytic efficiency of the samples has been checked in the slurry type batch reactor (Fig. 9). In this typical experiment, the slurry mixture consisting of dye and catalyst suspension was stirred magnetically.

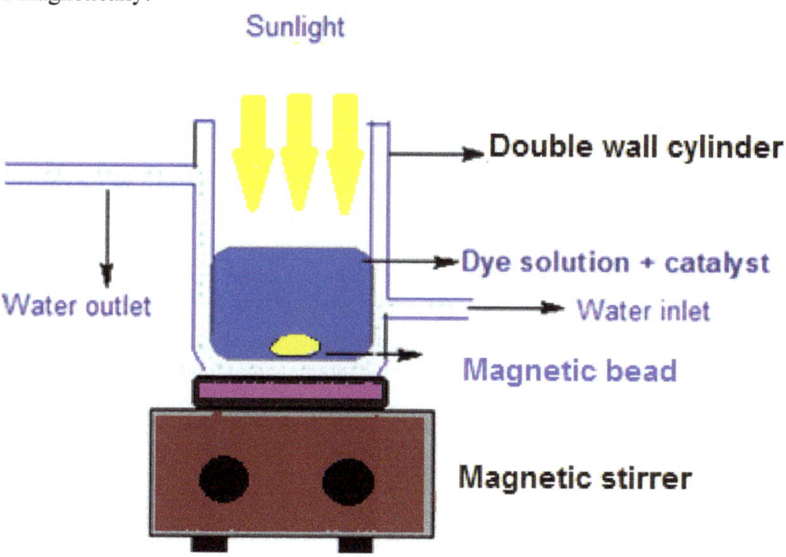

Figure 9 Arrangement for studying the degradation of Dye on illumination to natural sunlight.

The observed absorption spectrum of MB in the presence of Fe, Ni@ ZnO nanocomposites is presented in Fig. 10. The sample ZNF1 has higher degradation capacity in comparison to ZNF3 and ZNF5. The percentage of photodegradation of MB was calculated from the following equation:

$$\%\,Degradation = [1 - \frac{C_t}{C_0}] \times 100 \qquad (4)$$

where C_t is the absorbance after time t and C_0 is the initial concentration of dye. The calculated value of percentage degradation of dye is plotted with experiment length and is shown in Fig. 11(a-c). After 5 hours 81% degradation was achieved for the ZNF1 sample. While for the ZNF3 sample as a photocatalyst the degradation was 89%. However, ZNF5 decolorizes 79% of the dye during 5 hours. This implies that ZNF3 is a better photocatalyst than ZNF1 and ZNF5 samples. Earlier, we have observed that the band gap decreases from ZNF1 to ZNF5. This variation in band gap explains the photocatalytic behavior of ZNF1 and ZNF3. But for ZNF5 sample, it doesn't fit better. This can be explained in view of the fact, that the increase in iron in the form Fe^{3+}ions indicates the presence of more cation vacancies. The concentration of Ni^{2+} is fixed and it provides an almost equal number of charge carriers. The charge carries are sufficient for the recombination in the case of ZNF1 and ZNF3, but for ZNF5 it is no more possible which leads to decrease in photocatalytic efficiency. The bar depiction of dye degradation with time is presented in Fig. 11(d). The mechanism involved in the dye degradation is probably the interfacial electron transfer which takes place between donor states of oxygen vacancies and the Zn interstitials and dye. Since methylene blue (MB) is a cationic dye, it takes an electron from excited donor states and then decomposes into leucomethylene which is non-toxic, harmless and environment-friendly in nature.

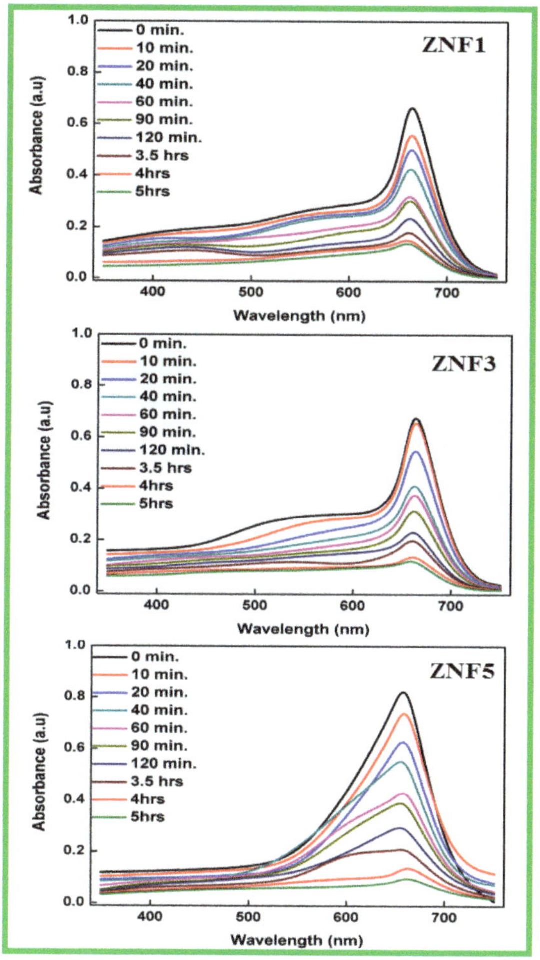

Figure 10 Absorption spectra of MB in presence of ZNF1, ZNF3, and ZNF5 samples at various times (initial concentration of MB = 10^{-5}M, pH=7, temperature = 30±0.5°C).

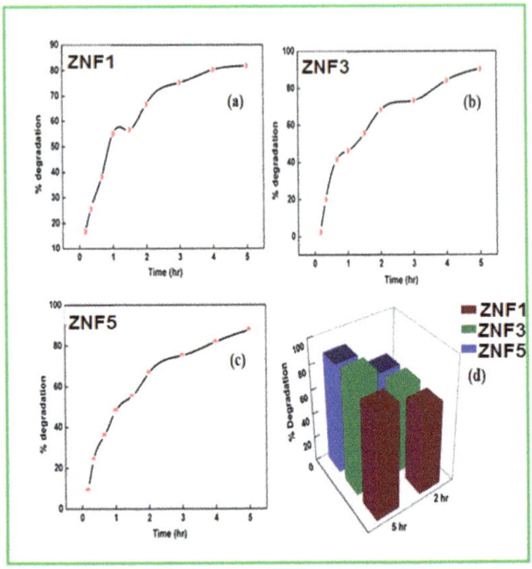

Figure 11 The extent of decomposition of MB with respect to time in presence of ZNF1, ZNF3 and ZNF5 samples (initial concentration of MB = $10^{-5}M$, pH=7, temperature = $30\pm0.5^{o}C$).

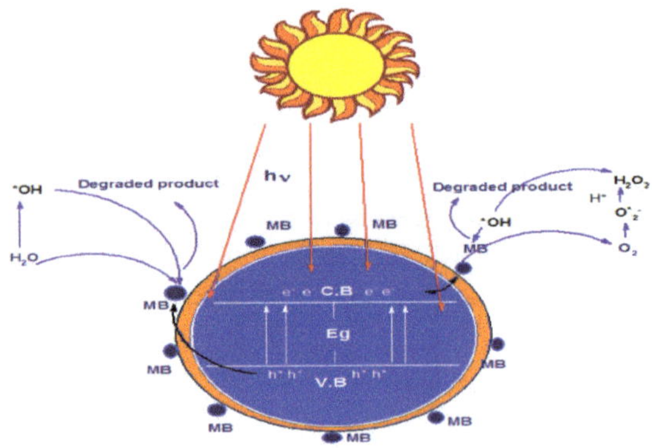

Figure 12 The depiction of dye removal by zinc oxide nanoparticles.

4. Conclusions

The solution combustion synthesized ZnO:Fe:Ninanocomposites were found to be highly effective photocatalyst for the decomposition of methylene blue dye via interfacial electron transfer mechanism. X-ray diffraction and transmission electron micrographs showed the crystalline nature of the samples without any impurity phase formation. X-ray photoelectron microscopy confirms the elemental presence, valence state and the composition of the samples. UV-Visible spectrophotometry studies demonstrate the decrease in band gap with an increase in Fe content.

Acknowledgement

Dr. Pooja Dhiman gratefully acknowledges USIC Facility, Himachal Pradesh University, Shimla for XRD measurements.

References

[1] H.Y. Zhu, R. Jiang, Y.Q. Fu, Y.J. Guan , J. Yao, L. Xiao, G.M. Zeng, Effective photocatalytic decolorization of methyl orange utilizing TiO_2/ZnO/ chitosan nanocomposite films under simulated solar irradiation. Desalination 286(2012)41–48. https://doi.org/10.1016/j.desal.2011.10.036

[2] X. Chen, Z. Wu, D. Liu, Z. Gao, Preparation of ZnO photocatalyst for the efficient and rapid photocatalytic degradation of azo dyes, Nanoscale Res. Lett.12 (2017)143. https://doi.org/10.1186/s11671-017-1904-4

[3] R. Kant, Textile dyeing industry an environmental hazard, Nat. Sci. 4 (2012) 22–26. https://doi.org/10.4236/ns.2012.41004

[4] U.G. Akpan, B.H. Hameed, Parameters affecting the photocatalytic degradation of dyes using TiO_2-based photocatalysts: Areview, J. Hazard. Mater. 170 (2009) 520–529. https://doi.org/10.1016/j.jhazmat.2009.05.039

[5] K. Kaviyarasu, L. Kotsedi, A. Simo, X. Fuku, G. T. Mola, J. Kennedy, M. Maaza, Photocatalytic activity of ZrO_2 doped lead dioxide nanocomposites: Investigation of structural and optical microscopy of RhB organic dye, Appl. Surf.Sci.,421 (2017) 234-239. https://doi.org/10.1016/j.apsusc.2016.11.149

[6] P. Dhiman, J. Chand, A. Kumar, R.K. Kotnala, K.M. Batoo, M. Singh, Synthesis and characterization of novel Fe@ZnO nanosystem, J. Alloys Compd. 578 (2013) 235–241. https://doi.org/10.1016/j.jallcom.2013.05.015

[7] S. Anandan, T. Selvamani, G.G. Prasad, A.M. Asiri, J.J. Wu, Magnetic and catalytic properties of inverse spinel $CuFe_2O_4$nanoparticles, J. Magn. Magn. Mater. 432 (2017) 437–443. https://doi.org/10.1016/j.jmmm.2017.02.026

[8] E.V. Casbeer, K. Sharma, X. Li, Synthesis and photocatalytic activity of ferrites under visible light: A review, Sep. Purif. Technol. 87 (2012) 1–14. https://doi.org/10.1016/j.seppur.2011.11.034

[9] K. Intarasuwan, P. Amornpitoksuk, S. Suwanboon, P. Graidist, Photocatalytic dye degradation by ZnO nanoparticles prepared from $X_2C_2O_4$ (X = H, Na and NH_4) and the cytotoxicity of the treated dye solutions, Sep. Purif. Technol. 177 (2017) 304–312. https://doi.org/10.1016/j.seppur.2016.12.040

[10] C. Aggelopoulos, M. Dimitropoulos, A. Govatsi, L. Sygellou, D. C. Tsakirogloua, N.S. Yannopoulos, Influence of the surface-to-bulk defects ratio of ZnO and TiO_2 on their UV-mediated photocatalytic activity, Appl. Catal. B: Environ. 205 (2017) 292–301. https://doi.org/10.1016/j.apcatb.2016.12.023

[11] M. Law, L.E. Greene, J.C. Johnson, R. Saykally, P. Yang, Nanowire dye sensitized solar cell, Nat. Mater. 4 (2005) 455–459. https://doi.org/10.1038/nmat1387

[12] Y. Liu, L.Yu, Y. Hu, C.F. Guo, F.M. Zhang, X.W. Lou, A magnetically separable photocatalyst based on nest-like γ-Fe_2O_3/ZnO double-shelled hollow structures with enhanced photocatalytic activity. Nanoscale 4 (2012)183–187. https://doi.org/10.1039/C1NR11114K

[13] S. Kant, D. Pathania, P. Singh, P. Dhiman, A. Kumar, Removal of malachite green and methylene blue by Fe0.01Ni0.01Zn0.98O/polyacrylamide nanocomposite using coupled adsorption and photocatalysis, Appl. Catal. B: Environ. 147 (2014) 340– 352. https://doi.org/10.1016/j.apcatb.2013.09.001

[14] G. Wang, B. Huang, Z. Li, Z.Lou,Z. Wang, Y. Dai, M.-H. Whangbo, Synthesis and characterization of ZnS with controlled amount of S vacancies for photocatalytic H_2 production under visible light, Sci. Rep. 5 (2015) 8544. https://doi.org/10.1038/srep08544

[15] G. Sharma, S. Bhogal, M. Naushad, Inamuddin, A. Kumar, F.J. Stadler, Microwave assisted fabrication of La/Cu/Zr/carbon dots trimetallic nanocomposites with their adsorptional vs photocatalytic efficiency for remediation of persistent organic pollutants, J. Photochem. Photobiol. A Chem. 347 (2017) 235–243. https://doi.org/10.1016/j.jphotochem.2017.07.001

[16] M. Wenxin, R. Baosheng, H. Zhen, C.Qifeng, C. Xiaofeng, G. Yanchuan, Mesostructured zinc oxide architectures with high photocatalytic activity, Mater. Cem. Phys. 186 (2017) 341–352. https://doi.org/10.1016/j.matchemphys.2016.11.005

[17] P. Ranjan, R.K. Singh,H.Suematsu,L.Phillip,R.Sarathi, Synthesis of nano-ZnO by wire explosion process and its photocatalytic Activity, J. Environ. Chem. Eng. 5 (2017) 1676-1684. https://doi.org/10.1016/j.jece.2017.02.036

[18] S. Wenzhong, L. Zhijie, H. Wang, Y. Liu, Q. Guo, Y. Zhang, J. Hazard. Mater. 152 (2008) 172–175. https://doi.org/10.1016/j.jhazmat.2007.06.082

[19] O. Mekasuwandumrong, P. Pawinrat, P. Praserthdam, J. Panpranot, Effects of synthesis conditions and annealing post-treatment on the photocatalytic activities of ZnO nanoparticles in the degradation of methylene blue dye, Chem. Eng. J. 164 (2010) 77-84. https://doi.org/10.1016/j.cej.2010.08.027

[20] R.Y. Hong, J.H. Li, L.L. Chen, D.Q. Liu, H.Z. Li, Y. Zheng, J. Ding, Synthesis, surface modification and photocatalytic property of ZnO nanoparticles, Powder Technol. 189 (2009) 426-432. https://doi.org/10.1016/j.powtec.2008.07.004

[21] A. Kumar, M. Naushad, A. Rana, Inamuddin, Preeti, G. Sharma, A.A. Ghfar, F.J. Stadler, M.R. Khan, ZnSe-WO3 nano-hetero-assembly stacked on Gum ghatti for photo-degradative removal of Bisphenol A: Symbiose of adsorption and photocatalysis, Int. J. Biol. Macromol. 104 (2017) 1172–1184. https://doi.org/10.1016/j.ijbiomac.2017.06.116

[22] J. Qin, X. Zhang, C. Yang, M. Cao, M. Ma, R. Liu, ZnO microspheres-reduced graphene oxide nanocomposite for photocatalytic degradation of methylene blue dye, Appl. Surf. Sci. 392 (2017) 196-203. https://doi.org/10.1016/j.apsusc.2016.09.043

[23] M. Ahmad, E.Ahmed, Z.L. Hong, W. Ahmed, A. Elhissi, N.R. Khalid, Photocatalytic, sonocatalytic and sonophotocatalytic degradation of Rhodamine B using ZnO/CNTscomposites photocatalysts, Ultrason. Sonochem. 21 (2014) 761-773. https://doi.org/10.1016/j.ultsonch.2013.08.014

[24] P. Dhiman, K. M. Batoo, R.K. Kotnala, J. Chand, M. Singh,Room temperature ferromagnetism and structural characterization of Fe,Ni co-doped ZnO nanocrystals, Appl. Surf. Sci. 287 (2013) 287– 292. https://doi.org/10.1016/j.apsusc.2013.09.144

[25] M.A. Ali, M.R. Idris, M.E. Quayum, Fabrication of ZnO nanoparticles bysolution-combustion method for the photocatalytic degradation of organic dye, J. Nanostruct. Chem. (2013) 3-36.

[26] C.J. Raj, R.K. Joshi, K.B.R. Varma, Synthesis from zinc oxalate, growth mechanism and optical properties of ZnO nano/microstructures, Cryst. Res. Technol. 46 (2011) 1181. https://doi.org/10.1002/crat.201100201

[27] X.Chuanhui, H. Chenguo, T. Yongshu, C.Peng, W. Buyong, X. Jing, Room-temperature ferromagnetic properties of Fe-doped ZnO rod arrays, Solid State Sci. 13 (2011) 388. https://doi.org/10.1016/j.solidstatesciences.2010.11.041

[28] A.P. Rambu, C. Doroftei, L. Ursu , F. Iacomi, Structure and gas sensing properties of nanocrystalline Fe-doped ZnO films prepared by spin coating method, J. Mater. Sci. 48 (2013) 4305. https://doi.org/10.1007/s10853-013-7245-5

[29] W.J. Liu, X.D. Tang, Z. Tang, W. Bai, N.Y. Tang, Oxygen defects mediated magnetism of Ni doped ZnO, Adv. Cond. Matter. Phys. 6(2013) 424398. https://doi.org/10.1155/2013/424398

[30] A. Yildiz, B. Kayhan, B. Yurduguzel, A. P. Rambu, F. Iacomi, S. Simon, Ni doping effect on electrical conductivity of ZnO nanocrystalline thin films, J Mater. Sci: Mater Electron. 22 (2011) 1473. https://doi.org/10.1007/s10854-011-0332-y

[31] J.K. Salem, T.M. Hammad , R. R. Harrison, Synthesis, structural and optical properties of Ni-doped ZnO micro-spheres, J. Mater. Sci.: Mater. Electron. 24 (2013) 1670. https://doi.org/10.1007/s10854-012-0994-0

[32] R. Elilarassi, G. Chandrasekaran, Structural, optical and electron paramagnetic resonance studies on Cu-doped ZnO nanoparticles synthesized using a novel auto-combustion method, J. Mater. Sci.: Mater. Electron. 24 (2013) 96. https://doi.org/10.1007/s10854-012-0893-4

[33] S. Kant, D. Pathania, P. Singh, P. Dhiman, A. Kumar, Removal of malachite green and methylene blue by $Fe_{0.01}Ni_{0.01}Zn_{0.98}O$/polyacrylamide nanocomposite using coupled adsorption and photocatalysis, Appl. Catal. B: Environ. 147 (2014) 340. https://doi.org/10.1016/j.apcatb.2013.09.001

[34] S. Wenzhong, L. Zhijie, W. Hui, L. Yihong, G. Qingjie, Z. Yuanli, Photocatalytic degradation for methylene blue using zinc oxide prepared by codeposition and sol–gel methods, J. Hazard. Mater. 152 (2008) 172–175. https://doi.org/10.1016/j.jhazmat.2007.06.082

[35] S. Kant, A. Kumar, A comparative analysis of structural, optical and photocatalytic properties of ZnO and Ni doped ZnO nanospheres prepared by sol gel method, Adv. Mat. Lett. 3 (2012) 350-354. https://doi.org/10.5185/amlett.2012.5344

[36] M.A. Mahmood, S. Baruah, J. Dutta, Enhanced visible light photocatalysis by manganese doping or rapid crystallization with ZnO nanoparticles, Mater. Chem. Phys. 130 (2011) 531-535. https://doi.org/10.1016/j.matchemphys.2011.07.018

[37] K. Rekha, M. Nirmala, M.G. Nair, A. Anukaliani, Structural, optical, photocatalytic and antibacterial activity of zinc oxide and manganese doped zinc oxide nanoparticles, Physica B Condensed Matter 405 (2010) 3180-3185. https://doi.org/10.1016/j.physb.2010.04.042

[38] Y. H Tong, J.Cheng, Y. L.Liu, G. G. Siu, Enhanced photocatalytic performance of ZnO hierarchical nanostructures synthesized via a two-temperature aqueous solution route, Scr. Mater. 60 (2009) 1093. https://doi.org/10.1016/j.scriptamat.2008.12.060

[39] X.L. Xu, X. Duan, Z.G. Yi, Z.W. Zhou, X.M. Fan, Y. Wang, Photocatalytic production of superoxide ion in the aqueous suspensions of two kinds of ZnO under simulated solar light, Catal. Commun. 12 (2010) 169-172. https://doi.org/10.1016/j.catcom.2010.09.006

[40] M.Y. Guo, A.M. Ching Ng, F. Liu, A.B. Djurisic, W.K. Chan, H. Su, K.S. Wong, Effect of native defects on photocatalytic properties of ZnO, J. Phys. Chem. C. 115 (2011) 11095-11101. https://doi.org/10.1021/jp200926u

Chapter 4

Use of Agricultural Solid Wastes as Adsorbents

Shilpa Nag[a], Akanksha Bhardwaj[a], Puneeta Pandey[a], Meenu Arora[b], J. Nagendra Babu[c,*]

[a]Centre for Environmental Science and Technology, School Earth and Environment Sciences, Central University of Punjab, Mansa Road, Bathinda, Punjab – 151001, INDIA

[b] Department of Applied Chemistry, Giani Zail Singh College of Engineering and Technology (GZSCET), Maharaja Ranjit Singh Punjab Technical University, Dabwali Road, Bathinda, Punjab – 151001, INDIA

[c]Centre for Chemical Sciences, School of Basic And Applied Chemistry, Central University of Punjab, Mansa Road Bathinda, Punjab – 151001, INDIA

*nagendra.rd@gmail.com

Abstract

Biosorption through agricultural lignocellulosic wastes and by-products has been identified as a viable substitute to current technologies applied to remove toxic metal ion and organic pollutants from water and wastewater. The present study emphasizes the use of agricultural and agro-industries based residues as low-cost biosorbents. The study aims to revisit the status of biosorption and various recent advances made in this arena. Biomasses are the main focuse of this study which requires substantial management. Further, this is supplemented with the physicochemical processing of such biomasses and their application in adsorption. The surge in biomass to energy applications in recent years has resulted in charred biomass production as a residual. These biochars have been used as adsorbents. The biosorbents have been divided into the following three groups: (i) raw biomass, (ii) processed biomass and (iii) charred biomass. The affinity of sorbents in the removal of organic and inorganic pollutants and their applications on water and wastewater have also been studied.

Keywords

Agricultural Waste, Heavy Metals, Organic Contaminants, Biomass, Biochar

Contents

1. Introduction

Agriculture and akin sectors are pivotal to sustainable growth and development. Agriculture derives its prominence from the fact that it has vital supply and demand links with the manufacturing sector. During the past few decades, the production and productivity of food grains, commercial crops, oilseeds, fruits, and vegetables in India have witnessed remarkable growth [1]. The Indian economy is driven by the agricultural sector and since independence, an exponential growth has been encouraged in this particular sector. It has been primarily an outcome of multi-pronged, government-driven program known as 'Green Revolution' which aimed at ensuring availability and use of quality seeds, irrigation facilities, chemical fertilizers, pesticides, farm machinery, equipment, etc.

With the onset of the green revolution, the surplus crop residue generation increased proportionally in developing and developed countries. About 4000 Mt/yr crop residue is produced worldwide from food crops [2]. The widely available agricultural residue is one of the most abundant renewable resources in the world. About 25% of nitrogen, 25% phosphorus, 50% of sulfur and 75% of potassium uptake by cereal crops are retained in residues, thus serving as valuable sources of nutrients. These residues have little or no economic value and often create a disposal problem. Globally, open agricultural residue burning is practiced for farm management, resulting in the emission of particulate matter, polycyclic aromatic hydrocarbons (PAH), soot and greenhouse gasses [3,4]. A Large proportion of the residues, about 140 Mt, is burned in open fields, to immediately clear straw after harvest of the previous crop [5]. Typically, the residue is burnt for rice, wheat, maize, sugarcane, cotton, jute, millet, rapeseed, and groundnut crops. Several endeavors have been started for the proper utilization of the waste viz: composting of residues for manure, farmyard manure (FYM), energy from waste, bioethanol, biomethanation, gasification of residues and pyrolysis to produce biochar [6,7]. Particularly, biochar technology is investigated for its carbon sequestration characteristics and is suitable for multifaceted applications in conservation agriculture [2,8] to combat the climate change scenario. Alternatively, biomass, its composites, and biochar have been studied for the biosorption of various organic and inorganic pollutants in water and wastewater treatment [9,10,11].

Biomass-based biosorption is one of the prominent processes for removal of organic and inorganic pollutants from water and wastewater at lower concentrations. In the recent past, several studies have been carried out, where the plant materials constituting lignocellulosic content have proven efficient and effective biosorbents for heavy metal ion removal. Heavy metals are hazardous for the environment and humans due to their toxic nature and non-biodegradability [12,13]. Both natural and anthropogenic sources contribute to heavy metals present in the environment. Natural sources are generally geogenic in origin. Volcanic eruptions, weathering of rocks, and leaching to water bodies act as a mode of entry of these contaminants to air, water and soil. Anthropogenic sources can be classified as industrial, agricultural, domestic effluents, mining, motor vehicle emissions etc. many of the inorganic pollutants remain unaffected by water and wastewater treatment processes because of their conservative nature. These include heavy metals from metal finishing and plating operations, and other industrial discharges [14]. Heavy metals are regarded as trace elements, whose bioavailability is influenced by many physical, chemical and biological factors. Toxicity of non-essential metals like Hg(I, II), Cd(II) and Pb(II) etc. mainly occurs through displacement of trace metals from their binding sites or through ligand interactions [15,16]. Long-term exposure to these metals may lead to carcinogenicity (through arsenic); dysfunctional nervous system (through mercury); hypertension and heart ailments (through cadmium); and interference with hemoglobin formation (through lead). Organic contaminants can be broadly classified as pesticides, dyes, antibiotics and polycyclic aromatic hydrocarbons (PAH). These contaminants originate from industries, agriculture, landfills, household wastes and through accidental leakage or spills leading to their entry in environmental matrices. Such contaminants can pose danger to living organisms by entering into the food chain. Therefore, there is a need to adopt proper disposal methods or treatment technologies to remove these harmful chemicals from the environment.

Several methods have been adopted to treat water and wastewater, which include biological, chemical and physical techniques. In the past few decades, the commonly used techniques are ion-exchange [17], electrolysis [18], flotation [19], membrane filtration [20], coagulation [21] and adsorption [22]. Treatment technologies available for removal of organic contaminants include ultrasound combined with photo-Fenton treatment [23], advanced oxidation processes [24], ozonation [25], electrodialysis membranes [26] etc. However, these methods have several shortcomings such as high operational and maintenance costs, generation of secondary pollutants etc. Compared to all these treatment techniques, adsorption has emerged to be an effective approach for the abatement of water and soil organic contaminants. The most commonly used adsorbent is activated carbon, which is used for removal of a variety of organics broadly classified as pesticides, dyes, drugs, pharmaceutical products, polycyclic aromatic hydrocarbons (PAH) etc. However, concerns related to its use include higher costs, regeneration capacity, and disposal.

Agro-wastes are energy efficient, low cost and effective substitute for activated carbon as an adsorbent. Agricultural wastes are lignocellulosic rich in organic content and contain many functional groups which help in binding the contaminants on the adsorbent surface. The basic components of the agricultural waste biomass include cellulose, hemicellulose, lignin, lipids, proteins, sugars, starch, polysaccharides, and pigments. These constituents have varied functional groups on their surface. In the present chapter, the physicochemical adsorption mechanism of the biomass, chemically modified biomass and thermochemical treated biomass has been explored for the removal of organic and inorganic pollutants.

1.1. Use of raw biomass as biosorbents

Agricultural based plant adsorbents are abundant, economical, and renewable and are mainly composed of cellulose and lignin, hence making them a viable option for wastewater treatment. The presence of structural-functional groups plays important role in efficacy of the biosorbents [27,28]. In the last few decades various researchers have explored the efficacy of agricultural wastes as biosorbents viz: husk [29,30], stalk [31,32], bagasse [33,34], shell [35,36] and straw etc. [37,38].

Table 1 Adsorption behavior and efficiency of raw biomass for heavy metal pollutants.

S.No.	Agricultural Waste	Heavy metal	Adsorbent dose (g/L)	pH	Reaction time (min)	Amount adsorbed (q_e) (mg/g)	Ref.
1.	Rice straw	Cd(II)	-	2-6	180	13.84	[39]
2.	Barley straw	Cu(II)	1	6-7	120	4.64	[37]
3.	Cashew nut shell straw	Ni(II)	3	5	30	18.86	[35]
4.	Rice shell	Cu(II)	10	6	180	2.95	[40]
5.	Wheat shell	Cu(II)	10	6	180	17.42	[40]
6.	Lentil shell	Cu(II)	10	6	180	9.59	[40]
7.	Wheat stem	Cd(II)	-	5	60	11.2	[41]
8.	Wheat bran	Cr(VI)	20	-	-	0.942	[42]
9.	*Picea smithiana*	Cr(VI)	2.0	4.5	1440	228	[43]
10.	*Pinus sylvestris*	Cr(VI)	1	7	120	0.201	[44]
11.	Sugarcane bagasse	Hg(II)	5	4	60	35.71	[45]
12.	Agave bagasse raw	Pb(II)	2	5	-	35.60	[34]
13.	Agave bagasse raw	Zn(II)	2	5	-	7.84	[34]
14.	Olive stone	Cu(II)	-	5.5-6	60	2.02	[46]
15.	Olive stone	Pb(II)	-	5.5-6	60	9.261	[46]
16.	Olive stone	Ni(II)	-	5.5-6	60	2.13	[46]
17.	Olive stone	Cd(II)	-	5.5-6	60	7.733	[46]
18.	Egyptian mandarin peel	Hg(II)	5	6.02	1440	19.01	[47]
19.	Sunflower hull	Cu(II)	2	5	180	57.14	[48]

Table 2 Agricultural wastes as adsorbent for organic pollutants.

S. No.	Agro-waste	Organic contaminant	Particle size	Adsorption conditions	Adsorbent dose (g/L)	Amount adsorbed(q_e) (mg/g)	Ref.
1.	*Phaseolus vulgaris* biomass	Reactive Red 198	<300μm	pH=2 Contact time=20min at 20°C	1.6	80	[49]
2.	Rice husk ash	2, 4-Dichloro phenoxy acetic acid	-	Contact time=12h at 30°C	0.1	1.425	[50]
3.	Pine leaves	Acid yellow 220 dye	63–500 μm	pH=2 Contact time=90 min	1	32.26	[51]
4.	Coir pith	Crystal violet	600 μm	300K	0.1	65.53	[52]
5.	Coir pith	Rhodamine B (RB)	600 μm	300K	0.1	55.54	[52]
6.	Sugarcane fiber	Crystal violet	600 μm	300K	0.1	10.44	[52]
7.	Sugarcane fiber	rhodamine B (RB	600 μm	300K	0.1	15.98	[52]
8.	Sugar beet pulp	Safranin	-	pH=10 Contact time=240 min	8	147.00	[53]
9.	Sugar beet pulp	Methylene blue	-	pH=10 Contact time=240 min	8	211	[53]
10.	Sunflower seed shells	Trifluralin	-	298K	12	2.29×10^{-2}	[54]
11.	Sugarcane bagasse	Indosol turquoise FBL dye	300 μm	pH=3 Contact time=30 min	1	22.72	[55]

12.	*Citrus sinensis*	Reactive yellow 42	<0.25	Contact time= 60min	20	13.99	[56]
13.	*Citrus sinensis*	Reactive red 45	<0.25	Contact time= 60min	20	15.21	[56]
14.	*Citrus sinensis*	Reactive blue 19	<0.25	Contact time= 60min	20	14.80	[56]
15.	*Citrus sinensis*	Reactive blue 49	<0.25	Contact time= 60min	20	27.41	[56]

1.2 Mechanism of biosorption

Biosorption is a multifaceted process affected by several factors. Mechanisms involved in the biosorption process includes chemisorption, microprecipitation, ion-exchange, complexation, adsorption–complexation on surface and pores, heavy metal hydroxide condensation onto the surface and surface adsorption [57]. Functional groups present on the surface of adsorbent play a pivotal role in immobilizing the metal adsorbed onto biomass material. Plant cell walls are the sites where all the functional groups are present which are helpful in the binding process, cell walls are generally composed of cellulose molecules, organized in microfibrils, surrounded with hemicellulose (glucomannans, mannans, galactans, arabinogalactans, xylans), lignin and pectin [58]. The arrangement of cellulose, hemicelluloses, and lignin is depicted in Fig. 1. The inner core is made of crystalline cellulose, stacked one above another by interlayer hydrogen bonding with few of the sections carrying amorphous cellulose [59]. These amorphous regions are better known for the adsorption of heavy metals based on the available functional groups. The cellulose layers are intertwined with the amorphous hemicellulose followed by the covalent binding with lignin in the shell surface. Lignin is a polyphenolic polymer comprising of three main phenol derivatives namely, coniferyl alcohol, coumaryl alcohol and sinapyl alcohol. These three components primarily have hydroxyl and carboxyl components, which are responsible for heavy metals sorption, whereas the crystalline structure with hydrophobic pores is essentially responsible for the adsorption of various organic adsorbates. Thus, the behavior of the substrate is highly dependent upon the porosity, crystallinity, surface area and degree of polymerization of fibers. The additional features of high ash content i.e. silica, N, P, K etc. in the crop residues result in enhancing adsorption behavior of the crop residues vis-a-vis the woody biomass [60].

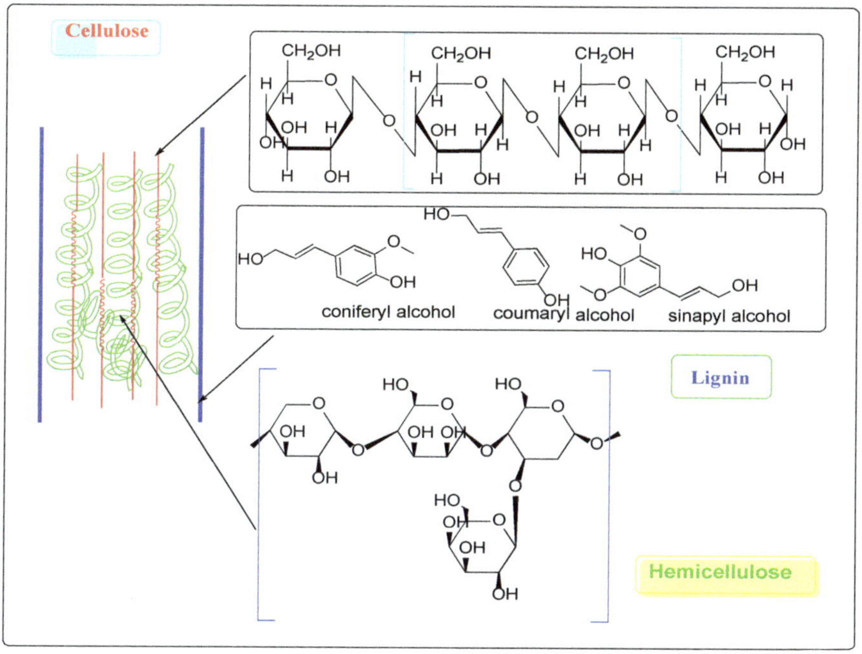

Figure 1 Lignocellulosic biomass compositions, structural arrangement and chemical characteristics of the components of biomass [2].

Laszlo and Dintzis, [61] have shown that lignocellulosic have sorption capacity. Lignocellulosic are hygroscopic in nature, where water permeates the non-crystalline portion of cellulose, hemicellulose, and lignin. Cellulose can thus adsorb heavy metals from a solution [62]. The molecular structure and supramolecular structure of cellulose have a strong influence on sorption properties. Adsorption of an aqueous solution by fibersresult in cellulose swelling. The more it swells, the higher is the amount of adsorption. Swelling also depends on the fibre structure, degree of crystallinity or amorphous and void regions [63].

2. Physicochemically treated biomass and their adsorption characteristics

As discussed earlier, agricultural solid waste materials are available in abundance and can be used as biosorbents due to their physicochemical characteristics. The utilization of agricultural solid wastes is of countless significance and can play a vital role in the national economy [63]. The major constituents of agricultural waste are lignin, hemicellulose, and cellulose that contain several functional groups such as aldehydes, alcohols, carboxylic acid, ketones, phenolic and ether linkages. These groups have resilient ability to bind with toxic metal ions by utilizing an electron pair to form different complexes in solution [64]. Recently, Zafar et al. [65] modified rice bran through various chemicals and utilized for the adsorption of nickel ions. Similarly, various results (Table 3) focus on using treated biomass for contaminant removal.

Table 3 Physicochemically processed agricultural residual biomass and their adsorption efficiency for metal ions.

S.No.	Agri waste	Heavy metal	Adsorbent dose (g/L)	pH	Reaction time (min)	Amount adsorbed (q_e) (mg/g)	Ref.
1.	Barley straw citric acid	Cu(II)	1	6-7	120	31.71	[37]
2.	Grapefruit $ZnCl_2$	Pb(II)	10	5.3-6.5	90	12.73	[66]
3.	Agave bagasse HCl	Cd(II)	1	5	-	12.50	[34]
4.	Agave bagasse HCl	Pb(II)	1	5	-	42.31	[34]
5.	Agave bagasse HCl	Zn(II)	1	5	-	12.40	[34]
6.	Agave bagasse HNO_3	Cd(II)	1	5	-	13.50	[34]
7.	Agave bagasse HNO_3	Pb(II)	1	5	-	54.29	[34]
8.	Agave bagasse HNO_3	Zn(II)	1	5	-	14.43	[34]
9.	Agave bagasse	Cd(II)	1	5	-	18.32	[34]

	NaOH						
10.	Agave bagasse NaOH	Pb(II)	1	5	-	50.12	[34]
11.	Agave bagasse NaOH	Zn(II)	1	5	-	20.54	[34]
12.	Orange peel citric acid	Cd(II)	4.3	6	120	0.90	[67]
13.	Egyptian mandarin peel NaOH	Hg(II)	5	6.02	1440	23.26	[68]
14.	Orange peel KCl	Cu(II)	5	5-5.5	120	59.77	[44]
15.	Orange peel KCl	Cd(II)	5	5-5.5	120	125.63	[44]
16.	Orange peel KCl	Pb(II)	5	5-5.5	120	142.94	[44]
17.	Orange peel KCl	Zn(II)	5	5-5.5	120	45.29	[44]
18.	Orange peel KCl	Ni(II)	5	5-5.5	120	49.14	[44]
19.	Orange peel K^+	Cu(II)	5	5-5.5	120	59.77	[69]
20.	Orange peel Mg^{2+}	Cu(II)	5	5-5.5	120	40.37	[69]
21.	Orange peel sulfured	Pb(II)	5	5	-	164.0	[70]
22.	Orange peel sulfured	Zn(II)	5	5	-	80.0	[70]

Various studies have reported the effect of pre-treatment on raw agricultural wastes (Table 4). Sadaf et al. [55] investigated the effect of chemical (acids, chelating agents, and organic solvents) and physical treatment (autoclaving and boiling) on sorption capacity of sugarcane bagasse and reported an increase in sorption capacity by these treatments. Treatment with HCl achieved the highest increase in the investigations. Increase in porosity, and high surface area, resulting in the availability of more active sites was the reason for higher sorption capacity.

Table 4 Physicochemically treated agricultural residual biomass and their adsorption characteristics for organic pollutants.

S. No.	Agricultural waste biomass	Organic contaminant	Particle size	Kinetic conditions	Loading (g/L)	Amount Adsorbed (mg/g)	Ref.
1.	Rice husk Mechanical	2, 4-Dicholoro phenoxyacetic acid	42.1 nm	pH=5 60min	1.5	76.92	[71]
2.	Rice hull Tartaric acid	Methylene blue	-	300min	5	25	[72]
3.	Sugarcane bagasse HCl	Indosol turquoise FBL dye	300 µm	pH=3 30 min	1	27.17	[55]
4.	Sugarcane bagasse Na-alginate-immobilized forms	Indosol turquoise FBL dye	300 µm	pH=3 1h	1	10.54	[55]
5.	Cotton stalks NaOH	Methylene blue	-	1440 min	2	-	[73]
6.	Palm shells H_2SO_4	Chloropyriphos	-	pH=6 240 min	-	52.63	[74]
7.	Palm shells H_2SO_4	Monocrotophos	-	pH=6 240 min	-	51.099	[74]
8.	*Citrus sinensis* Acetic acid	Reactive yellow 42	<0.25 mm	60 min	2	17.64	[56]
9.	*Citrus sinensis* Acetic acid	Reactive blue 19	<0.25 mm	60 min	2	23.31	[56]
10.	*Citrus sinensis* Acetic acid	Reactive blue 49	<0.25 mm	60 min	2	33.53	[56]

3. Use of charred biomass

While many adsorbents derived from agro-waste are widely available, but their adsorption capacity is low when compared with the processed forms. Many researchers have reported enhanced activity of pyrolyzed waste or activated char compared with the parent material in the raw form [75]. Biochar is gaining attention for its use as an adsorbent to remove organic compounds from water and soil environments. It's used as adsorbents environmental friendly, renewable and generates minimum waste production besides supporting, regeneration of adsorbent. Biochar is a solid material obtained from thermochemical conversion of biomass in an oxygen-limited environment. It is a source of renewable energy with the high potential to aid in environmental management. Biochar considered a form of black carbon is stable, recalcitrant organic, and produced from a variety of waste biomass materials such as crop residues, wood waste, garden waste, municipal solid waste, animal manure and food waste etc. under temperature and oxygen controlled conditions [76]. It has varied functional groups on its surface (Fig. 2). Slow pyrolysis leads to the generation of biochar and gas as major products. The key application of biochar is in sequestration of carbon and hence combating climate change. Beside this, when added to soil, biochar also acts as a soil conditioner which helps in holding carbon, improving microbial activity, increasing soil biodiversity by replenishing nutrient in deficient soil and thus enhance crop yield and food security. Biochar also helps in retaining nutrients and agrochemicals for a plant, hence resulting in reduced run-off and leaching into the underlying water. Due to all these functions, it has emerged as a very effective tool for environmental management [76]. Biochar was primarily introduced as a soil amendment, but the recent research has highlighted its ability to immobilize various organic contaminants in soil [75,77,78]. Biochar is known to enhance the sorption as well as the bioavailability of pesticide residues for biota. The two major processes governing the fate of organic contaminants in soil are sorption-desorption and degradation [79]. Biochar amended soils are reported to have high inorganic content, high cation exchange capacity (CEC), moisture content, adsorptive capacity, and pH. Biochar increases soil's sorption capacity for pesticides. Wheat and rice residue biochars were found to be 2500 times more effective than soil in sorbing diuron [80]. Highly carbonaceous and aromatic structure and high surface area play an important role for such high sorption [79].

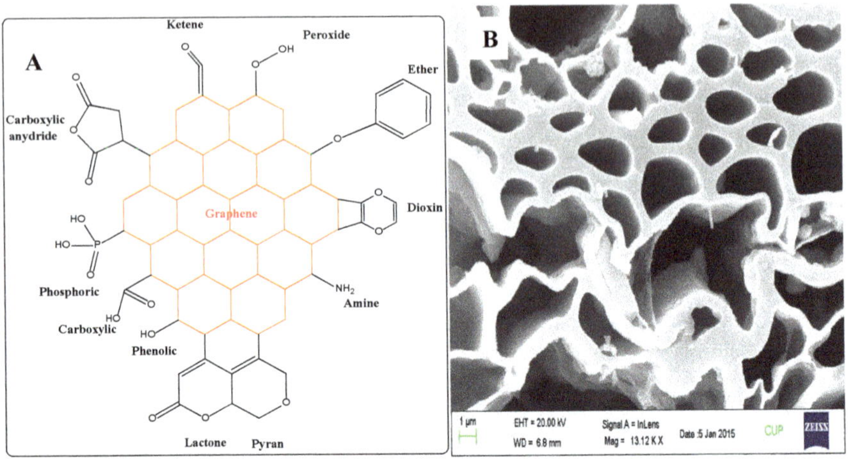

Figure2 (A) Structure of biochar with different functional groups present on its surface (Adapted and redrawn from [81]and [76]); and (B) SEM images of wheat straw biochar synthesized at 400°C.

3.1 Production of biochar

Biochar is produced through thermochemical conversion of biomass at high temperatures, usually ranging from 200-900°C with limited or no oxygen, by the process known as pyrolysis. Pyrolysis can be fast or slow. Fast pyrolysis results in more liquid fuel (bio-oil), with the lesser solid product (biochar), whereas slow pyrolysis yields a high amount of biochar and lesser liquid fuel. Residence time is <2s in fast pyrolysis, which yields about 75% bio-oil, and for slow pyrolysis, it varies from few minutes to days, with biochar as the major product (25-35%) [82,83].

Biochar yield generally depends on the temperature of pyrolysis, residence time, heating rate and type of feedstock. Sohi et al. [84] reported that high biochar yield is obtained from biomass with high lignin content. Biochars produced from crop residues and wood biomass has low carbon content and high molar H/C and O/C ratios and thus higher surface areas than those produced from animal litter and solid wastes. The heating rate is found to be the least influencing factor in determining biochar yield [85]. Pyrolysis temperature is known to affect biochar characteristics significantly. The increase in pyrolysis temperature enhances carbonization and lowers O and H contents. The surface area of biochar also increases with increasing pyrolysis temperatures as it removes volatile material which results in high micropore volume [86]. Keiluweit et al. [87] observed a decline in biochar yield at a temperature of less than 300°C due to initial dehydration reactions. With an increase in temperature, plant-based biomass undergoes dehydration and depolymerization into smaller dissociation products of lignin and cellulose. This change can be shown by Van krevelen diagram. McKendry et al. [88] compared biomass and fossil fuels in terms of their oxygen to carbon and hydrogen to carbon ratios and inferred that these ratios are inversely proportional to the energy content of material (Fig 3).

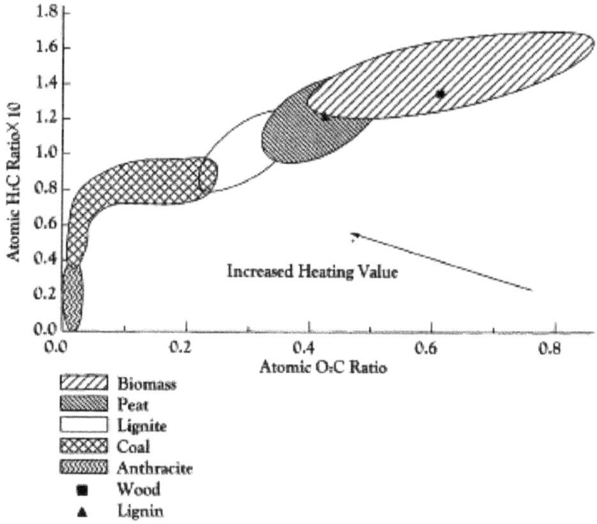

Figure 3 Van krevelen diagram depicting fuel characteristics (Source: McKendry et al. (2002).

Use of biochar as an adsorbent can be limited because of its relatively low surface area and high residual volatile matter. However, these shortcomings can be overcome by various pre- and post-treatment technologies (Fig 4). These modified biochars have large surface area and high porosity, which leads to their use as an alternative for activated carbon. Activation is suggested for enhancing biochar properties. Biochar properties can be modified by physical activation (steam activation and heat treatment) and chemical activation (acidic or alkaline modification and impregnation methods) [89]. These two methods are successfully used to enhance the surface area and porosity of biochar [90,91,92]. The biochars thus produced from these treatment types are called *activated biochars*. In physical activation by steam, biomass is subjected to steam at 800-900°C temperature after carbonization. Heat treatment involves heating of biochar at 800-900°C for 1-2 hours to provide more basic surface functional groups which are quite efficient for sorbing hydrocarbons [93,94]. Chemical activation, which yields better porous structure [95] involves impregnation of raw material with an activating agent under heat treatment in the inert atmosphere [96]. Different oxidants are used for acidic modification of biochars to increase their acidic property by removing mineral elements [94]. It is conducted by soaking biochars in acidic solutions at 120°C in biochar to acid ratio of 1:10 [96,97]. Alkaline modification helps to adsorb negatively charged species. It is carried out by soaking biochars in different basic concentrations at 25-100°C after which pyrolysis is done with limited or no oxygen [93,98]. In impregnation, biochars are mixed with metal salts or oxides to facilitate easy adsorption of metal ions. Application of these methodologies gives biochar different properties. Azargohar and Dalai [90] investigated the effects of steam and potassium hydroxide (KOH) as a means of physical and chemical activation of biochar. The BET surface area was found to have increased when the temperature for steam and chemical activation was increased. Biochars produced from fast pyrolysis were given steam treatment, which showed enhanced sorption capacity for the removal of contaminants by increasing the biochars' surface area and porosity [91]. Wheat straw was used as a precursor for the production of activated biochar by activating it using phosphoric acid, acid treated char was heated afterward to acquire desired properties for the removal of heptachlor from aqueous solution [99].

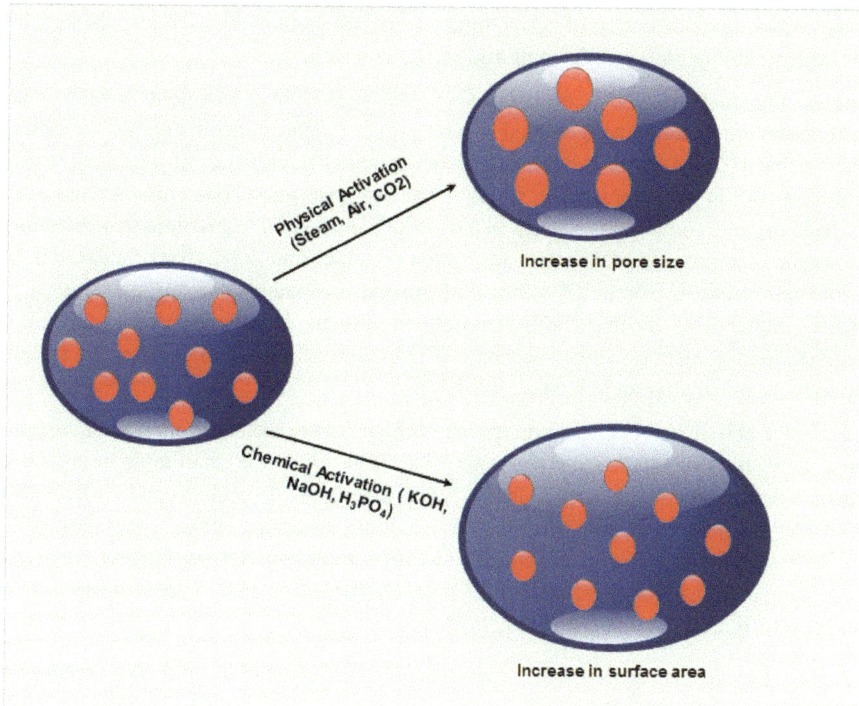

Figure 4 Physical and chemical activation of biochar leading to the different surface phenomenon.

3.2 Characteristics of biochar for adsorption

The pyrogenic conversion of waste materials into biochar is beneficial as it adds considerable economic value; helps in the reduction of waste disposal cost, and is a potentially reasonable alternative to the present commercial activated carbons. These waste materials have high adsorption capacity, considerable mechanical strength, and low ash content [100]. Pyrogenic carbon is an amorphous carbon with a high degree of porosity, which governs the way in which it performs the purifying role, and the large surface area provides multiple sites upon which the adsorption takes place [101]. The factors which favor selection of agricultural adsorbents include widespread presence, low cost, regenerability and organic composition.

Various mechanisms such as H-bonding, cation-bridging, covalent bonding, and hydrophobic interactions are involved in adsorption of anthropogenic organic compounds onto biochar [102]. Surface properties of biochar and characteristics of adsorbate play an important role in determining the reaction rate for adsorption. For various natural and synthetic organic compounds present in terrestrial environments, aromatic structures form the major components. Different adsorption mechanisms have been suggested for interactions between aromatic π-systems of organic compounds and sorption sites at the organic matter and biochar [87]. Aromatic π-systems have been considered to be involved in π-π electron donor-acceptor (EDA) interactions, and polar-π interactions between organic matter and biochar [103,104].

The surface activity of biochar particles may become attenuated with time as the external space available for adsorption reactions diminishes with the increasing presence of organic compounds at the surface [105]. Surface activity of biochar can be affected by aging process (like pore blockage, surface oxidation, swelling, surface coverage), which has been studied in the soil in detail [106-110].Various studies on biochar from agro waste have manifested a good adsorption capacity for heavy metal contaminants such as Cu, [111-116], Zn [111,116], Pb [117, 118], and Cr [119].

Table 5 Biochar generated under various pyrolysis conditions and their adsorption characteristics for heavy metal ions.

S. No.	Agri waste	Pyrolysis Temperature (°C)	Heavy metal	Adsorbent dose (g/L)	Optimum pH	Reaction time (h)	Amount Adsorbed (q_e) (mg/g)	Kinetics	Ref.
1.	Rice husk	300-400	Cu(II)	10	5	24	44.47		[120]
2.	Rice husk	300	Pb(II)		5	5	2.40	Pseudo second order	[117]
3.	Corn straw	600	Cu(II)	0.1	5	24	12.52	Pseudo-second-order	[111]
4.	Corn straw	600	Zn(II)	0.1	5	24	11.0	Pseudo-second-order	[111]
5.	Corn bran	600	Cr(VI)	-	2	-	86.49	-	[119]

6.	Corn cob	300-400	Cu(II)	10	5	24	2.70		[120]
7.	Banana peel	230	Pb(II)	0.01	2-7	3	359	Pseudo-second-order	[121]
8.	Switchgrass	300	Cu(II)	1	5	24	-	-	[113]
9.	Switchgrass	300	Cd(II)	1	5	24	-	-	[113]
10.	*Spartina alterniflora*	400	Cu(II)	0.5	6	-	48.8	Pseudo-second-order	[122]
11.	Eucalyptus	300-400	Cu(II)	10	5	24	3.48		[120]
12.	*Acacia*	300-400	Cu(II)	10	5	24	9.70		[120]
13.	Olive mill waste	300-400	Cu(II)	10	5	24	24.10		[120]
14.	Pine wood char	400-450	F^-	10	2-10	48	-	Pseudo second order	[118]
15.	Pine bark char	400-450	F^-	10	2-10	48	-	Pseudo second order	[118]
16.	Pine wood	300	Pb(II)	-	5	5	4.25	Pseudo second order	[117]
17.	Hardwood	450	Zn(II)	0.1	5	24	4.54	pseudo-second-order	[111]
18.	Hardwood	450	Cu(II)	0.1	5	24	6.79	Pseudo-second-order	[111]

Various studies have shown the effect of agro-wastes on pesticides and other organic contaminants in an aqueous environment (Table 6). In a study by Cederlund et al. [123], heat treatment was found to enhance adsorption and decrease desorption of bentazone and 2-methyl-4-chlorophenoxyacetic acid (MCPA). Chemical treatment using magnetite increased the adsorption of glyphosate, decreasing desorption of chlorpyrifos. Chlorpyrifos and diuron were adsorbed on untreated biochar owing to their high octanol-water partition coefficients, whereas glyphosate was least adsorbed. The adsorption capacity of biochar iron composite was found to be highest for all pesticides [123]. In another study, the efficiency of rice husk agro waste in nano-sorbent form for removal of 2,4 D was 96.87% removal of the contaminant at pH 5.0, temperature 30°C, the adsorbent dosage of 1.5 g/l attaining an equilibrium within an hour [71].

Sorption of pesticides was found to be more with biochar compared to soil or soil amended with raw material (straw) [78,79]. At intermediate and quasi-equilibrium conditions 48-55 and 66.6-72.4% sorptions were achieved respectively, for soil amended with biochar compared to control and straw for simazine [78].

Table 6 Selected reports on adsorption-desorption of contaminants with biochar amended media.

S. No.	Biochar	Organic contaminant	Pyrolysis temperature (°C)	Surface area (m^2/g)	Adsorption Conditions	Loading rate (g/L)	Amount adsorbed (mg/g)	Ref.
1.	Soybean stover	Trichloroethylene	300	6	pH=7 2880 min	0.3	9.85	[101]
2.	Soybean stover	Trichloroethylene	700	420	pH=7 2880 min	0.3	25.38	[101]
3.	Peanut shells	Trichloroethylene	300	3	pH=7 2880 min	0.3	7.79	[101]
4.	Peanut shells	Trichloroethylene	700	448	pH=7 2880 min	0.3	30.74	[101]
5.	Switchgrass	Metribuzin	425	1.1	pH 2 1440 min	1	223	[124]
6.	Olive kernel (Physical activation under steam)	Bromopropylate	800	600	75 min	5	0.094	[125]
7.	Corn cobs	Bromopropylate	800	630	135 min	5	0.086	[125]

	(Physical activatio n under steam)							
8.	Rapeseed stalks (Physical activatio n under steam)	Bromopro pylate	800	490	135 min	5	0.0893	[125]
9.	Soya stalks (Physical activatio n under steam)	Bromopro pylate	800	570	135 min	5	0.078	[125]
10.	Corn cob	Atrazine (A) and imidaclopr id (I)	600	242.1	1440 min	0.1	18.0–30.4%(A),(14.7–28.4%(I)	[126]
11.	Eucalypt us bark	Atrazine (A) and Imidaclopr id (I)	600	188.2	Conta ct time= 1440 min, 27°C	0.1	23.4–40.1%(A),5.9–20.1%(I)	[126]
12.	Rice husk	Atrazine (A) and Imidaclopr id (I)	600	159.1	1440 min	0.1	11.8–42.6%(A), 28.0–46.2%(I)	[126]
13.	Rice straw	Atrazine (A) and Imidaclopr id (I)	600	220.2	1440 min	0.1	37.5–70.7%(A),	[126]
14.	Corncob	Methylene blue	400	700	45min	2	28.65	[127]
15.	Corncob	Methylene blue	500	633	120mi n	2	17.57	[127]
16.	Corncob	Methylene blue	600	600	120 min	2	0.809	[127]
17.	Korean cabbage	Congo red	500	11.44	pH=9. 18, 1440	1	21.9	[128]

					min			
18.	Rice straw	Congo red	500	34.73	pH=8. 61 1440 min	1	13	[128]
19.	Rice straw	Crystal violet	500	34.73	pH=8. 61 1440 min	1	261.5	[128]
20.	Wood chips	Congo red	500	<0.01	pH=4. 71 1440 min	1	9.4	[128]
21.	Wood chips	Crystal violet	500	<0.01	pH=4. 71 1440 min	1	78.21	[128]
22.	*Jatropha curca* pods	Remazol,b rilliant blue R	600		pH=3	8	90%	[129]
23.	Wheat straw	Heptachlo r	500	176.4 ± 5.6, 0.631 cm3 g- µm	pH=7 180 min	2	2.218	[99]
24.	Groundn ut shell biochar	2, 4-dichloroph enoxyaceti c acid	650	43	pH=8. 5	40	3.02	[75]
25.	Groundn ut shell activated carbon	2, 4-dichloroph enoxyaceti c acid	800	709	pH=9. 7	0.6	250	[75]
26.	Rice straw	Phenol	550	71.35	1440 min	2	83.4	[130]
27.	Rice straw	Phenol	550	143.3	KOH	2	93.5	[130]
28.	Rice straw	Phenol	550	87.2	HNO_3	2	66.8	[130]
29.	Rice straw	Phenol	550	56.9	H_2SO_4	2	65.6	[130]
30.	Rice straw	Phenol	550	110.9	H_2O_2	2	80	[130]
31.	Rice straw	Phenol	550	87.75	$KMnO_4$	2	64.2	[130]

Yavari et al. [131] found that biochars produced from oil palm empty fruit bunches were more efficient for the sorption of polar pesticides like imazapic and imazapyr. This sorption behavior is accounted to the elemental composition and surface functional groups on the oil palm biochar as compared to rice husk derived biochars. It was observed that low pyrolysis temperature retains organic functional groups on biochar's surface, which leads to higher sorption of these polar pesticides.

The surface functional moieties on biochar had a significant influence on adsorption than the surface area of the biochar at elevated pyrolysis temperature. In contrast to this study, biochar obtained at high temperature was more effective at terbuthylazine (non-polar pesticide) adsorption than the one produced at lower temperature [132]. This affinity for non-polar molecules is accounted to the non-polar nature of surface as well as increased porosity of biochar produced at elevated pyrolysis temperature. Trivedi et al. [75] highlighted the effect of preparation conditions on adsorption of 2, 4-D on groundnut shell biochar prepared at 650 and 850°C. The surface area and organic carbon content increased with thermal activation of biomass, which enhanced the sorption capacity. Phosphorous treated and untreated rice husk biochar (RHBC) showed maximum herbicide sorption as compared with eucalyptus bark, corn cob, and bamboo chips, prepared under similar operating conditions. The cation exchange capacity, pore volume, and polarity of RHBC were highest among other biochars. The performance of RHBC was further enhanced by phosphoric acid treatment as it increased the functional groups on the surface of biochar pores. Apart from this, the surface area for RHBC was lowest. Pore size and pore volume show good correlation with adsorption onto biochar whereas no correlation was observed with a surface area of biochar [126]. Carbera et al. [133] related sorption of bentazone on the biochar-amended soil to surface area and dissolved organic carbon (DOC) content of biochar. Sorption was lesser on biochar with high DOC, as DOC has the tendency to get adsorbed to soil particles, hence competing with herbicides for adsorption sites. On the other hand, sorption of aminocyclopyrachlor increased in biochar with high surface area and low DOC due to the interactions between biochar DOC and mineral soil surfaces. Polar herbicides such as glyphosate and 2-methyl-4-chlorophenoxyacetic acid (MCPA) were weakly adsorbed on biochar owing to the negative charge on biochar surface which offered electrostatic repulsion [77]. soybean stover and peanut shells biochars produced at 700 and 300°C have studied for Trichloroethylene (TCE) adsorption [101]. Low adsorptive properties of the biochars produced at 300°C were attributed to high oxygen content or low carbon content at low temperature. Adsorption capacity was positively correlated with carbon content but negatively correlated with an oxygen content of biochars. This positive and negative correlation with C and O content of biochar, respectively, with adsorption of TCE, is attributed to the hydrophobicity of biochar and to the removal of acidic functional groups. The surface functional moieties on the char result in increased polarity of biochar produced at low pyrolysis temperature. This results in reducing the adsorption of TCE, as polar sites are known to hinder the removal of TCE due to the formation of water clusters. Low oxygen content in high-temperature biochars hindered the formation of water clusters and enhanced adsorption of TCE. Another reason for high adsorption with rising temperature was the formation of micropores which increases surface area leading to greater diffusion of TCE.

4. Conclusion

Lignocellulosic biomass-based agricultural residues have been extensively studied for remediation of heavy metal and organic contaminants in water and wastewater. The basic approach involves raw biomass physicochemically treated biomass and pyrolyzed biomass (biochar) for use as surface active adsorbents for heavy metals and organic contaminants. Limited studies on the physicochemically treated biomasses and their applications in adsorption of heavy metal or organic contaminants are available. There is a substantial promise in the usage of biochar as an adsorbent for heavy metal and pesticides.

One of the major concerns about adsorption is the disposal of the waste generated in the process. It is believed that lignocellulosic can be used in further thermochemical processes (combustion & pyrolysis) whereby the ash liberated would carry the adsorbate and be concentrated significantly, up to 50 times in case of heavy metals. This would have the added advantage that the thermochemical processes are catalyzed by the heavy metal adsorbates leading to a variety of value-added biorefinery products. On the other hand, the organic impurities volatilize and yield more toxic products. The other alternatives of regeneration and reuse are cost and environmentally ineffective so far. However, efforts are required in the management of these adsorption based wastes for the effective use of industrial wastewater and water treatment technologies.

References

[1] SOIA, (2016) http://eands.dacnet.nic.in/PDF/State_of_Indian_Agriculture, 2015-16.pdf.

[2] R. Singh, J.N. Babu, R. Kumar, P. Srivastava, P. Singh, A.S. Raghubanshi, Multifaceted application of crop residue biochar as a tool for sustainable agriculture: an ecological perspective, Ecol. Eng. 77 (2015) 324-347. https://doi.org/10.1016/j.ecoleng.2015.01.011

[3] P.J. Crutzen, M.O. Andreae, Biomass burning in the tropics: Impact on atmospheric chemistry and biogeochemical cycles, Science. 250 (1990) 1669-1679. https://doi.org/10.1126/science.250.4988.1669

[4] A.K. Jain, Z. Tao, X. Yang, C. Gillespie, Estimates of global biomass burning emissions for reactive greenhouse gases (CO, NMHCs, and NOx) and CO_2, J. Geophys. Res-Atmos. 111 (2006) 1-14. https://doi.org/10.1029/2005JD006237

[5] T.C. Thakur, A. Kumar, Machinery for zero-till surface managed crop residue systems progress and prospects, Conserv. Agri. 1 (2005) 33-42.

[6] J. Rass-Hansen, H. Falsig, B. Jørgensen, C.H. Christensen, Bioethanol: fuel or feedstock? J. Chem. Technol. Biotechnol. 82 (2007) 329-333. https://doi.org/10.1002/jctb.1665

[7] O. Erenstein, Cropping systems and crop residue management in the trans-gangetic plains: Issues and challenges for conservation agriculture from village surveys, Agric. Syst. 104 (2011) 54-62. https://doi.org/10.1016/j.agsy.2010.09.005

[8] R. Singh, P. Srivastava, S. Upadhyay, P. Singh, A.S. Raghubanshi, J.N. Babu, Integrating biochar as conservation agriculture tool under climate change mitigation scenario, In Proceedings of National Conference on Climate Change: Impacts, Adaptation, Mitigation Scenario and Future Challenges in Indian Perspective, 2015.

[9] O.S. Lawal, O.S. Ayanda, O.O. Rabiu, K.O. Adebowale, Application of black walnut (*Juglan nigra*) husk for the removal of lead (II) ion from aqueous solution, Water Sci. Technol. 75 (2017) 2454-2464. https://doi.org/10.2166/wst.2017.125

[10] N. Tahir, H.N. Bhatti, M. Iqbal, S. Noreen, Biopolymers composites with peanut hull waste biomass and application for crystal violet adsorption, Int. J. Biol. Macromol. 94 (2017) 210-220. https://doi.org/10.1016/j.ijbiomac.2016.10.013

[11] M. Essandoh, D. Wolgemuth, C.U. Pittman, D. Mohan, T. Mlsna, Phenoxy herbicide removal from aqueous solutions using fast pyrolysis switchgrass biochar, Chemosphere 174 (2017) 49-57. https://doi.org/10.1016/j.chemosphere.2017.01.105

[12] R. Ayyappan, A.C. Sophia, K. Swaminathan, S. Sandhya, Removal of Pb(II) from aqueous solution using carbon derived from agricultural wastes, Process Biochem. 40 (2005) 1293-1299. https://doi.org/10.1016/j.procbio.2004.05.007

[13] R.K. Gautam, M.C. Chattopadhyaya, S.K. Sharma, Biosorption of heavy metals: recent trends and challenges, In Wastewater Reuse and Management, Springer, Netherlands, 2013, pp. 305-322. https://doi.org/10.1007/978-94-007-4942-9_10

[14] B. Alyüz, S. Veli, Kinetics and equilibrium studies for the removal of nickel and zinc from aqueous solutions by ion exchange resins, J. Hazard. Mater. 167 (2009) 482-488. https://doi.org/10.1016/j.jhazmat.2009.01.006

[15] A.S. Mohammed, A. Kapri, R. Goel, Heavy metal pollution: source, impact, and remedies, In Biomanagement of metal-contaminated soils, Springer, Netherlands, 2011, pp. 1-28. https://doi.org/10.1007/978-94-007-1914-9_1

[16] D.H. Nies, Microbial heavy-metal resistance, Appl. Microbiol. Biotechnol. 51 (1999) 730-750. https://doi.org/10.1007/s002530051457

[17] M.R. Bruins, S. Kapil, F.W. Oehme, Microbial resistance to metals in the environment, Ecotoxicol. Environ. Saf. 45 (2000) 198-207. https://doi.org/10.1006/eesa.1999.1860

[18] A. Abou-Shady, C. Peng, J. Bi, H. Xu, Recovery of Pb(II) and removal of NO^{3-} from aqueous solutions using integrated electrodialysis, electrolysis, and adsorption process, Desalination 286 (2012) 304-315. https://doi.org/10.1016/j.desal.2011.11.041

[19] V. Mavrov, T. Erwe, C. Blöcher, H. Chmiel, Study of new integrated processes combining adsorption, membrane separation and flotation for heavy metal removal from wastewater, Desalination 157 (2003) 97-104. https://doi.org/10.1016/S0011-9164(03)00388-6

[20] H.F. Shaalan, M.H. Sorour, S.R. Tewfik, Simulation and optimization of a membrane system for chromium recovery from tanning wastes, Desalination 141 (2001) 315-324. https://doi.org/10.1016/S0011-9164(01)85008-6

[21] S. Song, A. Lopez-Valdivieso, D.J. Hernandez-Campos, C. Peng, M.G. Monroy-Fernandez, I. Razo-Soto, Arsenic removal from high-arsenic water by enhanced coagulation with ferric ions and coarse calcite, Water Res. 40 (2006) 364-372. https://doi.org/10.1016/j.watres.2005.09.046

[22] W.W. Ngah, M.A.K.M. Hanafiah, Removal of heavy metal ions from wastewater by chemically modified plant wastes as adsorbents: a review, Bioresour. Technol. 99 (2008) 3935-3948. https://doi.org/10.1016/j.biortech.2007.06.011

[23] H. Katsumata, T. Kobayashi, S. Kaneco, T. Suzuki, K. Ohta, Degradation of linuron by ultrasound combined with photo-Fenton treatment, Chem. Eng. J. 166 (2011) 468-473. https://doi.org/10.1016/j.cej.2010.10.073

[24] T. Zhou, T.T. Lim, S.S. Chin, A.G. Fane, Treatment of organics in reverse osmosis concentrate from a municipal wastewater reclamation plant: feasibility test of advanced oxidation processes with/without pretreatment, Chem. Eng. J. 166 (2011) 932-939. https://doi.org/10.1016/j.cej.2010.11.078

[25] M.I. Maldonado, S. Malato, L.A. Pérez-Estrada, W. Gernjak, I. Oller, X. Doménech, J. Peral, Partial degradation of five pesticides and an industrial

pollutant by ozonation in a pilot-plant scale reactor, J. Hazard. Mater. 138 (2006) 363-369. https://doi.org/10.1016/j.jhazmat.2006.05.058

[26] L.J. Banasiak, B. Van der Bruggen, A.I. Schäfer, Sorption of pesticide endosulfan by electrodialysis membranes, Chem. Eng. J. 166 (2011) 233-239. https://doi.org/10.1016/j.cej.2010.10.066

[27] A. Kumar, M. Naushad, A. Rana, Inamuddin, Preeti, G. Sharma, A.A. Ghfar, F.J. Stadler, M.R. Khan, ZnSe-WO3 nano-hetero-assembly stacked on Gum ghatti for photo-degradative removal of Bisphenol A: Symbiose of adsorption and photocatalysis, Int. J. Biol. Macromol. 104 (2017) 1172–1184. https://doi.org/10.1016/j.ijbiomac.2017.06.116

[28] S.O. Lesmana, N. Febriana, F.E. Soetaredjo, J. Sunarso, S. Ismadji, Studies on potential applications of biomass for the separation of heavy metals from water and wastewater, Biochem. Eng. J. 44 (2009) 19-41. https://doi.org/10.1016/j.bej.2008.12.009

[29] L.S. Oliveira, A.S. Franca, T.M. Alves, S.D. Rocha, Evaluation of untreated coffee husks as potential biosorbents for treatment of dye contaminated waters, J. Hazard. Mater.155 (2008) 507-512. https://doi.org/10.1016/j.jhazmat.2007.11.093

[30] M. Imamoglu, Adsorption of Cd (II) ions onto activated carbon prepared from hazelnut husks, J. Dispersion Sci. Technol. 34 (2013) 1183-1187. https://doi.org/10.1080/01932691.2012.739869

[31] L. Zheng, Z. Dang, X. Yi, H. Zhang, Equilibrium and kinetic studies of adsorption of Cd (II) from aqueous solution using modified corn stalk, J. Hazard. Mater. 176 (2010) 650-656. https://doi.org/10.1016/j.jhazmat.2009.11.081

[32] M. Jalali, F. Aboulghazi, Sunflower stalk, an agricultural waste, as an adsorbent for the removal of lead and cadmium from aqueous solutions, J. Mater. Cycles Waste Manage. 15 (2013) 548-555. https://doi.org/10.1007/s10163-012-0096-3

[33] K.K. Krishnani, X. Meng, L. Dupont, Metal ions binding onto lignocellulosic biosorbent, J. Environ. Sci. Health Part A. 44 (2009) 688-699. https://doi.org/10.1080/10934520902847810

[34] L.H. Velazquez-Jimenez, J. Andrea-Pavlick, R. Rangel-Mendez, Chemical characterization of raw and treated agave bagasse and its potential as adsorbent of metal cations from water, Ind. Crops Prod. 43 (2013) 200-206. https://doi.org/10.1016/j.indcrop.2012.06.049

[35] P.S. Kumar, S. Ramalingam, S.D. Kirupha, A. Murugesan, T. Vidhyadevi, S. Sivanesan, Adsorption behavior of nickel (II) onto cashew nut shell: Equilibrium, thermodynamics, kinetics, mechanism and process design, Chem. Eng. J. 167 (2011) 122-131. https://doi.org/10.1016/j.cej.2010.12.010

[36] C.P.J. Isaac, A. Sivakumar, Removal of lead and cadmium ions from water using *Annona squamosa* shell: kinetic and equilibrium studies, Desalin. Water Treat. 51 (2013) 7700-7709. https://doi.org/10.1080/19443994.2013.778218

[37] E. Pehlivan, T. Altun, Ş. Parlayici. Modified barley straw as a potential biosorbent for removal of copper ions from aqueous solution, Food Chem. 135 (2012) 2229-2234. https://doi.org/10.1016/j.foodchem.2012.07.017

[38] U. Farooq, M.A. Khan, M. Athar, J.A. Kozinski, Effect of modification of environmentally friendly biosorbent wheat (*Triticum aestivum*) on the biosorptive removal of cadmium (II) ions from aqueous solution, Chem. Eng. J. 171 (2011) 400-410. https://doi.org/10.1016/j.cej.2011.03.094

[39] Y. Ding, J. Debing, G. Huili, Z. Lianbi, Y. Xiaosong, Biosorption of aquatic cadmium (II) by unmodified rice straw, Bioresour. Technol. 114 (2012) 20-25. https://doi.org/10.1016/j.biortech.2012.01.110

[40] H. Aydın, B. Yasemin, Y. Çiğdem, Removal of copper(II) from aqueous solution by adsorption onto low-cost adsorbents, J. Environ. Manag. 87 (2008) 37-45. https://doi.org/10.1016/j.jenvman.2007.01.005

[41] G. Tan, X. Dan, Adsorption of cadmium ion from aqueous solution by ground wheat stems, J. Hazard. Mater. 164 (2009) 1359-1363. https://doi.org/10.1016/j.jhazmat.2008.09.082

[42] M. Nameni, M.A. Moghadam, M. Arami, Adsorption of hexavalent chromium from aqueous solutions by wheat bran, Int. J. Environ. Sci. Technol. 5 (2008) 161-168. https://doi.org/10.1007/BF03326009

[43] S. Mittal, U. Vaid, G. Nabi Najar, J.N. Babu, Removal of hexavalent chromium from aqueous solution: a comparative study of cone biomass of "*Picea smithiana*" and activated charcoal, Desalin. Water Treat. 57 (2016) 11081-11095. https://doi.org/10.1080/19443994.2015.1038594

[44] H. Ucun, Y.K. Bayhan, Y. Kaya, A. Cakici, O.F. Algur, Biosorption of chromium (VI) from aqueous solution by cone biomass of *Pinus sylvestris*, Bioresour. Technol. 85 (2002) 155-158. https://doi.org/10.1016/S0960-8524(02)00086-X

[45] E. Khoramzadeh, B. Nasernejad, R. Halladj, Mercury biosorption from aqueous solutions by sugarcane bagasse, J. Taiwan Inst. Chem. E. 44 (2013) 266-269. https://doi.org/10.1016/j.jtice.2012.09.004

[46] N. Fiol, I. Villaescusa, M. Martínez, N. Miralles, J. Poch, J. Serarols, Sorption of Pb(II), Ni(II), Cu(II) and Cd(II) from aqueous solution by olive stone waste, Sep. Purif. Technol. 50 (2006) 132-140. https://doi.org/10.1016/j.seppur.2005.11.016

[47] D.Z. Husein, Adsorption and removal of mercury ions from aqueous solution using raw and chemically modified Egyptian mandarin peel, Desalin. Water Treat. 51 (2013) 6761-6769. https://doi.org/10.1080/19443994.2013.801793

[48] A. Witek-Krowiak, Analysis of temperature-dependent biosorption of Cu^{2+} ions on sunflower hulls: kinetics, equilibrium and mechanism of the process, Chem. Eng. J. 192 (2012) 13-20. https://doi.org/10.1016/j.cej.2012.03.075

[49] S.T. Akar, A.S. Özcan, T. Akar, A. Özcan, Z. Kaynak, Biosorption of a reactive textile dye from aqueous solutions utilizing an agro-waste, Desalination. 249 (2009) 757-761. https://doi.org/10.1016/j.desal.2008.09.012

[50] S.K. Deokar, S.A. Mandavgane, Rice husk ash for fast removal of 2,4-dichlorophenoxyacetic acid from aqueous solution, Adsorpt. Sci. Technol. 33 (2015) 429-440. https://doi.org/10.1260/0263-6174.33.5.429

[51] F. Deniz, S. Karaman, Removal of an azo-metal complex textile dye from colored aqueous solutions using an agro-residue, Microch. J. 99 (2011) 296-302. https://doi.org/10.1016/j.microc.2011.05.021

[52] H. Parab, M. Sudersanan, N. Shenoy, T. Pathare, B. Vaze, Use of agro-industrial wastes for removal of basic dyes from aqueous solutions, CLEAN–Soil, Air, Water. 37 (2009) 963-969. https://doi.org/10.1002/clen.200900158

[53] M.R. Malekbala, S. Hosseini, S.K. Yazdi, S.M. Soltani, M.R. Malekbala, The study of the potential capability of sugar beet pulp on the removal efficiency of two cationic dyes, Chem. Eng. Res. Des. 90 (2012) 704-712. https://doi.org/10.1016/j.cherd.2011.09.010

[54] R. Rojas, J. Morillo, J. Usero, E. Vanderlinden, H. El Bakouri, Adsorption study of low-cost and locally available organic substances and a soil to remove pesticides from aqueous solutions, J. Hydrol. 520 (2015) 461-472. https://doi.org/10.1016/j.jhydrol.2014.10.046

[55] S. Sadaf, H.N. Bhatti, S. Ali, K.U. Rehman, Removal of Indosol Turquoise FBL dye from aqueous solution by bagasse, a low cost agricultural waste: batch and column study, Desalin. Water Treat. 52 (2014) 184-198. https://doi.org/10.1080/19443994.2013.780985

[56] M. Asgher, H.N. Bhatti, Evaluation of thermodynamics and effect of chemical treatments on sorption potential of Citrus waste biomass for removal of anionic dyes from aqueous solutions, Ecol. Eng. 38 (2012) 79-85. https://doi.org/10.1016/j.ecoleng.2011.10.004

[57] J.L. Gardea-Torresdey, J.R. Peralta-Videa, M. Montes, G. De la Rosa, B. Corral-Diaz, Bioaccumulation of cadmium, chromium and copper by *Convolvulus arvensis* L.: impact on plant growth and uptake of nutritional elements, Bioresour. Technol. 92 (2004) 229-235. https://doi.org/10.1016/j.biortech.2003.10.002

[58] P.S. Nobel, Achievable productivities of certain CAM plants: basis for high values compared with C3 and C4 plants, New Phytol. 119 (1991) 183-205. https://doi.org/10.1111/j.1469-8137.1991.tb01022.x

[59] H. Chen, Chemical composition and structure of natural lignocellulose. In Biotechnology of lignocellulose, Springer, Netherlands, 2014, pp. 25-71. https://doi.org/10.1007/978-94-007-6898-7_2

[60] Y. Zhou, L. Zhang, Z. Cheng, Removal of organic pollutants from aqueous solution using agricultural wastes: A review, J. Mol. Liq. 212 (2015) 739-762. https://doi.org/10.1016/j.molliq.2015.10.023

[61] J.A. Laszlo, F.R. Dintzis, Crop resides as Ion-exchange materials. Treatment of soybean hull and sugar beet fiber (pulp) with epichlorohydrin to improve cation-exchange capacity and physical stability, J. Appl. Polym. Sci. 52 (1994) 531-538. https://doi.org/10.1002/app.1994.070520408

[62] B. Acemioglu, M.H. Alma, Equilibrium studies on adsorption of Cu (II) from aqueous solution onto cellulose, J. Colloid Interface Sci. 243 (2001) 81-84. https://doi.org/10.1006/jcis.2001.7873

[63] A. Demirbas, Heavy metal adsorption onto agro-based waste materials: a review, J. Hazard. Mater. 157 (2008) 220-229. https://doi.org/10.1016/j.jhazmat.2008.01.024

[64] A. Bhatnagar, A.K. Minocha, M. Sillanpää, Adsorptive removal of cobalt from aqueous solution by utilizing lemon peel as biosorbent, Biochem. Eng. J. 48 (2010) 181-186. https://doi.org/10.1016/j.bej.2009.10.005

[65] M.N. Zafar, I. Aslam, R. Nadeem, S. Munir, U.A. Rana, S.U.D. Khan, Characterization of chemically modified biosorbents from rice bran for biosorption of Ni (II), J. Taiwan Inst. Chem. Eng. 46 (2015) 82-88. https://doi.org/10.1016/j.jtice.2014.08.034

[66] Y.Y. Pei, J.Y. Liu, J.Y. Wang, M.W. Huang, J. Lan, H.M. Cai, Experimental study on adsorption of lead ions in water using grapefruit husk [J], Guangdong Agri. Sci. 16 (2011) 054.

[67] X. Li, Y. Tang, Z. Xuan, Y. Liu, F. Luo, Study on the preparation of orange peel cellulose adsorbents and biosorption of Cd^{2+} from aqueous solution, Sep. Purif. Technol. 55 (2007) 69-75. https://doi.org/10.1016/j.seppur.2006.10.025

[68] N.C. Feng, X.Y. Guo, Characterization of adsorptive capacity and mechanisms on adsorption of copper, lead and zinc by modified orange peel, Trans. Nonferrous Met. Soc. China 22 (2012) 1224-1231. https://doi.org/10.1016/S1003-6326(11)61309-5

[69] S. Liang, X.Y. Guo, N.C. Feng, Q.H. Tian, Effective removal of heavy metals from aqueous solutions by orange peel xanthate, Trans. Nonferrous Met. Soc. China 20 (2010) 187-191. https://doi.org/10.1016/S1003-6326(10)60037-4

[70] S. Liang, X. Guo, Q. Tian, Adsorption of Pb^{2+} and Zn^{2+} from aqueous solutions by sulfured orange peel, Desalination 275 (2011) 212-216. https://doi.org/10.1016/j.desal.2011.03.001

[71] R. Chidambaram, Rice husk as a low cost nano sorbent for 2,4-dichlorophenoxyacetic acid removal from aqueous solutions, Ecol. Eng. 92 (2016) 97-105. https://doi.org/10.1016/j.ecoleng.2016.03.020

[72] O. SiewTeng, K. PeiSin, C. AiWen, L. SiewLing, Y.T. Hung, Tartaric acid modified rice hull as a sorbent for methylene blue removal, Am. J. Environ. Sci. 6 (2010) 244-248. https://doi.org/10.3844/ajessp.2010.244.248

[73] Z. Ding, X. Hu, A.R. Zimmerman, B. Gao, Sorption and cosorption of lead(II) and methylene blue on chemically modified biomass, Bioresour. Technol. 167 (2014) 569-573. https://doi.org/10.1016/j.biortech.2014.06.043

[74] S. Kushwaha, G. Sreelatha, P. Padmaja, Evaluation of acid-treated palm shell powder for its effectiveness in the adsorption of organophosphorus pesticides: isotherm, kinetics, and thermodynamics, J. Chem. Eng. Data 56 (2011) 2407-2415. https://doi.org/10.1021/je1013334

[75] N.S. Trivedi, R.A. Kharkar, S.A. Mandavgane, 2, 4-Dichlorophenoxyacetic acid adsorption on adsorbent prepared from groundnut shell: Effect of preparation conditions on equilibrium adsorption capacity: Submitted to Arabian Journal of Chemistry (2016).

[76] J. Lehmann, S. Joseph, Biochar for environmental management: science, technology and implementation, IInd Ed., Routledge, 2015.

[77] H. Cederlund, E. Börjesson, J. Stenström, Effects of a wood-based biochar on the leaching of pesticides chlorpyrifos, diuron, glyphosate and MCPA, J. Environ. Manage. 191 (2017) 28-34. https://doi.org/10.1016/j.jenvman.2017.01.004

[78] H. Cheng, D.L. Jones, P. Hill, M.S. Bastami, Biochar concomitantly increases simazine sorption in sandy loam soil and lowers its dissipation, Arch. Agron. Soil Sci. (2016) 1-11.

[79] R.S. Kookana, The role of biochar in modifying the environmental fate, bioavailability, and efficacy of pesticides in soils: a review, Soil Res. 48 (2010) 627-637. https://doi.org/10.1071/SR10007

[80] H.F. Shaalan, M.H. Sorour, S.R. Tewfik, Simulation and optimization of a membrane system for chromium recovery from tanning wastes, Desalination 141 (2001) 315-324. https://doi.org/10.1016/S0011-9164(01)85008-6

[81] J.K. Brennan, T.J. Bandosz, K.T. Thomson, K.E. Gubbins, Water in porous carbons, Colloids Surf. A. 187 (2001) 539-568. https://doi.org/10.1016/S0927-7757(01)00644-6

[82] D. Mohan, C.U. Pittman, P.H. Steele, Pyrolysis of wood/biomass for bio-oil: a critical review, Energy Fuels 20 (2006) 848-889. https://doi.org/10.1021/ef0502397

[83] R. Brown, Biochar production technology, in: J. Lehmann, S. Joseph (Eds.), Biochar for Environmental Management Science and Technology, Earthscans, UK, 2009, pp. 127-146.

[84] S.P. Sohi, E. Krull, E. Lopez-Capel, R. Bol, A review of biochar and its use and function in soil, Adv. Agron. 105 (2010) 47-82. https://doi.org/10.1016/S0065-2113(10)05002-9

[85] M. Ahmad, A.U. Rajapaksha, J.E. Lim, M. Zhang, N. Bolan, D. Mohan, M. Vithanage, S.S. Lee, Y.S. Ok, Biochar as a sorbent for contaminant management in soil and water: a review, Chemosphere. 99 (2014) 19-33. https://doi.org/10.1016/j.chemosphere.2013.10.071

[86] J.W. Lee, M. Kidder, B.R. Evans, S. Paik, A.C. Buchanan III, C.T. Garten, R.C. Brown, Characterization of biochars produced from corn stovers for soil amendment, Environ. Sci. Technol. 44 (2010) 7970–7974. https://doi.org/10.1021/es101337x

[87] M. Keiluweit, M. Kleber, Molecular-level interactions in soils and sediments: the role of aromatic π-systems, Environ. Sci. Technol. 43 (2009) 3421-3429. https://doi.org/10.1021/es8033044

[88] P. McKendry, Energy production from biomass (part 2): conversion technologies, Bioresour. Technol. 83 (2002) 47-54. https://doi.org/10.1016/S0960-8524(01)00119-5

[89] M.B. Ahmed, J.L. Zhou, H.H. Ngo, W. Guo, M. Chen, Progress in the preparation and application of modified biochar for improved contaminant removal from water and wastewater, Bioresour. Technol. 214 (2016) 836-851. https://doi.org/10.1016/j.biortech.2016.05.057

[90] R. Azargohar, A.K. Dalai, Steam and KOH activation of biochar: Experimental and modeling studies, Microporous Mesoporous Mater. 110 (2008) 413-421. https://doi.org/10.1016/j.micromeso.2007.06.047

[91] I.M. Lima, A.A. Boateng, K.T. Klasson, Physicochemical and adsorptive properties of fast-pyrolysis bio-chars and their steam activated counterparts, J. Chem. Technol. Biotechnol. 85 (2010) 1515-1521. https://doi.org/10.1002/jctb.2461

[92] B. Wang, Y.S. Jiang, F.Y. Li, D.Y. Yang, Preparation of biochar by simultaneous carbonization, magnetization and activation for norfloxacin removal in water, Bioresour. Technol. 233 (2017) 159-165. https://doi.org/10.1016/j.biortech.2017.02.103

[93] Y. Li, J. Shao, X. Wang, Y. Deng, H. Yang, H. Chen, Characterization of modified biochars derived from bamboo pyrolysis and their utilization for target component (furfural) adsorption, Energy Fuels 28 (2014) 5119–5127. https://doi.org/10.1021/ef500725c

[94] W. Shen, Z. Li, Y. Liu, Surface chemical functional groups modification of porous carbon, Recent Patents Chem. Eng. 1 (2008) 27–40. https://doi.org/10.2174/2211334710801010027

[95] D. Angin, E. Altintig, T.E. Köse, Influence of process parameters on the surface and chemical properties of activated carbon obtained from biochar by chemical activation, Bioresour. Technol. 148 (2013) 542-549. https://doi.org/10.1016/j.biortech.2013.08.164

[96] Y. Zhou, B. Gao, A.R. Zimmerman, J. Fang, Y. Sun, X. Cao, Sorption of heavy metals on chitosan-modified biochars and its biological effects, Chem. Eng. J. 231 (2013) 512-518. https://doi.org/10.1016/j.cej.2013.07.036

[97] M.M. Zhang, Y.G. Liu, T.T. Li, W.H. Xu, B.H. Zheng, X.F. Tan, H. Wang, Y.M. Guo, F.Y. Guo, S.F. Wang, Chitosan modification of magnetic biochar produced from *Eichhornia crassipes* for enhanced sorption of Cr(VI) from aqueous solution, RSC Adv. 5 (2015) 46955–46964. https://doi.org/10.1039/C5RA02388B

[98] Y. Ma, W.J. Liu, N. Zhang, Y.S. Li, H. Jiang, G.P. Sheng, Polyethylenimine modified biochar adsorbent for hexavalent chromium removal from the aqueous solution, Bioresour. Technol. 169 (2014) 403–408. https://doi.org/10.1016/j.biortech.2014.07.014

[99] B. Seyhi, P. Drogui, P. Gortares-Moroyoqui, M.I. Estrada-Alvarado, L.H. Alvarez, Adsorption of an organochlorine pesticide using activated carbon produced from an agro-waste material, J. Chem. Technol. Biotechnol. 89 (2014) 1811-1816. https://doi.org/10.1002/jctb.4256

[100] Y. Yang, G. Sheng, Enhanced pesticide sorption by soils containing particulate matter from crop residue burns, Environ. Sci. Technol. 37 (2003) 3635-3639. https://doi.org/10.1021/es034006a

[101] M. Ahmad, S.S. Lee, X. Dou, D. Mohan, J.K. Sung, J.E. Yang, Y.S. Ok, Effects of pyrolysis temperature on soybean stover-and peanut shell-derived biochar properties and TCE adsorption in water, Bioresour. Technol. 118 (2012) 536-544. https://doi.org/10.1016/j.biortech.2012.05.042

[102] S.D. Joseph, M. Camps-Arbestain, Y. Lin, P. Munroe, C.H. Chia, J. Hook, L. Van Zwieten, S. Kimber, A. Cowie, B.P. Singh, J. Lehmann, An investigation into the reactions of biochar in soil, Soil Res. 48 (2010) 501-515. https://doi.org/10.1071/SR10009

[103] W. Chen, L. Duan, L. Wang, D. Zhu, Adsorption of hydroxyl-and amino-substituted aromatics to carbon nanotubes, Environ. Sci. Technol. 42 (2008) 6862-6868. https://doi.org/10.1021/es8013612

[104] X. Qu, L. Xiao, D. Zhu, Site-specific adsorption of 1,3-Dinitrobenzene to bacterial surfaces: A mechanism of–π electron-donor-acceptor interactions, J. Environ. Qual. 37 (2008) 824-829. https://doi.org/10.2134/jeq2007.0236

[105] J.J. Pignatello, S. Kwon, Y. Lu, Effect of natural organic substances on the surface and adsorptive properties of environmental black carbon (char): attenuation of surface activity by humic and fulvic acids, Environ. Sci. Technol. 40 (2006) 7757-7763. https://doi.org/10.1021/es061307m

[106] S. Kwon, J.J. Pignatello, Effect of natural organic substances on the surface and adsorptive properties of environmental black carbon (char): pseudo pore blockage by model lipid components and its implications for N_2-probed surface properties of natural sorbents, Environ. Sci. Technol. 39 (2005) 7932-7939. https://doi.org/10.1021/es050976h

[107] G. Cornelissen, Ö. Gustafsson, Sorption of phenanthrene to environmental black carbon in sediment with and without organic matter and native sorbates, Environ. Sci. Technol. 38 (2004) 148-155. https://doi.org/10.1021/es034776m

[108] Q. Li, V.L. Snoeyink, B.J. Mariãas, C. Campos, Elucidating competitive adsorption mechanisms of atrazine and NOM using model compounds, Water Res. 37 (2003) 773-784. https://doi.org/10.1016/S0043-1354(02)00390-1

[109] W.J. Braida, J.J. Pignatello, Y. Lu, P.I. Ravikovitch, A.V. Neimark, B. Xing, Sorption hysteresis of benzene in charcoal particles, Environ. Sci. Technol. 37 (2003) 409-417. https://doi.org/10.1021/es020660z

[110] A.R. Chughtai, G.R. Williams, M.M.O. Atteya, N.J. Miller, D.M. Smith, Carbonaceous particle hydration, Atmos. Environ. 33 (1999) 2679-2687. https://doi.org/10.1016/S1352-2310(98)00329-X

[111] X. Chen, G. Chen, L. Chen, Y. Chen, J. Lehmann, M.B. McBride, A.G. Hay, Adsorption of copper and zinc by biochars produced from pyrolysis of hardwood and corn straw in aqueous solution, Bioresour. Technol. 102 (2011) 8877-8884. https://doi.org/10.1016/j.biortech.2011.06.078

[112] X.J. Tong, J.Y. Li, J.H. Yuan, R.K. Xu, Adsorption of Cu(II) by biochars generated from three crop straws, Chem. Eng. J. 172 (2011) 828-834. https://doi.org/10.1016/j.cej.2011.06.069

[113] P. Regmi, J.L. García-Moscoso, S. Kumar, X. Cao, J. Mao, G. Schafran, Removal of copper and cadmium from aqueous solution using switchgrass biochar produced via hydrothermal carbonization process, J. Environ. Manage. 109 (2012) 61-69. https://doi.org/10.1016/j.jenvman.2012.04.047

[114] J. Jiang, R.K. Xu, Application of crop straw derived biochars to Cu(II) contaminated Ultisol: evaluating role of alkali and organic functional groups in Cu(II) immobilization, Bioresour. Technol. 133 (2013) 537-545. https://doi.org/10.1016/j.biortech.2013.01.161

[115] L. Trakal, R. Šigut, H. Šillerová, D. Faturíková, M. Komárek, Copper removal from aqueous solution using biochar: effect of chemical activation, Arab. J. Chem. 7 (2014) 43-52. https://doi.org/10.1016/j.arabjc.2013.08.001

[116] A. Bogusz, P. Oleszczuk, R. Dobrowolski, Application of laboratory prepared and commercially available biochars to adsorption of cadmium, copper and zinc ions from water, Bioresour. Technol. 196 (2015) 540-549. https://doi.org/10.1016/j.biortech.2015.08.006

[117] Z. Liu, F.S. Zhang, Removal of lead from water using biochars prepared from hydrothermal liquefaction of biomass, J. Hazard. Mater. 167 (2009) 933-939. https://doi.org/10.1016/j.jhazmat.2009.01.085

[118] D. Mohan, H. Kumar, A. Sarswat, M. Alexandre-Franco, C.U. Pittman Jr, Cadmium and lead remediation using magnetic oak Wood and oak bark fast pyrolysis bio-chars, Chem. Eng. J. 236 (2014) 513-528. https://doi.org/10.1016/j.cej.2013.09.057

[119] J. Zhang, P. Zheng, A preliminary investigation of the mechanism of hexavalent chromium removal by corn-bran residue and derived chars, RSC Adv. 5 (2015) 17768-17774. https://doi.org/10.1039/C4RA12351D

[120] D. Arán, J. Antelo, S. Fiol, F. Macías, Influence of feedstock on the copper removal capacity of waste-derived biochars, Bioresour. Technol. 212 (2016) 199-206. https://doi.org/10.1016/j.biortech.2016.04.043

[121] N. Zhou, H. Chen, J. Xi, D.Yao, Z. Zhou, Y. Tian, X. Lu, Biochars with excellent Pb (II) adsorption property produced from fresh and dehydrated banana peels via hydrothermal carbonization, Bioresour. Technol. 232 (2017) 204-210. https://doi.org/10.1016/j.biortech.2017.01.074

[122] M. Li, Q. Liu, L. Guo, Y. Zhang, Z. Lou, Y. Wang, G. Qian, Cu(II) removal from aqueous solution by *Spartina alterniflora* derived biochar, Bioresour. Technol. 141 (2013) 83-88. https://doi.org/10.1016/j.biortech.2012.12.096

[123] H. Cederlund, E. Börjesson, D. Lundberg, J. Stenström, Adsorption of pesticides with different chemical properties to a wood biochar treated with heat and iron, Water Air Soil Pollut. 227 (2016) 1-12. https://doi.org/10.1007/s11270-016-2894-z

[124] M. Essandoh, D. Wolgemuth, C.U. Pittman, D. Mohan, T. Mlsna, Adsorption of metribuzin from aqueous solution using magnetic and nonmagnetic sustainable

low-cost biochar adsorbents, Environ. Sci. Pollut. Res. 24 (2016) 4577-4590. https://doi.org/10.1007/s11356-016-8188-6

[125] O.A. Ioannidou, A.A. Zabaniotou, G.G. Stavropoulos, M.A. Islam, T.A. Albanis, Preparation of activated carbons from agricultural residues for pesticide adsorption, Chemosphere 80 (2010) 1328-1336. https://doi.org/10.1016/j.chemosphere.2010.06.044

[126] A. Mandal, N. Singh, T.J. Purakayastha, Characterization of pesticide sorption behaviour of slow pyrolysis biochars as low cost adsorbent for atrazine and imidacloprid removal, Sci. Total Environ. 577 (2017) 376-385. https://doi.org/10.1016/j.scitotenv.2016.10.204

[127] G.O. El-Sayed, M.M. Yehia, A.A. Asaad, Assessment of activated carbon prepared from corncob by chemical activation with phosphoric acid, Water Resour. Ind. 7 (2014) 66-75. https://doi.org/10.1016/j.wri.2014.10.001

[128] D.D. Sewu, P. Boakye, S.H. Woo, Highly efficient adsorption of cationic dye by biochar produced with Korean cabbage waste, Bioresour. Technol. 224 (2017) 206-213. https://doi.org/10.1016/j.biortech.2016.11.009

[129] P. Sathishkumar, M. Arulkumar, T. Palvannan, Utilization of agro-industrial waste *Jatropha curcas* pods as an activated carbon for the adsorption of reactive dye Remazol Brilliant Blue R (RBBR), J. Cleaner Prod. 22 (2012) 67-75. https://doi.org/10.1016/j.jclepro.2011.09.017

[130] S.M. Yakout, Monitoring the changes of chemical properties of rice straw-derived biochars modified by different oxidizing agents and their adsorptive performance for organics, Biorem. J.19 (2015) 171-182. https://doi.org/10.1080/10889868.2015.1029115

[131] S. Yavari, A. Malakahmad, N.B. Sapari, S. Yavari, Sorption-desorption mechanisms of imazapic and imazapyr herbicides on biochars produced from agricultural wastes, J. Environ. Chem. Eng. 4 (2016) 3981-3989. https://doi.org/10.1016/j.jece.2016.09.003

[132] B. Wang, Y.S. Jiang, F.Y. Li, D.Y. Yang, Preparation of biochar by simultaneous carbonization, magnetization and activation for norfloxacin removal in water, Bioresour. Technol. 233 (2017) 159-165. https://doi.org/10.1016/j.biortech.2017.02.103

[133] A. Cabrera, L. Cox, K.U.R.T. Spokas, M.C. Hermosín, J. Cornejo, W.C. Koskinen, Influence of biochar amendments on the sorption–desorption of aminocyclopyrachlor, bentazone and pyraclostrobin pesticides to an agricultural soil, Sci. Total Environ. 470 (2014) 470438-443. https://doi.org/10.1016/j.scitotenv.2013.09.080

Chapter 5

Carbon Quantum Dot Composites for Photocatalytic Degradation of Organic Pollutants

Gaurav Sharma[1], Shweta Sharma[1], Amit Kumar[1], Pooja Dhiman[2], Deepika Jamwal[1], Dolly Rana[1], Ram Prakash Dwivedi[3], Genene Tessema Mola[4]

[1]School of Chemistry, Shoolini University, Solan -173212, Himachal Pradesh, India

[2]Department of Physics, IEC University, Baddi, Himachal Pradesh, India

[3]School of Electrical and Computer Science Engineering, Shoolini University, Solan (H.P.) India

[4]School of Chemistry & Physics, University of KwaZulu-Natal, Pietermaritzburg Campus, Private Bag X01, Scottsville 3209, South Africa

Abstract

In the present chapter, basic chemistry of carbon quantum dots (CQDs) has been presented. CQDs actually represent a class of materials comprising of carbon atoms in the nanometric range. These are more important as compared to other quantum dots. The CQDs can be synthesized using different methods such as hydrothermal treatment, microwave method, electrochemical and synthetic route, etc. CQDs are of immense importance due to their eco-friendly nature, easy availability and cost-effectiveness. Different composites based on CQDs have been synthesized in the recent years and being used for a number of applications such as biosensing, bioimaging, photocatalysis and chemical sensing, etc. CQDs based composites have been used as catalysts due to their outstanding electrochemical catalytic activity.

Keywords

Carbon Quantum Dots, Photocatalytic Degradation, Hydrothermal Treatment, Quantum Dots, Fluorescence

Contents

1. Introduction to carbon quantum dots

Quantum dots (QDs) are the minute fragments of matter; even so small that it can be concentrated into a single point. These are zero-dimensional particles. Due to their extremely small size, the particles carrying electricity get restricted and have well-defined energy levels according to the laws of quantum theory [1-4]. QDs are few nanometer wide crystals which can be made from semiconductors, organic materials as well as from conductors. In spite of the small size, these have the tendency of acting as individual atoms. And although QDs are crystals, they behave more like individual atoms that is why they are also known as artificial atoms [5]. Exceptional quantum size effect of QDs enhances their properties such as wide excitation spectrum, exceptional photostability, high quantum yield of fluorescence and maximum photobleaching.

Increasing environmental issues have made the researchers to discover some particles or materials that are not harmful to nature. Carbon is the skeleton of all living tissues and is also non-toxic in nature. Carbon-based quantum dots (CQDs, C-dots or CDs) offer an alternative to the traditionally used toxic materials. CQDs were first accidentally discovered during the purification of single-walled carbon nanotubes which were prepared through electrophoresis. Later on, these were also prepared through laser ablation of graphite powder and cement in 2006. CQDs have gained huge attention because of their chemical stability, biocompatibility, and low toxicity in comparison to other quantum dots [6-9]. In addition, CQDs are also abundant and inexpensive in nature which increases their utility range [10]. Carbon is basically a black material and normally deliberated to have low solubility in water and feeble fluorescence. Extensive consideration has been concentrated on CQDs due of their noble solubility and strong luminescence; that is why they are also referred to as carbon nano lights [11]. In the past few years, much progress has been achieved in the synthesis, properties, and applications of carbon-based quantum dots and much has been reported by Baker et al., Lee et al., and Zhu et al. Compared to traditional semiconductor quantum dots and organic dyes, photoluminescent CQDs are superior in terms of high (aqueous) solubility, robust chemical inertness, facile modification and high resistance to photobleaching [12-17]. The superior biological properties of carbon quantum dots, such as low toxicity and good biocompatibility, assign them with probable applications in bioimaging, biosensor and drug delivery [18,19]. The high electronic properties of CQDs as electron donor and acceptor, causing chemiluminescence and electrochemical luminescence, bestow them with extensive potentials in optronics, catalysis, and sensors.

2. Structure of carbon quantum dots

Carbon quantum dots are generally spherical in shape with a diameter of approximately 10 nm. CQDs comprise only a few molecules or atoms of nanostructured materials. Their molecular weight ranges from only a few thousand to tens of thousands. The presence of $-COOH$, $-OH$ and $-NH_2$ groups in the structure, gave them the ability to attach to a number of organic, inorganic and biological molecules or groups [20-23]. These groups also make them an efficient tool for a number of applications such as targeted drug delivery, photoluminescence, and sensors, etc.

3. Synthesis of carbon quantum dots

A number of techniques have been proposed for the synthesis of carbon quantum dots. Basically, their synthesis techniques are categorized into two: top- down and bottom- up approaches [24-26]. The top-down approach includes cutting, oxidation and the bigger molecules are broken down to the nanometer size particles. However, the bottom-up approach involves chemical reactions of the precursor molecules to form the nano range products [27,28].

3.1 Top-down approach

Top-down approaches for CQDs are the commonly used preparation measures which include arc-discharge approach [30], electrochemical method [31] and laser ablation [32]. In 2004, during the synthesis of single-walled carbon nanotubes by arc discharge method, Xu et al. accidentally discovered CQDs. They further introduced nitryl onto the CQDs by arc discharge fuming nitric acid so as to enhance its hydrophilicity [33,34]. After that, NaOH solution having pH 8.4 was utilized to extract the precipitate to address a black suspension. Three groups of fluorescent carbon nanoparticles with different molecular weights were obtained after further electrophoretic separation [35-37]. These three groups were known to exhibit a characteristic color of green, yellow and orange, respectively.

3.1.1 Electrochemical method

In the starting stage, electrochemical methodologies were employed mostly for the CQDs synthesis. S.T. Lee et al., utilized graphite electrodes as the carbon source [38]. Particularly, researchers employed graphite rods as both cathode and anode in an electrochemical setup, with NaOH/ethanol as the electrolyte solution. The flow of electric current in this electrochemical cell helps in the production of CQDs. These CQDs show different colors exhibiting their luminescence capacity. Fig. 1 shows the images of CQDs prepared from the electrochemical method. In spite of its high production ability, the electrochemical method still suffers from certain drawbacks such as CQDs produced do not have a uniform size distribution, same surface morphologies, and uniform crystallinity, etc.

3.2 Bottom-up approach

This approach includes the combination of atoms or molecules so as to form a nanoscale structure through chemical reactions and polymerization methods. It basically includes hydrothermal method [39-41], simple pyrolysis [42,43], microwave-based pyrolysis [44,45] and acid dehydration method [46]. Among these, hydrothermal and microwave-based pyrolysis methods are the most extensively used methods for CQD preparation.

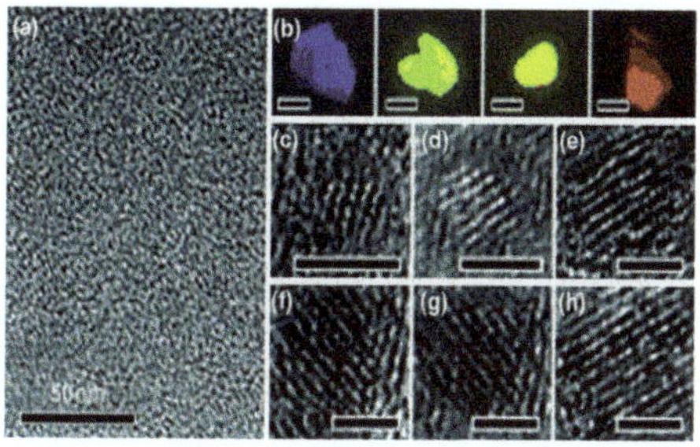

Figure 1 CQDs produced through an electrochemical method. (a) TEM image of the CQDs (exhibiting small < 5 nm diameters); (b) photoluminescence images recorded upon excitation at different wavelengths; (c–h) high-resolution TEM images showing the crystalline graphite lattice planes. Scale bars correspond to 2 nm [38].

3.2.1 Hydrothermal method

Hydrothermal method is the most commonly used method for the synthesis of carbon quantum dots due to its simple mode of operation and versatile nature. It is actually a heterogeneous process which mineralizes under high pressure and temperature conditions. High temperature and pressure conditions help in dissolving the materials which are not under ordinary conditions and produce a nanomaterial with uniform size distribution, surface morphologies, and high purity content, etc. As compared to others, this method is quite eco-friendly and cost-effective w.r.t. instrumentation and maintenance cost [47]. It can be hybridized with many other processes such as microwave, ultrasound, and electrochemistry.

In 2007, the hydrothermal method was used for the first time for the preparation of carbon quantum dots by Liu and his coworkers. Candle ash was chosen as the carbon source mixed with nitric acid through heat reflux and produced a black solution. It was purified through a number of processes such as centrifugation, dialysis, and electrophoresis. This method resulted in fluorescent CQDs having uniform size distribution [48]. S.P. Lau and his colleagues also synthesized CQDs through hydrothermal treatment. They used glucose as the carbon precursor. In this experiment, the CQDs were accumulated by the nucleation process which leads to the development of "self- passivated" layers having different groups such as –OH, -COR and –CH [47] as shown in Fig. 2.

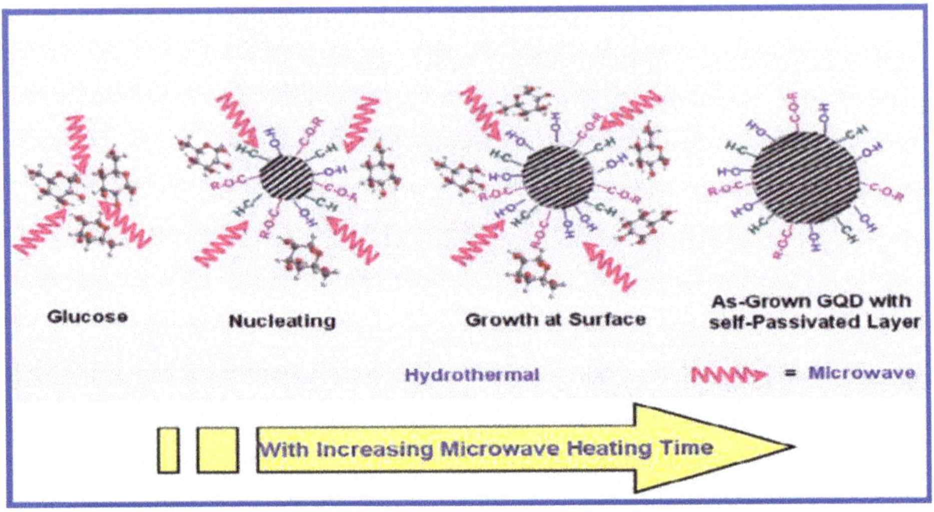

Figure 2 Hydrothermal synthesis of carbon dots using glucose as a carbon source. A model for nucleation and growth of carbon dots induced by microwave heating of glucose solution [47].

3.2.2 Microwave method

In addition to the hydrothermal method, microwave method for the synthesis of carbon quantum dots is also quite in usage due to its fast processing and easy mode of use. Zhu with his coworkers synthesized fluorescent CQDs using the simple microwave method. Carbohydrate and PEG200 was chosen as the carbon precursor and coating agent, respectively [49]. Under 500W microwave power radiation for 2-10 min, the reaction system progressively changed from colorless to a dark brown solution. The dark brown solution was further diluted with water and fluorescent CQDs were obtained. They also demonstrated how the particle size and yield of CQDs depend upon the reaction time. Wang et al. also used this method for the synthesis of CQDs. They used eggshells as the carbon source and converted them into ashes. These ashes were then thoroughly mixed with NaOH aqueous solution. The resultant solution was subjected to the microwave radiation for about 4-5 min. This radiation treatment yielded CQDs with high purity content. The fluorescent quantum yield was noted to be about 14% [50].

Another experiment which demonstrated the synthesis of CQDs was carried out by Chandra and his group [51]. They used sucrose as the carbon source in the presence of phosphoric acid environment under the microwave radiation. The reaction time was adjusted between 3-4 min and the obtained fluorescent CQDs were bright green colored. Yang et al. also used microwave method for the CQD preparation using chitosan as the carbon source. They chemically modified them with amino groups and the resulting particles were used for the adenocarcinoma A549 cell imaging [52].

3.3 Synthetic methods

Researchers have suggested several synthetic methods for the assembly of carbon quantum dots having scarce luminescent properties. H. Lin with his coworkers chemically modified the carbon quantum dots with –N containing functional groups such as amines, amides, and other carbon-nitrogen units. These N- doped CQDs unveiled a fascinating properties as these showed strong emission around 600nm. This is one of the rare observations since a maximum of the CQDs showed luminescence properties at shorter wavelengths (green- blue).

One of the most interesting experiments was the use of waste frying oil for the production of CQDs which were doped with sulfur atoms to improve properties, as demonstrated by J.S. Yu and his colleagues [53]. These sulfur-doped carbon quantum dots were found to show a wide range of properties and were used mainly for the cell imaging. These properties were found to be dependent on the pH of the solution. Fig. 3 is the schematic representation of a synthetic method for the preparation of sulfur-doped carbon quantum dots using waste frying oil as the carbon precursor [53].

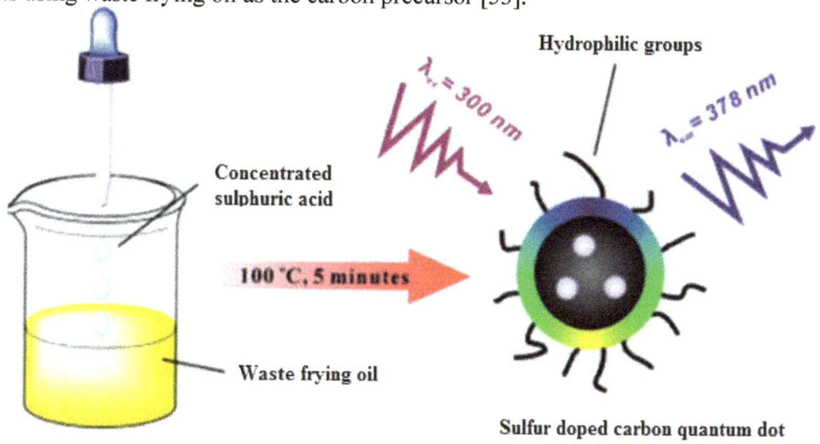

Figure 3 Preparation of sulfur-doped carbon- dots from waste frying oil [53].

The additional experiment includes the use of folic acid as the carbon precursor for the preparation of CQDs. The hydrothermal method was used for their preparation as synthesized CQDs showed high sensitivity towards the folate receptors. These receptors are mostly found in the cancerous cells in the large amount [54]. Fig. 4 shows general route for the synthesis of CQDs using folic acid as a carbon precursor.

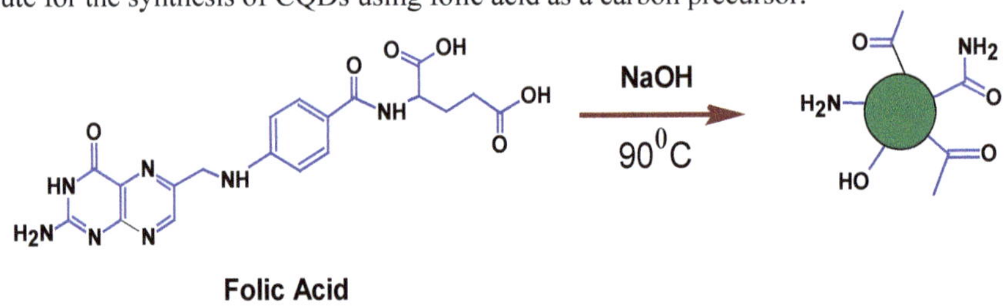

Figure 4 Carbon- dot preparation from folic acid. Scheme of the thermal process for generation of the Carbon dots in basic aqueous solution [54].

Table 1 Different synthesis methods of CQDs from different carbon precursors with their major applications.

S. No.	Carbon source	Method	Application	Ref.
1.	Glucose	Hydrothermal method	Bioimaging	55
2.	Sucrose	Microwave method	Bioimaging	56
3.	Chitosan	Hydrothermal method	Bioimaging	57
4.	Citric acid and ethylenediamine	Hydrothermal treatment	Fe^{3+} sensing	58
5.	Grass	Hydrothermal method	Cu^{2+} sensing	59
6.	Ascorbic acid	Hydrothermal method	pH sensing	60
7.	PEG and saccharide	Microwave method	-	61
8.	Watermelon peels	Carbonization	Bioimaging	62
9.	Chicken egg	Plasma irradiation	Printing	63
10.	Organogel	Topochemical polymerization	-	64

4. Properties of carbon quantum dots

4.1 Optical

CQDs have lucrative absorption in the ultraviolet range, which can also outspread to the visible range. Modification with the help of some passivating agents, the redshift region may be obtained unceasingly. Photoluminescence and electrochemical luminescence are the two main types of luminescence properties shown by CQDs. However, photoluminescence is the most important one [65]. As a pertinent role in all areas of fluorescent nanomaterials, the brilliant optical properties of CQDs chiefly include steep fluorescence durability, nonblinking, tunable excitation, and emission wavelengths. Nevertheless, the complete mechanism of emission of CQDs is still not clear. The intensive quantum study needs an aspiring establishment. Few researchers speculated that the emitting mechanisms of CQDs are complex due to quantum confinement, surface blockade, or exciton recombination radiation process [66].

4.2 Biologically compatible

Carbon quantum dots have quite lower toxicity value and extremely small size which make their smooth passage through the cell walls. Their easy penetrating nature makes them to be used in the biological field. In addition, the amount of CQDs contains a predestine functional groups and thus, can be modified mutually with organic, inorganic, polymer, and other substances endowing different pragmatic properties.

5. Applications of carbon quantum dots

5.1 Chemical sensing

Carbon quantum dots have been used for a number of applications such as chemical sensing, targeted drug delivery, and bioimaging, etc. Fluorescent CQDs have exceptional optical properties, chemical stability, and high water solubility. For this reason, they have been used for metal ions as well as anion recognition and biomolecule detection [67,68]. CQDs have the ability to alter the proficiency of recombination between the surfaces of electron-hole pairs by interaction with the analyte. This phenomenon occurred during the quenching treatment so as to acquire qualitative or quantitative analysis of the object.

CQDs have been used for the detection of toxic heavy metals such as mercury, lead, and chromium, etc. Mercury is among the most toxic metal ions present in the environment having serious carcinogenic effects. Researchers have suggested many different methods for the removal of Hg^{2+} from the environment by the use of CQDs [69-71].

Carbon quantum dots synthesized by the hydrothermal method using grapefruit peel as the carbon precursor has received much attention due to high detection ability towards Hg^{2+}. The detection limit was calculated to be about 0.23nM. Liu et al. acquired CQDs with exceptional luminous stability by the polyethylene glycol (PEG) refluxed with NaOH. They used these CQDs as a sensor that can precisely detect Hg^{2+} in solution and the detection limit was calculated to be 1fM. This test method efficaciously detected Hg^{2+} in the rivers, lakes, and tap water samples and the sensitivity noted was very high [72,73]. In addition, anion and smaller molecules can also be detected using carbon quantum dots based materials.

5.2 Photocatalysis

With the rising urge for low-carbon materials, more emphasis is laid on how to make the full use of solar energy and other relevant cheap energy sources [74]. The visualization of photocatalytic technology caused its expansion, involving solar energy, photovoltaic cells, self-cleaning materials, environmental atomic waste control and multiple other familiar fields. However, the commonly used photocatalytic materials such as TiO_2 and ZnO consistent with absorption range in ultraviolet region; visible light cannot be taken as a perfect advantage and have been greatly isolated in prudent applications [75-79]. Thus, formulating nanomaterials with high stability is of great significance in deciphering environmental problems and energy concerns.

TiO_2 is among the most commonly used photocatalysts. It has been used for the decomposition of organic pollutants and also to generate H_2 from H_2O. But, due to the marginal gap between the valence and conduction band, it shows absorption peaks in shorter wavelengths only (ultraviolet). Thus, to increase its absorption range to higher wavelengths, researchers formulated that the composites of TiO_2 nanoparticles and CQDs can meritoriously enlarge the range of the optical response of the composite structure and intensify the consumption of solar energy and transformation [80-85]. Fig. 5 shows a general mechanism of CQD sensitization. For instance, Zhang et al. utilized CQDs-TiO_2 composites to get H_2 from the water decomposition under visible light irradiation, wherein the introduction of CQDs obviously widens the light response range of TiO_2. Consequently, based on the upconversion property of CQDs and the synergistic effect between the CdSe QDs and CQDs, the CdSe quantum dots were aided to amend the TiO_2-CQD nanocomposites, and the photocatalytic efficiency was greatly enhanced [86].

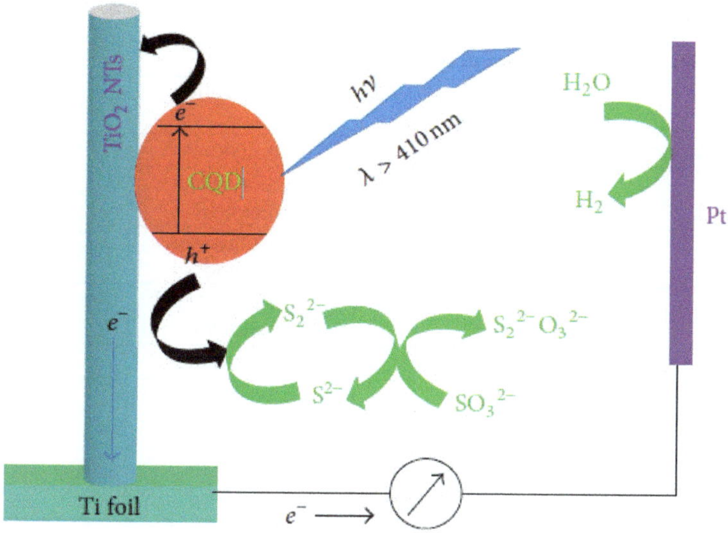

Figure 5 Illustration of the sensitization mechanism of CQDs [86].

5.3 Bioimaging

We know that the CQDs possess great advantages as compared to the traditionally used semiconductor QDs due to their exceptional optical and steady chemical properties. Moreover, CQDs are eco-friendly and low noxious nanomaterials [87]. These characteristics make the CQDs to replace the semiconductor quantum dots in bioimaging.

As an experiment, Yang et al. injected CQDs into the 1 mice body in three different ways: subcutaneous injection, intradermal injection, and intravenous injection. They got effective *in-vivo* imaging under light irradiation of 470 and 545nm. They found that CQDs subcutaneously injected into the mice of the forelimb migrate along the forearm to the axillary lymph nodes, while CQDs of intravenous injection in the systemic circulation concentrated at the bladder area. They also notified that fluorescence images of the urine can also be detected after 3h of injection. After 4h, the fluorescent CQDs exist mainly in the kidney of mice. They speculated that polyethylene glycol could diminish the effective surface of CQDs on the affinity to protein and makes CQDs more difficult in uptake by the liver [88]. Bioimaging applications of different CQDs having different carbon precursor with their method of preparation have been presented in Table 2.

Table 2 Bioimaging applications of different CQDs having different carbon precursors with their method of synthesis.

S. No.	Carbon precursor	Method	Ref.
1.	Phenol/formaldehyde resin	Carbonization	89
2.	Candle soot	HNO_3 oxidation	90
3.	Orange juice	Hydrothermal treatment	91
4.	Sugarcane juice	Hydrothermal method	92
5.	Gelatine	Hydrothermal treatment	93
6.	Hair fibre	H_2SO_4 treatment	94
7.	3-(3,4-Dihydroxyphenyl)-L-alanine, L-histidine, and L-arginine	Carbonization	95
8.	Acetic acid	Carbonization	96
9.	Carbohydrate	Amine passivation	97
10.	Chitosan	Microwave oven	98

5.4 Drug delivery

Traditionally used drug carrier materials do not offer the ability of their observability and traceability. Researchers have suggested the use of fluorescent nanomaterials to the drug carrier system due to their easy detectability [99]. Due to low toxicity, admirable biocompatibility, and modifiable surface functional groups, CQDs have become a center of attraction of the drug delivery research.

Lai et al. prepared CQDs by pyrolysis using glycerol as the carbon precursor. The synthesized CQDs were modified using polyethylene glycol (PEG) and the particle size was controlled through mesoporous nanoparticles to acquire a uniform and excellent optical properties of CQDs (CQDs @ $mSiO_2$-PEG). They used CQDs @ $mSiO_2$-PEG as an anticancer drug carrier, doxorubicin (DOX) and studied the release rate of DOX in HeLa cells. The results revealed that CQDs and DOX in the cytoplasm emitted blue fluorescence. However, red fluorescence was observed in the nucleus [100]. According to the proportions of the two fluorescent colors, this method acts as an active pharmaceutical carrier. They also reported that these can act as a traceability agent in the amount of drug released in the cells.

Zheng et al. imprinted amino groups on CQDs surface with oxaliplatin through the chemical coupling reaction integrating the two substances as one. The reformed CQDs can enter into the interior of cancer cells via endocytosis. Due to the hasty changes in the interior environment of the cancer cells, the drug can be easily released from the surface of the CQDs. The results showed that by monitoring the fluorescent CQD signals in cancer cells, dispersal of oxaliplatin in the cancer cells can be easily detected, which will deliver great help to doctors to grasp the exact time and drug dose injection.

6. Degradation of organic pollutants using carbon quantum dot based composites

Rapid growth in industrialization has led to increasing of effluents in the water system. Industries are the major source for the addition of organic pollutants, dyes and pharmaceutical effluents, etc. to the water. These pollutants have many adverse effects including eutrophication and aquatic life degradation. For this reason, their removal is of immense importance. Researchers have suggested that solar energy can be used in this context due to its renewable, cheap and abundant nature. Degradation of pollutants in the presence of solar light has been proved to be a simple and the most effective method.

The probable utilization of CQDs in photocatalytic mechanisms has been extensively revealed, where they have the tendency to absorb both UV and visible light and directly help in transferring photoexcited electrons. CQDs are highly used for photodegradation processes due to their low cost, easy production, and non-toxicity.

A number of CQD based composites have been prepared and used for the photocatalytic degradation of the organic pollutants. Zhang with his coworkers synthesized CQDs using an electrochemical method which was further modified and CQDs/Ag_3SO_4 & CQDs/Ag/Ag_3SO_4 composite catalysts were prepared. They studied the ability of these composite catalysts for the photocatalytic degradation of methylene blue under visible light radiation. Their experimental results showed that the photocatalytic degradation capability of CDs/Ag/Ag_3SO_4 is about 5.5 times greater than that of CDs/Ag_3SO_4. B.Y. Yu with his co-workers synthesized carbon quantum dots embedded with mesoporous hematite nanospheres with the solvent-thermal process in aqueous solution. The mesoporous hematite nanospheres provide high surface area and the CQDs act as a catalyst, promoting the rate of photocatalytic degradation. Their results showed the high photocatalytic degradation ability having up to 97% retention capacity [101].

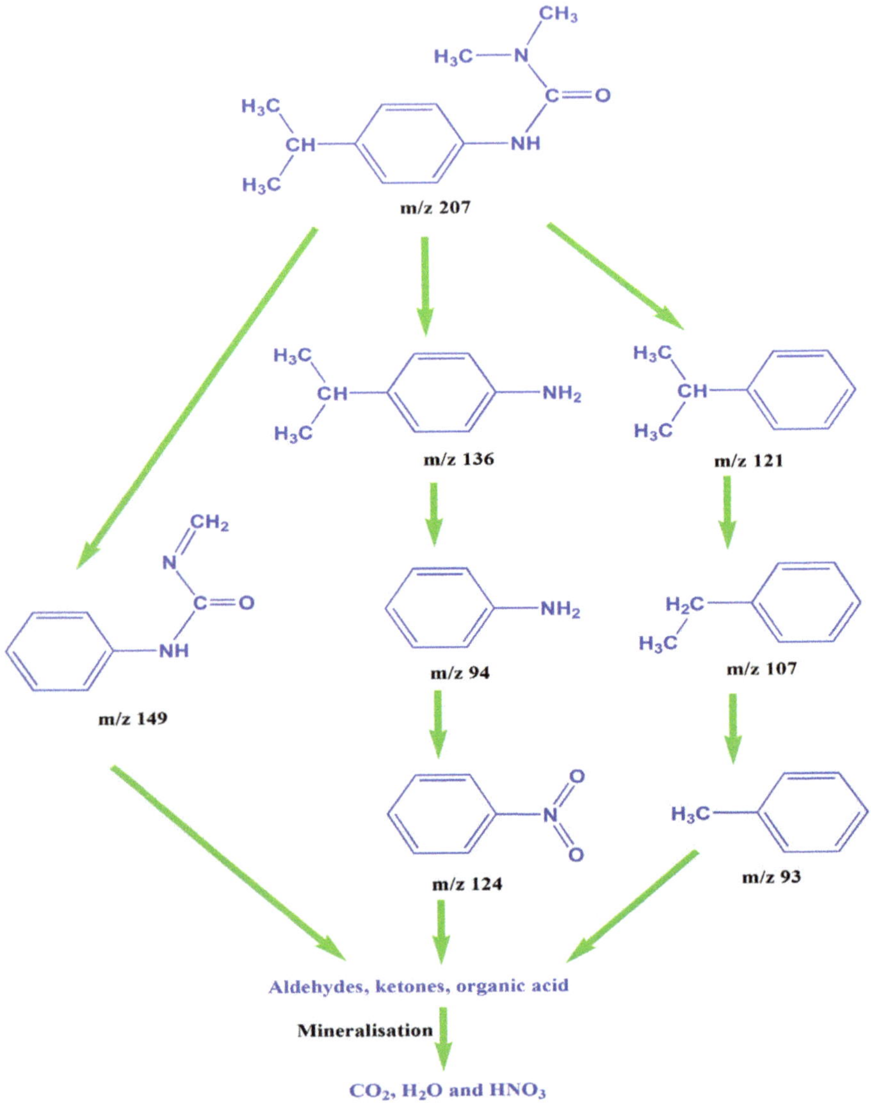

Figure 6 Photocatalytic degradation pathway of isoproturon [102].

For complete exploitation of solar energy, R. Xie, and his co-workers synthesized a novel carbon quantum dots (CQDs) modified $Bi_{20}TiO_{32}$ photocatalysts with varying carbon quantum dot content. Results revealed the uniform distribution of CQDs on the surface of $Bi_{20}TiO_{32}$. Prepared photocatalyst was capable of absorbing light both under visible and near- infrared region due to the presence of CQDs which were acting as photosensitizers. As prepared CQDs/$Bi_{20}TiO_{32}$ composite was used for the degradation of herbicide isoproturon and a maximum of 98.1% degradation was recorded with only 1% carbon quantum dot content. Active species in the photodegradation process were h^+ and $^{\cdot}O_2^{-}$. The degradation rate was about 4.5 times greater than pure $Bi_{20}TiO_{32}$ indicating the high catalytic ability of catalytic ability of CQDs. Fig. 6 shows the photocatalytic degradation pathway of isoproturon [102].

Rhodamine B is an organic dye used for the detection of rate of reaction and the direction of flow. It is a fluorescent dye so quite easily detectable but its presence in high amount in water is a serious matter of concern. M. Ji et al. [103], tried to remove it from the wastewater so to protect the aquatic life. They utilized (CQDs)/$Bi_4O_5I_2$ which were synthesized by solvothermal method. The CQDs, imprinted successfully on the $Bi_4O_5I_2$ acted as transfer sites for the electrons generated during photocatalysis and adsorption site for the organic pollutant. A multipurpose CQDs/ N-ZnO photocatalyst was capable of degrading three commercial dyes namely, malachite green, methylene blue, and fluorescein dyes, within 30–45 min of daylight irradiation. Due to the photo-corrosion property of the synthesized CQDs/N-ZnO photocatalyst, it was used up to four times without significant deactivation [103].

7. Conclusions

Fluorescent CQDs have excellent optical properties as well as good biocompatibility due to which they show a wide range of applications such as bioimaging, biosensing, chemical sensing, drug delivery and photocatalysis, etc. Hydrothermal and microwave methods have been proved to be the best preparative techniques of CQDs with high purity level and maximum yield. Modification helps in injecting a number of functional groups on CQDs which makes it a highly effective sensor and catalyst. The addition of functional groups helps in the generation of electron-hole pairs which further enhance the rate of degradation of the organic pollutants from the wastewater.

References

[1] T. Yang, M. Lu, X. Mao, W. Liu, L. Wan, S. Miao, J. Xu, Synthesis of CdS quantum dots (QDs) via a hot-bubbling route and co-sensitized solar cells

assembly, Chem. Eng. J. 225 (2013) 776-783. https://doi.org/10.1016/j.cej.2013.04.028

[2] Q. Xiao, S. Huang, W. Su, P. Li, J. Ma, F. Luo, J. Chen, Y. Liu, Y. Systematically investigations of conformation and thermodynamics of HSA adsorbed to different sizes of CdTe quantum dots, Colloids Surf. B: Biointerfaces 102 (2013) 76-82. https://doi.org/10.1016/j.colsurfb.2012.08.028

[3] V.K. Gupta, T.A. Saleh, D. Pathania, B.S. Rathore, G. Sharma, A cellulose acetate based nanocomposite for photocatalytic degradation of methylene blue dye under solar light, Ionics 21 (2015) 1787-1793. https://doi.org/10.1007/s11581-014-1323-9

[4] S.N. Baker, G.A. Baker, Luminescent Carbon Nanodots: Emergent Nano lights. Angew. Chem. Int. Ed. 49 (2010) 6726- 6744. https://doi.org/10.1002/anie.200906623

[5] Z. Yang, Z. Li, M. Xu, Y. Ma, J. Zhang, Y. Su, F. Gao, H. Wei, L. Zhang, Controllable Synthesis of Fluorescent Carbon Dots and Their Detection Application as Nanoprobes. Nano-Micro Letters 5 (2013) 247-259. https://doi.org/10.1007/BF03353756

[6] A. Fu, W. Gu, C. Larabell, A.P. Alivisatos, Semiconductor nanocrystals for biological imaging, Curr. Opin. Neurobio. 15 (2005) 568-575. https://doi.org/10.1016/j.conb.2005.08.004

[7] J.E. Halpert, V.J. Porter, J.P. Zimmer, M.G. Bawendi, Synthesis of CdSe/CdTe nanobarbells. J. Am. Chem. Soc. 128 (2006) 12590-12591. https://doi.org/10.1021/ja0616534

[8] P. Reiss, S. Carayon, J. Bleuse, A. Pron, Low polydispersity core/shell nanocrystals of CdSe/ ZnSe and CdSe/ZnSe/ZnS type: preparation and optical studies, Synth. Met. 139 (2003) 649- 652. https://doi.org/10.1016/S0379-6779(03)00335-7

[9] J. Chen, J.L. Song, X.W. Sun, W.Q. Deng, C.Y. Jiang, W. Lei, J.H. Huang, R.S. Liu, An oleic acid capped CdSe quantum-dot sensitized solar cell, Appl. Phys. Lett. 94 (2009) 153115(1-3). https://doi.org/10.1063/1.3117221

[10] P. Reiss, J. Bleuse, A. Pron, Highly luminescent CdSe/ZnSe core/shell nanocrystals of low size dispersion, Nano Lett. 2 (2002) 781-784. https://doi.org/10.1021/nl025596y

[11] Y. Jin-nouchi, T. Hattori, Y. Sumida, M. Fujishima, H. Tada, PbS quantum dot-sensitized photoelectrochemical cell for hydrogen production from water under illumination of simulated sunlight, Chemphyschem 11 (2010) 3592-3595. https://doi.org/10.1002/cphc.201000593

[12] Y. Tachibana, H.Y. Akiyama, Y. Ohtsuka, T. Torimoto, S. Kuwabata, CdS quantum dots sensitized TiO_2 sandwich type photoelectrochemical solar cells, Chem. Lett. 36 (2007) 88-89. https://doi.org/10.1246/cl.2007.88

[13] Y. Song, D. Feng, W. Shi, X. Li, H. Ma, Parallel comparative studies on the toxic effects of unmodified CdTe quantum dots, gold nanoparticles, and carbon nanodots on live cells as well as green gram sprouts, Talanta 116 (2013) 237-244. https://doi.org/10.1016/j.talanta.2013.05.022

[14] Y. Liu, C.Y. Liu, Z.Y. Zhang, Synthesis and surface photochemistry of graphitized carbon quantum dots, J. Colloid Interface Sci. 356 (2011) 416-421. https://doi.org/10.1016/j.jcis.2011.01.065

[15] Y. Dong, N. Zhou, X. Lin, J. Lin, Y. Chi, G. Chen, Extraction of Electrochemiluminescent Oxidized Carbon Quantum Dots from Activated Carbon, Chem. Mater. 22 (2010) 5895-5899. https://doi.org/10.1021/cm1018844

[16] Y. Liu, Y.X. Yu, W.D. Zhang, Carbon quantum dots-doped CdS microspheres with enhanced photocatalytic performance, J. Alloys Compd. 569 (2013) 102-110. https://doi.org/10.1016/j.jallcom.2013.03.202

[17] Y.Q. Dong, C.Q. Chen, J.P. Lin, N.N. Zhou, Y.W. Chi, G.N. Chen, Electrochemiluminescence emission from carbon quantum dot-sulfite coreactant system, Carbon 56 (2013) 12-17. https://doi.org/10.1016/j.carbon.2012.12.086

[18] X. Zhang, H. Ming, R.H. Liu, X. Han, Z.H. Kang, Y. Liu, Y.L. Zhang, Highly sensitive humidity sensing properties of carbon quantum dots films, Mater. Res. Bull. 48 (2013) 790-794. https://doi.org/10.1016/j.materresbull.2012.11.056

[19] K. Hola, Y. Zhang, Y. Wang, E.P. Giannelis, R. Zboril, A.L. Rogach, Carbon dots—Emerging light emitters for bioimaging, cancer therapy and optoelectronics, Nano Today 9 (2014) 590-603. https://doi.org/10.1016/j.nantod.2014.09.004

[20] L. Cao, X. Wang, M.J. Meziani, F. Lu, H. Wang, P.G. Luo, Y. Lin, B.A. Harruff, L.M. Veca, D. Murray, S.Y. Xie, Carbon dots for multiphoton bioimaging, J. Am. Chem. Soc. 129 (2007) 11318-9. https://doi.org/10.1021/ja073527l

[21] H. Zhang, Y. Chen, M. Liang, L. Xu, S. Qi, H. Chen, X. Chen, Solid-phase synthesis of highly fluorescent nitrogen-doped carbon dots for sensitive and

selective probing ferric ions in living cells, Anal. Chem. 86 (2014) 9846-52. https://doi.org/10.1021/ac502446m

[22] S.Y. Park, H.U. Lee, E.S. Park, S.C. Lee, J.W. Lee, S.W. Jeong, C.H. Kim, Y.C. Lee, Y.S. Huh, J. Lee, Photoluminescent green carbon nanodots from food-waste-derived sources: large-scale synthesis, properties, and biomedical applications, ACS Appl. Mater. Interfaces 6 (2014) 3365-70. https://doi.org/10.1021/am500159p

[23] S.L. Hu, K.Y. Niu, J. Sun, J. Yang, N.Q. Zhao, X.W. Du, One-step synthesis of fluorescent carbon nanoparticles by laser irradiation, J. Mater. Chem. 19 (2009) 484–488. https://doi.org/10.1039/B812943F

[24] H. Liu, T. Ye, C. Mao, Fluorescent carbon nanoparticles derived from candle soot, Angew. Chem. Int. Ed. 46 (2007) 6473–6475. https://doi.org/10.1002/anie.200701271

[25] L. Wang, S.J. Zhu, H.Y. Wang, S.N. Qu, Y.L. Zhang, J.H. Zhang, Q.D. Chen, H.L. Xu, W. Han, B. Yang, H.B. Sun, Common origin of green luminescence in carbon nanodots and graphene quantum dots, ACS Nano 8 (2014) 2541-2547. https://doi.org/10.1021/nn500368m

[26] H. Li, Z. Kang, Y. Liu, S.T. Lee, Carbon nanodots: synthesis, properties and applications, J. Mater. Chem. 22 (2012) 24230–24253. https://doi.org/10.1039/c2jm34690g

[27] X. Jia, J. Li, E. Wang, One-pot green synthesis of optically pH-sensitive carbon dots with upconversion luminescence, Nanoscale 4 (2012) 5572–5575. https://doi.org/10.1039/c2nr31319g

[28] V. Polshettiwar, M.N. Nadagouda, R.S. Varma, Microwave-assisted chemistry: A rapid and sustainable route to synthesis of organics and nanomaterials. Aust. J. Chem. 62 (2009) 16-26. https://doi.org/10.1071/CH08404

[29] X. Xu, R. Ray, Y. Gu, H.J. Ploehn, L. Gearheart, K. Raker, W.A. Scrivens, Electrophoretic analysis and purification of fluorescent single-walled carbon nanotube fragments, J. Am. Chem. Soc. 126 (2004) 12736-1227. https://doi.org/10.1021/ja040082h

[30] H. Jiang, F. Chen, M.G. Lagally, F.S. Denes, Newstrategy for synthesis and functionalization of carbon nanoparticles, Langmuir 26 (2010) 1991–1995. https://doi.org/10.1021/la9022163

[31] J. Zhou, C. Booker, R. Li, X. Zhou, T.K. Sham, X. Sun, Z. Ding, An electrochemical avenue to blue luminescent nanocrystals from multiwalled carbon nanotubes (MWCNTs), J. Am. Chem. Soc. 129 (2007) 744-745. https://doi.org/10.1021/ja0669070

[32] H.M.R. Gonc¸alves, A.J. Duarte, J.C.G. Esteves da Silva, Optical fiber sensor for Hg(II) based on carbon dots, Biosens. Bioelectron. 26 (2010) 1302–1306. https://doi.org/10.1016/j.bios.2010.07.018

[33] S. Hu, J. Liu, J. Yang, Y. Wang, S. Cao, Laser synthesis and size tailor of carbon quantum dots, J. Nanopart. Res. 13 (2011) 7247–7252. https://doi.org/10.1007/s11051-011-0638-y

[34] Y.P. Sun, B. Zhou, Y. Y. Lin, W. Wang, K.S. Fernando, P. Pathak, M.J. Meziani, B.A. Harruff, X. Wang, H. Wang, P.G. Luo, Quantum-sized carbon dots for bright and colorful photoluminescence, J. Am. Chem. Soc. 128 (2006) 7756–7757. https://doi.org/10.1021/ja062677d

[35] X. Zheng, A. Ananthanarayanan, Q. Luo, P. Chen, Glowing graphene quantum dots and carbon dots: properties, syntheses, and biological applications, Small 11 (2015) 1620-1636. https://doi.org/10.1002/smll.201402648

[36] K.A.S. Fernando, S. Sahu, Y. Liu, W.K. Lewis, E. Guliants, A. Jafariyan, P. Wang, C. Bunker, Y. Sun, Carbon quantum dots and applications in photocatalytic energy conversion, ACS Appl. Mater. Interfaces 7 (2015) 8363-8376. https://doi.org/10.1021/acsami.5b00448

[37] Q. Xiang, J. Yu, M. Jaroniec, Graphene-based semiconductor photocatalysts, Chem. Soc. Rev. 41 (2012) 782-796. https://doi.org/10.1039/C1CS15172J

[38] H. Li, X. He, Z. Kang, H. Huang, Y. Liu, J. Liu, S. Lian, C.H.A. Tsang, X. Yang, S.T. Lee, Water-soluble fluorescent carbon quantum dots and photocatalyst design, Angew. Chem. Int. Ed. 49 (2010) 4430–4434. https://doi.org/10.1002/anie.200906154

[39] H. Dong, A. Kuzmanoski, D. M.G¨oßl, R. Popescu, D. Gerthsen, C. Feldmann, Polyol-mediated C-dot formation showing efficient Tb^{3+}/Eu^{3+} emission, Chem. Commun. 50 (2014) 7503–7506. https://doi.org/10.1039/C4CC01715C

[40] V.K. Gupta, G. Sharma, D. Pathania, N.C. Kothiyal, Nanocomposite pectin Zr(IV) selenotungstophosphate for adsorptional/photocatalytic remediation of methylene blue and malachite green dyes from aqueous system, J. Ind. Eng. Chem. Res. 21 (2014) 957-964. https://doi.org/10.1016/j.jiec.2014.05.001

[41] S.C. Ray, A. Saha, N.R. Jana, R. Sarkar, Fluorescent carbon nanoparticles: synthesis, characterization, and bioimaging application, J. Phys. Chem. C 113 (2009) 18546–18551. https://doi.org/10.1021/jp905912n

[42] G. Sharma, M. Naushad, A. Kumar, Shilpa, M.R. Khan, Lanthanum/cadmium/polyaniline bimetallic nanocomposite for the photodegradation of organic pollutant, Iran Polym. J. 24 (2015) 1003-1013. https://doi.org/10.1007/s13726-015-0388-2

[43] A.B. Bourlinos, A. Stassinopoulos, D. Anglos, R. Zboril, M. Karakassides, E.P. Giannelis, Surface functionalized carbogenic quantum dots, Small 4 (2008) 455–458. https://doi.org/10.1002/smll.200700578

[44] H. Zhu, X. Wang, Y. Li, Z. Wang, F. Yang, X. Yang, Microwave synthesis of fluorescent carbon nanoparticles with electrochemiluminescence properties, Chem. Commun. 34 (2009) 5118–5120. https://doi.org/10.1039/b907612c

[45] Q. Wang, X. Liu, L. Zhang, Y. Lv, Microwave-assisted synthesis of carbon nanodots through an eggshell membrane and their fluorescent application, The Analyst 137 (2012) 5392–5397. https://doi.org/10.1039/c2an36059d

[46] G.E. LeCroy, S.K. Sonkar, F. Yang, L.M. Veca, P. Wang, K.N. Tackett, J.J. Yu, E. Vasile, H. Qian, Y. Liu, P. Luo, Toward structurally defined carbon dots as ultracompact fluorescent probes, ACS Nano 8 (2014) 4522-9. https://doi.org/10.1021/nn406628s

[47] L. Tang, R. Ji, X. Cao, J. Lin, H. Jiang, X. Li, K.S. Teng, C.M. Luk, S. Zeng, J. Hao, S.P. Lau, Deep ultraviolet photoluminescence of water-soluble self-passivated graphene quantum dots, ACS Nano 6 (2012) 5102-10. https://doi.org/10.1021/nn300760g

[48] M.X. Gao, C.F. Liu, Z.L. Wu, Q.L. Zeng, X.X. Yang, W.B. Wu, Y.F. Li, C.Z. Huang, A surfactant-assisted redox hydrothermal route to prepare highly photoluminescent carbon quantum dots with aggregation-induced emission enhancement properties, Chem. Commun. 49 (2013) 8015-8017. https://doi.org/10.1039/c3cc44624g

[49] M. Xu, G. He, Z. Li, F. He, F. Gao, Y. Su, L. Zhang, Z. Yang, Y. Zhang, A green heterogeneous synthesis of N-doped carbon dots and their photoluminescence applications in solid and aqueous states, Nanoscale 6 (2014) 10307-10315. https://doi.org/10.1039/C4NR02792B

[50] D. Pathania, R. Katwal, G. Sharma, M. Naushad, M.R. Khan, A.H. Al-Muhtaseb, Novel guar gum/Al_2O_3 nanocomposite as an effective photocatalyst for the degradation of malachite green dye, Int. J. Biol. Macromol. 87 (2016) 366–374. https://doi.org/10.1016/j.ijbiomac.2016.02.073

[51] S. Chandra, P. Das, S. Bag, D. Laha, P. Pramanik, Synthesis, functionalization and bioimaging applications of highly fluorescent carbon nanoparticles, Nanoscale 3 (2011) 1533–1540. https://doi.org/10.1039/c0nr00735h

[52] V.K. Gupta, D. Pathania, T.A. Saleh, G. Sharma, Liquid phase synthesis of pectin–cadmium sulfide nanocomposite and its photocatalytic and antibacterial activity, J. Mol. Liq. 196 (2014) 107-112. https://doi.org/10.1016/j.molliq.2014.03.021

[53] Y. Hu, J. Yang, J. Tian, L. Jia, J.S. Yu, Waste frying oil as a precursor for one-step synthesis of sulfur-doped carbon dots with pH-sensitive photoluminescence, Carbon 77 (2014) 775–782. https://doi.org/10.1016/j.carbon.2014.05.081

[54] S.K. Bhunia, A.R. Maity, S. Nandi, D. Stepensky, R. Jelinek, Imaging cancer cells expressing the folate receptor with carbon dots produced from folic acid, Chem. Bio. Chem. 17 (2016) 614-619. https://doi.org/10.1002/cbic.201500694

[55] Z.C. Yang, M. Wang, A.M. Yong, S.Y. Wong, X.H. Zhang, H. Tan, A.Y. Chang, X. Li, J. Wang, Intrinsically fluorescent carbon dots with tunable emission derived from hydrothermal treatment of glucose in the presence of monopotassium phosphate, Chem. Commun. 47 (2011) 11615-11617. https://doi.org/10.1039/c1cc14860e

[56] W. Wang, Y.R. Ni, Z.Z. Xu, One-step uniformly hybrid carbon quantum dots with high-reactive TiO_2 for photocatalytic application, J. Alloys Compd. 622 (2015) 303–308. https://doi.org/10.1016/j.jallcom.2014.10.076

[57] Y. Yang, J. Cui, M. Zheng, C. Hu, S. Tan, Y. Xiao, Q. Yang, Y. Liu, One-step synthesis of amino-functionalized fluorescent carbon nanoparticles by hydrothermal carbonization of chitosan, Chem. Commun. 48 (2012) 380-382. https://doi.org/10.1039/C1CC15678K

[58] S. Zhu, Q. Meng, L. Wang, J. Zhang, Y. Song, H. Jin, K. Zhang, H. Sun, H. Wang, B. Yang, Highly photoluminescent carbon dots for multicolor patterning, sensors, and bioimaging, Angew. Chem. Int. Ed. 52 (2013) 3953-3957. https://doi.org/10.1002/anie.201300519

[59] S. Liu, J. Tian, L. Wang, Y. Zhang, X. Qin, Y. Luo, A.M. Asiri, A.O. Al-Youbi, X. Sun, Hydrothermal treatment of grass: A low-cost, green route to nitrogen-

doped, carbon-rich, photoluminescent polymer nanodots as an effective fluorescent sensing platform for label-free detection of Cu(II) ions, Adv. Mater. 24 (2014) 2037-2041. https://doi.org/10.1002/adma.201200164

[60] H. Wu, C. Mi, H. Huang, B. Han, J. Li, S. Xu, Solvothermal synthesis of green-fluorescent carbon nanoparticles and their application, J. Lumin. 132 (2012) 1603-1607. https://doi.org/10.1016/j.jlumin.2011.12.077

[61] Z.A. Qiao, Y. Wang, Y. Gao, H. Li, T. Dai, Y. Liu, Q. Huo, Commercially activated carbon as the source for producing multicolor photoluminescent carbon dots by chemical oxidation, Chem. Commun. 46 (2010) 8812-8814. https://doi.org/10.1039/c0cc02724c

[62] J. Zhou, Z. Sheng, H. Han, M. Zou, C. Li, Facile synthesis of fluorescent carbon dots using watermelon peel as a carbon source, Mater. Lett. 66 (2012) 222-4. https://doi.org/10.1016/j.matlet.2011.08.081

[63] J. Wang, C.F. Wang, S. Chen, Amphiphilic egg-derived carbon dots: Rapid plasma fabrication, pyrolysis process, and multicolor printing patterns, Angew. Chem. Int. Ed. 124 (2010) 9431-9435. https://doi.org/10.1002/ange.201204381

[64] J.R. Néabo, C. Vigier-Carrière, S. Rondeau-Gagné, J.F. Morin, Room-temperature synthesis of soluble, fluorescent carbon nanoparticles from organogel precursors, Chem. Commun. 48 (2012) 10144-10146. https://doi.org/10.1039/c2cc35087d

[65] S.L. Hu, K.Y. Niu, J. Sun, J. Yang, N.Q. Zhao, X.W. Du, One-step synthesis of fluorescent carbon nanoparticles by laser irradiation, J. Mater. Chem. 19 (2009) 484–488. https://doi.org/10.1039/B812943F

[66] H. Peng, J. Travas-Sejdic, Simple aqueous solution route to luminescent carbogenic dots fromcarbohydrates, Chem. Mater. 21 (2009) 5563–5565. https://doi.org/10.1021/cm901593y

[67] H.M. Goncalves, A.J. Duarte, J.C. da Silva, Optical fiber sensor for Hg(II) based on carbon dots, Biosens. Bioelectron. 26 (2010) 1302-1306. https://doi.org/10.1016/j.bios.2010.07.018

[68] S. Barman, M. Sadhukhan, Facile bulk production of highly blue fluorescent graphitic carbon nitride quantum dots and their application as highly selective and sensitive sensors for the detection of mercuric and iodide ions in aqueous media, J. Mater. Chem. 22 (2012) 21832-21837. https://doi.org/10.1039/c2jm35501a

[69] Y. Liu, C.Y. Liu, Z.Y. Zhang, Synthesis of highly luminescent graphitized carbon dots and the application in the Hg^{2+} detection, Appl. Surf. Sci. 263 (2012) 481-485. https://doi.org/10.1016/j.apsusc.2012.09.088

[70] F. Yan, Y. Zou, M. Wang, X. Mu, N. Yang, L. Chen, Highly photoluminescent carbon dots-based fluorescent chemosensors for sensitive and selective detection of mercury ions and application of imaging in living cells, Sens. Actuators, B 192 (2014) 488-495. https://doi.org/10.1016/j.snb.2013.11.041

[71] H. Huang, J.J. Lv, D.L. Zhou, N. Bao, Y. Xu, A.J. Wang, J.J. Feng, One-pot green synthesis of nitrogen-doped carbon nanoparticles as fluorescent probes for mercury ions, RSC Adv. 3 (2013) 21691-21696. https://doi.org/10.1039/c3ra43452d

[72] S.Y. Lim, W. Shen, Z. Gao, Carbon quantum dots and their applications, Chem. Soc. Rev. 44 (2015) 362-381. https://doi.org/10.1039/C4CS00269E

[73] H. Gonçalves, P.A. Jorge, J.R. Fernandes, J.C. da Silva, Hg (II) sensing based on functionalized carbon dots obtained by direct laser ablation, Sens. Actuators, B 145 (2010) 702-717. https://doi.org/10.1016/j.snb.2010.01.031

[74] A.S. Kumar, T. Ye, T. Takami, B.C. Yu, A.K. Flatt, J.M. Tour, P.S. Weiss, Reversible photo-switching of single azobenzene molecules in controlled nanoscale environments, Nano Letters 8 (2008) 1644-1648. https://doi.org/10.1021/nl080323+

[75] M. Zhang, Q. Wang, C. Chen, L. Zang, W. Ma, J. Zhao, Oxygen atom transfer in the photocatalytic oxidation of alcohols by TiO_2: oxygen isotope studies, Angew. Chem. Int. Ed. 121 (2009) 6197-6200. https://doi.org/10.1002/ange.200900322

[76] Q. Wang, M. Zhang, C. Chen, W. Ma, J. Zhao, Photocatalytic aerobic oxidation of alcohols on TiO2: the acceleration effect of a Brønsted acid, Angew. Chem. Int. Ed. 49 (2010) 7976-7979. https://doi.org/10.1002/anie.201001533

[77] Y. Zhang, N. Zhang, Z.R. Tang, Y.J. Xu, Improving the photocatalytic performance of graphene–TiO 2 nanocomposites via a combined strategy of decreasing defects of graphene and increasing interfacial contact, Ber. Bunsen-Ges. Phys. Chem. 14 (2012) 9167-9175.

[78] H. Li, R. Liu, S. Lian, Y. Liu, H. Huang, Z. Kang, Near-infrared light controlled photocatalytic activity of carbon quantum dots for highly selective oxidation reaction, Nanoscale 5 (2013) 3289-3297. https://doi.org/10.1039/c3nr00092c

[79] Z. Ma, H. Ming, H. Huang, Y. Liu, Z. Kang, One-step ultrasonic synthesis of fluorescent N-doped carbon dots from glucose and their visible-light sensitive photocatalytic ability, New J. Chem. 36 (2012) 861-864. https://doi.org/10.1039/c2nj20942j

[80] S. Hu, R. Tian, Y. Dong, J. Yang, J. Liu, Q. Chang, Modulation and effects of surface groups on photoluminescence and photocatalytic activity of carbon dots, Nanoscale 5 (2013) 11665-11671. https://doi.org/10.1039/c3nr03893a

[81] S. Hu, R. Tian, L. Wu, Q. Zhao, J. Yang, J. Liu, S. Cao, Chemical regulation of carbon quantum dots from synthesis to photocatalytic activity, Chem. Asian J. 8 (2013) 1035-1041. https://doi.org/10.1002/asia.201300076

[82] H. Li, R. Liu, W. Kong, J. Liu, Y. Liu, L. Zhou, X. Zhang, S.T. Lee, Z. Kang, Carbon quantum dots with photo-generated proton property as efficient visible light controlled acid catalyst, Nanoscale 6 (2014) 867-873. https://doi.org/10.1039/C3NR03996J

[83] J. Di, J. Xia, Y. Ge, H. Li, H. Ji, H. Xu, Q. Zhang, H. Li, M. Li, Novel visible-light-driven CQDs/Bi 2 WO 6 hybrid materials with enhanced photocatalytic activity toward organic pollutants degradation and mechanism insight, Appl. Catal. B 168 (2015) 51-61. https://doi.org/10.1016/j.apcatb.2014.11.057

[84] X. Chen, S.S. Mao, Titanium dioxide nanomaterials: synthesis, properties, modifications, and applications, Chem. Rev. 107 (2007) 2891-2895. https://doi.org/10.1021/cr0500535

[85] Z. Kang, Y. Liu, C.H. Tsang, D.D. Ma, X. Fan, N.B. Wong, S.T. Lee, Water-soluble silicon quantum dots with wavelength-tunable photoluminescence, Adv. Mater. 21 (2009) 661-664. https://doi.org/10.1002/adma.200801642

[86] X. Zhang, F. Wang, H. Huang, H. Li, X. Han, Y. Liu, Z. Kang, Carbon quantum dot sensitized TiO_2 nanotube arrays for photoelectrochemical hydrogen generation under visible light, Nanoscale 5 (2013) 2274– 2278. https://doi.org/10.1039/c3nr34142a

[87] V.N. Mehta, S. Jha, S. K. Kailasa, One-pot green synthesis of carbon dots by using Saccharum officinarum juice for fluorescent imaging of bacteria (Escherichia coli) andyeast (Saccharomyces cerevisiae) cells, Mater. Sci. Eng. C 38 (2014) 20–27. https://doi.org/10.1016/j.msec.2014.01.038

[88] S.T. Yang, L. Cao, P.G. Luo, F. Lu, X. Wang, H. Wang, M.J. Meziani, Y. Liu, G. Qi, Y.P. Sun, Carbon dots for optical imaging in vivo, J. Amer. Chem. Soc. 131 (2009) 11308–11309. https://doi.org/10.1021/ja904843x

[89] R. Liu, D. Wu, S. Liu, K. Koynov, W. Knoll, Q. Li, An aqueous route to multicolor photoluminescent carbon dots using silica spheres as carriers, Angew. Chem. Int. Ed. 121 (2009) 4668-4671. https://doi.org/10.1002/ange.200900652

[90] S.C. Ray, A. Saha, N.R. Jana, R. Sarkar, Fluorescent carbon nanoparticles: synthesis, characterization, and bioimaging application, J. Phys. Chem. 113 (2009) 18546-18551. https://doi.org/10.1021/jp905912n

[91] S. Sahu, B. Behera, T.K. Maiti, S. Mohapatra, Simple one-step synthesis of highly luminescent carbon dots from orange juice: application as excellent bio-imaging agents, Chem. Commun. 48 (2012) 8835-8837. https://doi.org/10.1039/c2cc33796g

[92] V.N. Mehta, S. Jha, S.K. Kailasa, One-pot green synthesis of carbon dots by using Saccharum officinarum juice for fluorescent imaging of bacteria (Escherichia coli) and yeast (Saccharomyces cerevisiae) cells, Mater. Sci. Eng. C 38 (2014) 20-27. https://doi.org/10.1016/j.msec.2014.01.038

[93] Q. Liang, W. Ma, Y. Shi, Z. Li, X. Yang, Easy synthesis of highly fluorescent carbon quantum dots from gelatin and their luminescent properties and applications, Carbon 60 (2013) 421-428. https://doi.org/10.1016/j.carbon.2013.04.055

[94] D. Sun, R. Ban, P.H. Zhang, G.H. Wu, J.R. Zhang, J.J. Zhu, Hair fiber as a precursor for synthesizing of sulfur-and nitrogen-co-doped carbon dots with tunable luminescence properties, Carbon 64 (2013) 424-434. https://doi.org/10.1016/j.carbon.2013.07.095

[95] Y. Xu, M. Wu, Y. Liu, X.Z. Feng, X.B. Yin, X.W. He, Y.K. Zhang, Nitrogen-doped carbon dots: A facile and general preparation method, Photoluminescence investigation, and imaging applications, Chem. Eur. J. 19 (2013) 2276-2283. https://doi.org/10.1002/chem.201203641

[96] P. Miao, K. Han, Y. Tang, B. Wang, T. Lin, W. Cheng, Recent advances in carbon nanodots: synthesis, properties and biomedical applications, Nanoscale 7 (2015) 1586-1595. https://doi.org/10.1039/C4NR05712K

[97] H. Peng, J. Travas-Sejdic, Simple aqueous solution route to luminescent carbogenic dots from carbohydrates, Chem. Mater. 21 (2009) 5563-5565. https://doi.org/10.1021/cm901593y

[98] D. Chowdhury, N. Gogoi, G. Majumdar, Fluorescent carbon dots obtained from chitosan gel, RSC Adv. 2 (2012) 12156-12159. https://doi.org/10.1039/c2ra21705h

[99] C.W. Lai, Y.H. Hsiao, Y.K. Peng, P.T. Chou, Facile synthesis of highly emissive carbon dots from pyrolysis of glycerol; gram scale production of carbon dots/mSiO2 for cell imaging and drug release, J. Mater. Chem. 22 (2012) 14403–14409. https://doi.org/10.1039/c2jm32206d

[100] D. Pathania, D. Gupta, A.H. Al-Muhtaseb, G. Sharma, A. Kumar, M. Naushad, T. Ahamad, S.M. Alshehri, Photocatalytic degradation of highly toxic dyes using chitosan-g-poly(acrylamide)/ZnS in presence of solar irradiation, J. Photochem. Photobiol. A 329 (2016) 61-68. https://doi.org/10.1016/j.jphotochem.2016.06.019

[101] B.Y. Yu, S.Y. Kwak, Carbon quantum dots embedded with mesoporous hematite nanospheres as efficient visible light-active photocatalysts, J. Mater. Chem. 22 (2012) 8345- 8353. https://doi.org/10.1039/c2jm16931b

[102] R. Xie, L. Zhang, H. Xu, Y. Zhong, X. Sui, Z. Mao, Construction of up-converting fluorescent carbon quantum dots/$Bi_{20}TiO_{32}$ composites with enhanced photocatalytic properties under visible light, Chem. Eng. J. 310 (2017) 79–90. https://doi.org/10.1016/j.cej.2016.10.089

[103] S. Muthulingam, I.H. Lee, P. Uthirakumar, Highly efficient degradation of dyes by carbon quantum dots/N-doped zinc oxide(CQD/N-ZnO) photocatalyst and its compatibility on three different commercial dyes under day light, J. Colloid Interface Sci. 455 (2015) 101–109. https://doi.org/10.1016/j.jcis.2015.05.046

Chapter 6

Methods for the Detection, Determination and Removal of Phenolic Compounds from Wastewater

Qudsia Kanwal[1], Dure Najaf Iqbal[1], Munawar Iqbal[1,*]

[1] Department of Chemistry, University of Lahore, Lahore, Pakistan

* bosalvee@yahoo.com

Abstract

Phenolic compounds are pollutants of main concern, which at very low concentration are highly toxic. In this chapter, we have discussed the comparison of the applicability and efficiency of conventional as well as advanced methods of treatment of wastewater for removal of phenols. Conventional treatments like extraction, adsorption, electrochemical as well as chemical oxidation, and distillation, have been effectively used for many phenolic compounds, but advanced treatments like ozonation, Fenton processes, photochemical treatment, and wet air oxidation have received little attention as compared to that of conventional treatment methods. Compared to physicochemical treatment, biological treatment is energy saving and environment-friendly as compared to that of physicochemical treatment. But it is not a very effective treatment method for pollutants with high concentration. Enzymatic treatment is the best treatment method for the removal of phenols with a number of enzymes such as peroxidases, laccases, and tyrosinases under gentle conditions.

Keywords

Wastewater Treatment, Phenol Toxicity, Degradation, Bioremediation

Contents

1. Introduction

At present, the most crucial problem faced by humans is the environmental pollution. Groundwater and soil quality is affected by larger production/leakage of toxic compounds into the environment as an effluent [1, 2]. For the past 20 years, the immense increase in industrialization in the region of the world has a big role in environmental pollution. Moreover, the chlorinated phenolic compounds and derivatives are produced by the natural decomposition of plants by fungi e.g. tannins are produced within the plants as secondary metabolites. Different human activities are responsible for various sources of phenols such as phenolic herbicides and biocides such as 2, 4-dichlorophenoxyacetic acid (2, 4-D) and pentachlorophenol (PCP reside in the environment due to human activities [3]. Dissolved phenols and polycyclic aromatic hydrocarbons (PAHs) are present in industrial effluents which may pollute the underground water table and cause severe drinking water pollution problems. These chemicals, when exposed to the human body, can affect the respiratory system, the central nervous system, blood system, and kidney [4, 5]. Phenols have been classified as the 45th top hazardous compounds of main concern by ATSDR (Agency for Toxic Substances and Disease Registry), USA, harmful compounds which must be treated on an urgent basis before their exposure to the environment [5, 6].

Chronic and *acute* exposure are the two types of phenol toxicity, the acute poisoning of phenol in humans can be seen through symptoms such as dryness in throat and mouth, and dark colors of urine excreted due to the occurring of lipid peroxidation [7]. Chronic exposure causes several symptoms such as anderoxia, muscle pain, headache, gastrointestinal pain which will then lead to cancer [8]. Phenol is easy to be absorbed after being exposed via inhalation because it can be present in the atmosphere as a gaseous or particulate form or directly dermal contact. Phenol has the ability to irritate skin, and long-term of dermal contact can lead to severe skin damage [9]. Oral accidental exposure to phenol may cause severe damage to the liver and kidney due to accidental ingestion of 1 g phenol which is reported to cause fatality to human [10] (Table 1).

Toxicity of phenols is related with two causes i.e. *Organic free radical species* generation and *hydrophobicity of the phenolic compound* [3]. First, the logarithm of the partition coefficient (log P_{ow}) of phenolic compounds in water and n-octanol correlates with the hydrophobicity of solvent. The toxicity of phenols is influenced by the log P_{ow} value. The value between 1.5-4.0 is toxic to living cells especially microorganism. Log P_{ow} value of phenol solvent within that range has the ability to disturb the structure of membrane hence damaging its vital function [11]. Secondly, Different substitution and other chemical reactions of the reactive hydroxyl group of phenols enhance the toxicity of phenols like chlorination of phenols. Different chemical reactions can be carried out on the benzene ring of phenol such as nitrification, alkylation, and halogenations. For example, the presence of chlorine in phenolic structure because of water treatment brings about the increase of toxicity of phenolic compound and resistance in the atmosphere. The ability of chemical groups attached to phenol, their properties and position on a ring structure are the deciding factor for their residence in the environment [3].

Due to the toxic effects of phenol and derivatives to human and animals, it is necessary to remove such toxic compounds from wastewater. Therefore, it has become a challenging problem to remove such organic compounds to decrease their concentrations up to the permitted levels before releasing waters to the environment. For the removal of phenolics from wastewater, a number of methods have been introduced, which includes chemical coagulation [12], electrochemical oxidation [[13], membrane separation [14], solvent extraction [15] biodegrading, photocatalytic degradation [9] bioremediation [16] and using many synthetic and natural adsorbents. For the remediation of organic compounds from wastewater, different techniques have been developed. Bioremediation of organic phenolic compounds is a most desirable technique due to its easy implementation but it is not very suitable for high molecular weight phenolic compounds having 5-6 aromatic rings [16].

This chapter is related to removal of phenol from wastewater, to ensure and provide a safe place for the next generation in future especially water sources. Literature included in this chapter is from 1983 to 2016.

Table 1 Harmful Effects of Phenols [17].

HUMAN AND ANIMALS	AGRICULTURE
Short-term effects: Respiratory infection, headaches, burning eyes, skin rashes.	**Effects on soil:** Porosity of soil decreases, flocculation
Chronic effects (high exposure): Muscle pain, weakness, weight loss, anorexia, fatigue.	**Effects on plants:** Reduced amount of germination of seeds
Long-term effects (low exposure): Heart disease, respiratory cancer, immune system weakness.	**Effect on groundwater:** Groundwater near hazardous site is contaminated by phenol which seeps through the soil.

2. Methods of determination of Phenols

Generally, a spectrophotometric method is used for the determination of phenolic contents in wastewater. For determination of specific phenolic compounds, at very low concentration, a gas chromatographic (GC) method with flame ionization detector or (FID) or electron capture detection (ECD) is used. A liquid-liquid extraction gas chromatographic/mass spectrometric (GC/MS) method can also be used for the determination of phenols at a little higher concentration.

A highly selective spectrophotometric method can be used for the determination of phenolic in industrial and municipal wastewater samples (Table 3). For this purpose, aminoantipyrine is an extensively used reagent. Another reagent N-hydroxyl-N, N`-diphenyl benzamidine (HODPBA) is used for determination of phenols by spectrophotometric method. In this method phenol reacts with HODPBA in the presence of vanadium and a complex of blue-colour is formed which is extractable into $CHCl_3$.

2.1 Physical methods

2.1.1 Solvent extraction

Solvent extraction is the widely used technique for the removal of phenols from different effluents in wastewater [15]. Main types of MSE (membrane solvent extraction) are non-porous MSE and porous MSE. A driving force like the potential difference is responsible for the transportation of solute across the membrane. With low mass transfer resistance due to its pore size range from 0.1-10 µ, the microfiltration membrane has be extensively used for MSE amongst porous membranes. For coal gasification wastewater treatment method methyl isobutyl ketone was used as extracting solvent. By this method, about 93% of phenol was recuperated as a by-product from wastewater, which was an additional economic advantage of this method [18].

2.1.2 Liquid-liquid extraction

It is the process in which components of a solution are separated on the basis of their relative solubility (Fig. 1). It is a very simple method for removal of phenols using different solvents. Piava *et al* worked on the batch removal of phenol by liquid-liquid extraction from methyl isobutyl ketone (14.4%) as a solvent, in a bench-scale mixed vessel [19]. The temperature range was 10 to 40 °C, along with NaOH concentration between 5.5 and 6.5% and the range of rotational speed was between 400 to 800 rpm. The removal efficiency was noted to be 94.0-97.6%. Different solvents used in the removal of phenol are summarized in Table 2.

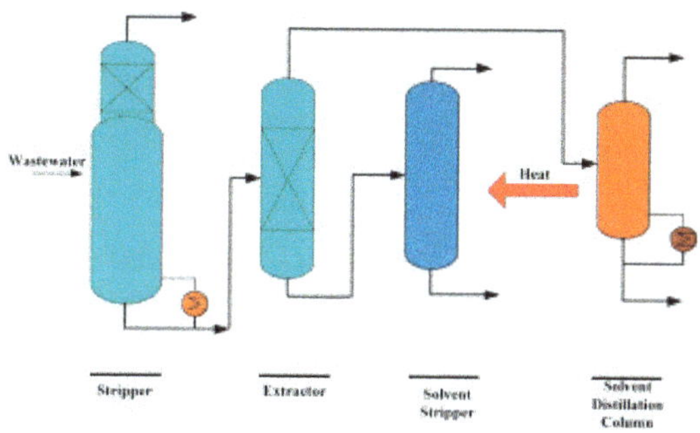

Figure 1 Continuous liquid-liquid extractors with solvent-dispersing frits [20].

Table 2 Different solvents used for removal of phenol.

Solvents used	**%Removal**	**Reference**
Amide and its compounds.	92%	[21]
Imidazole and its homolog compounds	90%	[21]
Mixed solvents: 20% TBP, 20% n-octanol and 60% Cyclohexane	99.3%	[21]
Green based liquid organic solvent: Palm oil	83%	[22]
Methyl-ter-butyl ether (MTBE)	98%	[23]
1-decanol	98%	[24]
Methyl iso-butyl ketone	94-97%	[19]
Kerosene	95%	[25]
Aliquat-336	75-96%	[26]

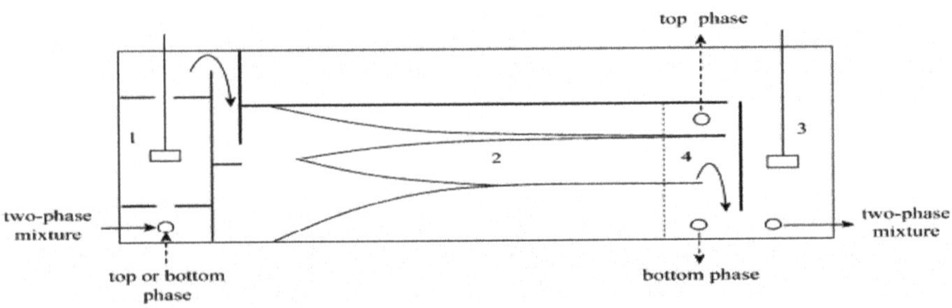

Figure 2 Sketch structure of the proposed three-liquid-phase extractor: 1—three-phase mixer; 2—three-phase settler; 3—two-phase mixer; 4—liquid-flow control zone [27].

2.1.3 Three Phase Liquid System

The three phase liquid system is composed of organic extractant, high molecular polymer, and salt. Pinhua Yu et al [20] investigated separation of P-Nitro phenol and O-Nitro phenol by using a three-phase extraction system. *O*- nitrophenol was collected at top phase, *p*-nitrophenol in the middle and mixture of *o& p* nitrophenol was at the bottom phase [Fig. 2& 3]. About 85% of *o-* nitro phenol and 90% of *P-* nitro phenol were recovered using this method at pH 4.

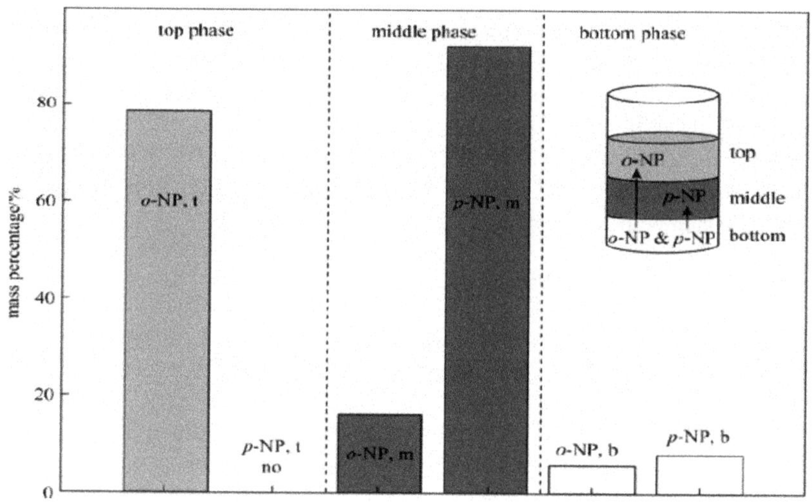

Figure 3 The partition of two phenols in three-layered liquid phases (Agitating at 300rmin^{-1} for 15 min for the extraction reaches final equilibrium. O- NP: ortho-nitrophenol, p-NP: p- nitrophenol, t: top phase, m: middle phase, b: bottom phase. V_t: V_m: V_b (1: 2.1:1.7) [20]

Table 3 Various industries and its associated phenols [3].

Industries	Type of Phenols
Textile	Phenol, Chloro phenol, Alkyl phenols, Catechol, Chloro catechol, Nitro phenol.
Wood processing	Phenol, Chloro phenol, Alkyl phenols.
Pharmaceutical	Catechol, Chloro catechol, Chloro phenol, Methyl phenol, Buthyl hydroxyl toluene, Buthyl hydroxyanisole.
Rubber	Petrochemical Phenol, Methyl phenol.
.	
Cosmetics	Chlorocetachols, Methyl phenol, Buthyl hydroxyl toluene, Buthyl hydroxyanisole.
Coal-tar production	Phenol, Nitrophenols, Methyl phenols.

2.1.4 Pervaporation

Pervaporation is a separation process wherein the compounds are separated from the solution by partial vaporization through a non-porous or porous membrane. The two steps involved in this process are permeation and evaporation. The membrane used acts as a selective barrier that allows the desired component of the liquid feed to pass through it by vaporization [Fig. 4]. Kujawski et al [23] worked on pervaporation and adsorption for phenol removal. Various membranes like poly (ether-block-amide) (PEBA), pervaporation 1060 (PERVAP1060) and PERVAP1070 were used. All the membranes were found to be selective for phenols. PEBA membranes showed the highest selectivity for phenol removal but, these are not commercially available, PERVAP1060 and PERVAP1070 membranes are commonly used.

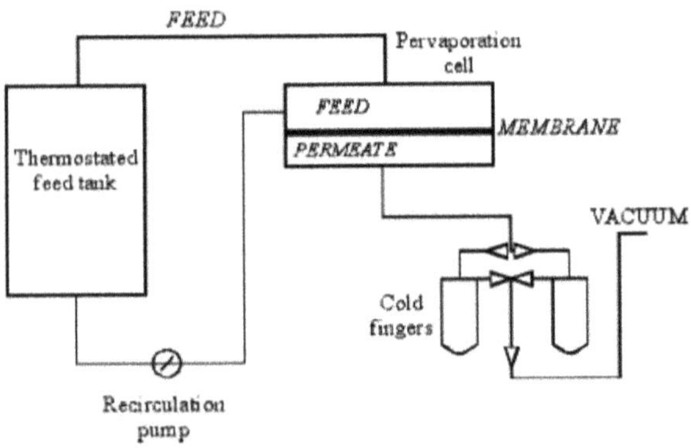

Figure 4. The schematic representation of laboratory scale of pervaporation setup [28].

2.1.5 Membrane filtration

Subha, *et al*, [29] reported different membrane filtration methods for the removal of phenols from wastewater these are as follows.

2.1.5.1 Reverse osmosis separation methods

Reverse osmosis (RO) is a very effective method for the removal of phenols from wastewater. It can easily be operated at room temperature and this is an energy saving process. It has drawn attention due to its energy saving nature. This method is widely used for the separation of organics from water. Depending on the mode of application, operating pressure remains in the range of 15-50 bars. RO rejects the organic molecules having a molecular weight of more than 50 along with monovalent ions. This method can be applied commonly to sea water and brackish water desalination. [30, 31]. For the characterization of reverse osmosis and nanofiltration membrane, two models: *the combined film theory-(Spiegler-Kedem)* and *combined film theory (solution diffusion)* were used. Both models used commercial polyamide composite thin-film reverse osmosis membrane for phenol rejection. In the phenol-water system, the combined film theory of Spiegler-Kedem model more accurately calculated rejection of phenols in water than the other model.

2.1.5.2 Ultrafiltration

The principle of ultrafiltration is alike to that of reverse osmosis, except that in ultrafiltration the pore size of the membrane (0.002-0.003 mm) is large and it can be operated at reduced pressure. The rejection of organic molecules with molecular weight more than 800 by ultrafiltration membranes frequently operated at pressures below 5 bar. The surfactant micelles i.e. charged particles containing these organic molecules are large enough so that they cannot passe through the pores of the membrane. This fact makes the ultrafiltration membrane more efficient for phenol removal from a stream of water [32-35].

2.1.5.3 Nanofiltration

Nanofiltration membranes have properties between the membranes of ultrafiltration and reverse osmosis with pore sizes ranging from 1-100 nm. Nanofiltration can be used for both, the removal of disinfectants from water and for water softening. Nanofiltration membranes permit the movement of monovalent ions through them, for instance, sodium/potassium. Organic molecules with molecular weight > 200 and high concentration of divalent ions like calcium & magnesium are rejected by the membranes. Normally the operating pressure is kept close to 5 bars. Nanofiltration may be effective for the removal of color and organic compounds [30]. For removal of phenol from its aqueous solution Bodalo *et al.* [30] used many nanofiltration membranes (NF- 99 DSS-HR98PP and NF-97) under different experimental conditions at operating pressures: 10, 15 and 20 x 105 N/m^2 with constant feed flow at 2.78 x 10-5 m^3/s, T= 25°C and pH=8 with variation of concentration of phenol from 50 -200 x 10-3 x kg/m^3. Different rejection percentages were obtained with a variation of concentration. It was concluded that molecular properties like solubility, hydrogen bonding and acidity etc., were responsible for a rejection rate of organic compounds in nanofiltration membranes [30].

Table 4 Phenol removal by adsorption using different plant sources.

Adsorbent	pH	Time (h)	Temperature (°C)	Adsorbant dose (g/L)	Phenol removal	Ref.
Date pit activated Carbon	7	24	25	4	The adsorption ability of low-cost date seed adsorbent was evaluated at elevated 2,4- dinitrophenol concentration	[36]
Mango peel	7	6	30	0.25	Mango peel and agricultural waste was used as adsorbent for the removal of chlorophenols from	[37, 38]
Chestnut shells	2.5	8	50	-	Chestnut shell was used as adsorbent for removal of phenolics from wastewater of paint industry	[38]
Pineapple peel	7	48	25	-		[39]
Rice husk and sugarcane bagasse	5	5	25	1g	Activated carbon prepared from Rice husk and sugar -cane bagasse impregnated with $ZnCl_2$ used as adsorbent for the removal of arsenic, phenols and other landfill leachates from municipal wastewater and solid waste	[40]
Activated tea waste	2.4	5	25	1.5	Synthetic wastewater was treated with activated tea leaves at different concentrations, PH, pollutant dose etc.	[41]
Potato peel	1-2	1	55	-	Polyphenol oxidase from Potato peel waste was used for the degradation of phenols from	[42]
Olive mill waste	-	30	--	1	Onion solid byproduct was used to produce peroxidase for the removal of phenolic from olive mill wastewater	[43]
Tamarind bean	-	50	-	-		[44]

2.1.6 Adsorption

It is a surface phenomenon. There are two types of adsorption *physical-sorption* (operated by van der Waals forces) and *chemisorption* (formation of covalent bonding), both of these depend on the forces between adsorbent and adsorbate. It can be normally illustrated by an isotherm at constant pressure. A thin film of adsorbate formed on the surface of adsorbent is the function of concentration (in case of liquid)/ pressure (in case of gas). To explain this phenomenon a variety of isotherm models like Langmuir, Freundlich constant, Henry adsorption constant, adsorption enthalpy, BET theory etc, are available. Adsorption is one of the most investigated techniques used for Phenol removal from wastewater of industries. Table 4 shows various synthetic sorbents and microbial, agricultural, synthetic and chemical sources for adsorption and respectively. They have been tested for their adsorption capacities of organic pollutants, including phenols. Due to diverse sources, adsorption can be performed in fluidized bed reactor, fixed bed reactor, trickle bed reactors and much more. Owing to its versatility and ease of operation, it is one of the most commonly used wastewater treatment methods for removal of phenols and phenolic compounds.

2.1.7 Cloud point extraction method

It is an advanced physical method for phenol extraction from wastewater. In this technique, the analyte gets partitioned itself between two isotropic phases i.e. aqueous phase and surfactant-rich phase. Surfactant solution (water-based solution without toxic organic solvents) is used as a solvent. This extraction technique is very efficient, low cost and less time consuming; environmentally benign separation approach frequently used as an alternative to conventional extraction systems. This technique has very good ability to concentrate the solutes with high recoveries, need very small amount of surfactant which is non-volatile and inflammable and the surfactants can be disposed of easily [45]. In this process, the aqueous solutions of some surfactants show the cloud point phenomenon. Due to a decrease in solubility of the surfactant in water, the surfactant solution becomes turbid suddenly. The clouding phenomenon is generated with an increase in temperature. Consequently, an isotropic layer is formed which further split into two isotropic phases discussed above [46].

It was concluded that the isotropic phase rich in surfactant which is separated under the cloud point condition is able to pre−concentrate a wide range of inorganic and organic compounds and to extract them from the aqueous phase. The solute present in the aqueous solution of the surfactant is distributed between the two phases above the cloud point temperature.

2.2 Chemical methods

2.2.1 Advanced oxidation process

Advanced oxidation process (AOP) can be broadly defined as aqueous phase oxidation methods based on the intermediacy of highly reactive species such as hydroxyl radicals in the mechanism leading to the destruction of the target pollutant. AOPs are gaining much importance in research and development due to their areas of prospective application and variety of technologies.

Over the past 30 years, AOPs has played a great part in research and development due to two main reasons namely the diversity of technologies involved. The key AOP processes include heterogeneous and homogeneous photocatalysis based on near ultraviolet (UV) or solar visible irradiation, electrolysis, ozonation, the Fenton's reagent, ultrasound and wet air oxidation. The processes like ionizing radiation, microwaves, pulsed plasma and the ferrate reagent are also emerging photodegradation processes, responsible for phenol removal. Depending on the properties of the waste stream to be treated and the treatment objective itself, AOPs can be employed either alone or coupled with other physicochemical and biological processes.

2.2.2 Air oxidation

For the estimation of phenol decomposition, byproduct accumulation, and final product accumulation, a new technique named as the differential neural network was introduced by Chairez and coworkers [47, 48]. The effect of raising pH on the dynamics of decomposition of phenol , 4-chlorophenol and 2, 4- Di chloroPhenol was very important. The total time for decomposition of phenol was reduced by the factor of ten. The presence of the chloro species in the phenol molecules reduces the degradation time during ozonation [37, 38].

2.2.3 Catalytic wet air oxidation of phenol

Wang *et al.* [49] used continuous packed bubble column reactor for wet air catalytic oxidation of phenol with pellets of catalyst (CeO_2-Ru/ZrO_2) ZrO_2 was introduced into Ru/CeO_2 to increase the adsorption capacity, specified surface area, and mechanical strength of this pelletized catalyst in this process, removal of phenol was stabilized about 100 % [39].

2.2.4 Photodecomposition

Photodegradation by UV radiations is used for the removal of phenol from wastewater. It is used by itself as well as with alternative adsorbents. In this process, UV radiation is used to irradiate the sample for the removal of contaminants. The UV light is absorbed by contaminant compound which undergoes decomposition from its excited state. The UV photoreactors are highly efficient because the solution which is to be analyzed surround the photo reactor's lamps. Generally, a medium pressure mercury lamp is used as a radiation source which has 400 W output. UV lamps radiate primarily UV radiation of 365 – 366 nm with radiations at 334, 313, 303, 297 and 265 nm in lesser amounts. The mechanism of phenol degradation is heterogeneous which follows zero-order kinetics. Heterogeneous degradation of phenols results in 70% conversion. Phenol degradation can be catalyzed in the presence of silver and TiO_2. In the presence of silver, degradation is carried out by direct electron transfer in the presence of TiO_2 in an aqueous medium at pH 5, phenol removal up to 99% has been observed after 4 hours at pH 5. Under the same conditions for 2 chloro phenol the percentage removal of other phenolic compounds was 92% and for 2, 4 dimethyl phenol was 83%. Phenol absorbs UV radiations from the light source and an electron is removed and taken up by the aqueous solvent solution which results in the formation of radicals OH˙& phenoxide radicals. The phenoxide radicals are being attacked by hydroxyl radical at ortho/para positions and after a further attack of ˙OH convert this pyrocatechol to maleic acid followed by the formation of formic acid which decomposes to give CO_2 and H_2O (Fig. 5).

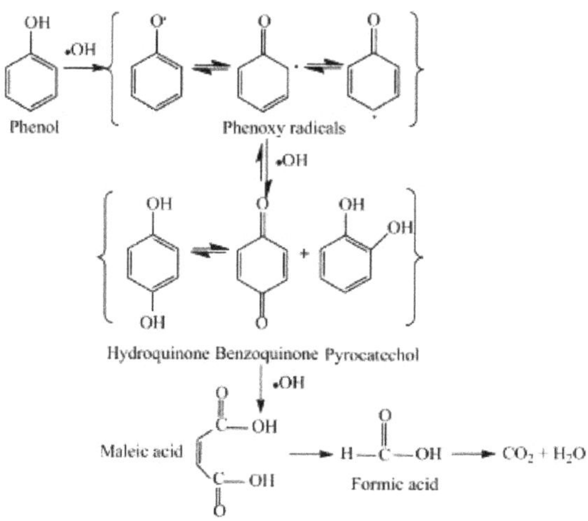

Figure 5 A schematic mechanism of photodegradation of phenol [50].

Udom and co-workers [51] worked on photodegradation of phenol using near UV radiation in the presence of aqueous FeO/TiO_2/ red mud / Nickle mud [mixture of Fe_2O_3 (31.80 wt. %), TiO_2 (22.60 wt. %), MgO (0.20 wt. %), Al_2O_3 (20.10 wt. %), CaO (4.78 wt. %), SiO_2 (6.10 wt. %), Na_2O (4.70 wt. %), K_2O (0.03 wt. %)] annealed on a glass surface. The surface of the glass wool was covered by black or red nickel mud and annealed in a muffle furnace (at 300°C) so as to reach the finest accessible rate for the adsorbent fixation, for the removal of possible organic impurities of adsorbent and increasing its adsorption capacity by converting the adsorbent particles into fiber through heating. These layers of glass 57 wool containing catalysts were kept in a UV photoreactor in order to increase the phenol removal efficiency. The influence of catalysts was monitored.

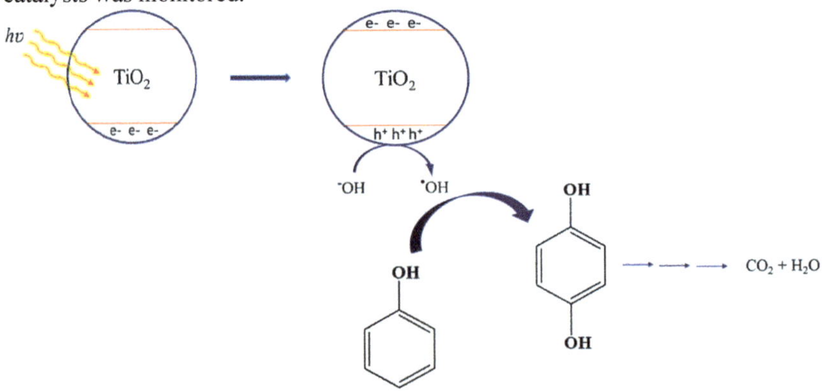

Figure 6: A schematic mechanism of photodegradation of phenol by TiO_2 [52].

Photodegradation of phenols in wastewater depends upon different parameters like light intensity, irradiation time, pH, the concentration of phenol and amount of TiO_2. Photo-degradation process catalyzed by TiO_2 is a highly efficient method for the elimination of phenol from wastewaters. The effectiveness of this process is strongly dependent on the experimental reaction conditions. It depends mainly upon pH, TiO_2 concentration, and UV light intensity. By increasing intensity of UV light, the rate of phenol degradation can be increased. Optimum pH value for the quick OH radicals' generation is 7. The generation of hydroxyl radicals produces a positive charge on the surface of TiO_2, which attracts the phenol having a negative charge. This results in efficient degradation of phenol into CO_2 and H_2O (Fig. 6) at loading range of TiO_2 0 to 0.20 wt.%, the degradation rate of phenol was observed to be increased with increased TiO_2 concentration. At loading rate of above 0.20 wt %, the phenol degradation was observed to remain constant with enhanced concentration of TiO_2. The degradation rate of phenol was best described by Pseudo-first order according to kinetic of reaction. The rate of degradation of phenol was more efficient as compared to sunlight under UV light.

2.2.5 Coupled ultrasound and Fenton's reagent

Coupled Fenton's reagent and sonochemistry technique has been used to study the phenol degradation in aqueous medium equilibrated with air for a variety of experimental conditions. Extra OH radicals produced by Fenton's reagent through the reaction of Fe(II) with H_2O_2 contributes to the degradation of phenol in wastewater. This production of ·OH radicals by coupled ultrasound and Fenton's reagent further enhance the rate of decomposition of phenols. In this technique, the rate of decomposition of aqueous phenol was extremely dependent upon the initial concentration and pH of reactants [53].

2.2.6 Three phase electrode system

The three-phase electrode system is an advanced method to remove phenol from wastewater. Ya Xiong and coworkers [54] worked on the performance of the three-phase three-dimensional electrode reactor which was highly effective for COD reduction in phenol-containing wastewater and its performance was compared with granulated activated carbon adsorption bed and 3D electrode system.

(v) Electrocoagulation

In this method, phenol is removed by electrocoagulation in which the sacrificed anodes form active coagulant that is then used to remove pollutant *in situ* by flotation and precipitation. Electrochemical cells containing an electrode arrangement in contact with polluted water are employed during electrocoagulation (Fig. 7). In 2013 Ashtoukhy and his team [55] worked on removal of phenolic compounds by electrocoagulation from petrochemical wastewater using a fixed bed reactor.

Parameters taken into consideration for electrocoagulation are pH, Retention time, initial phenol, current density, and NaCl, as well as alginic acid concentrations. For assessment of wastewater, the samples were taken up from reactors at different time periods and were centrifuged at 4000 rpm for 30 min. All analyses were carried out under conditions consistent with standard methods before analysis [44, 56]. The analysis of remaining wastewater was done for the concentration of phenols on the basis of color developed by the reaction of phenols with 4-aminoantipyrine. The phenol concentration was determined by spectrophotometer at λ_{max} 500 nm. Phenol removal was carried out by increasing concentration of NaCl and current density at pH 7.

Phenols are converted to phenoxide by removal of the electron under acidic conditions and these phenoxide ions combine with other phenol molecules present in solution under the effect of the electric field in an electrochemical cell. These aggregates thus formed and removed from water (Fig.7& 8).

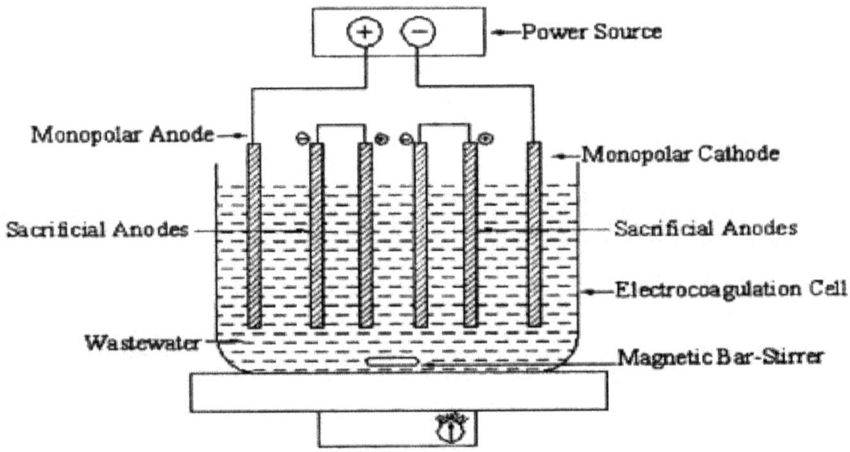

Figure 7. Schematic diagram for electrocoagulation [57].

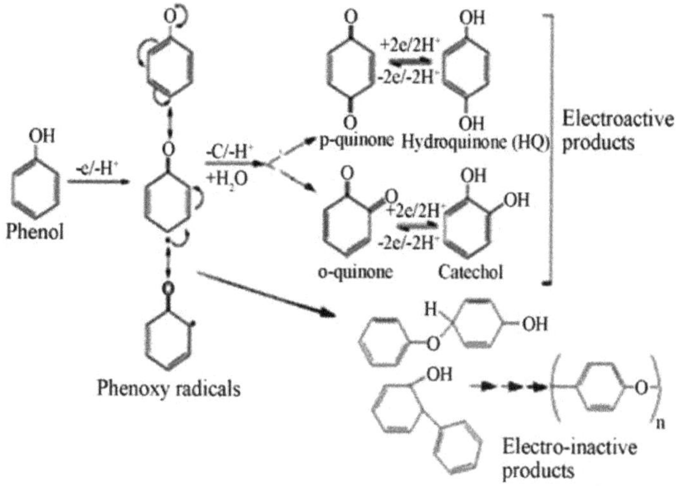

Figure 8. Proposed mechanism of Electrodegradation of phenols [58].

2.3 Biological Methods

Biodegradation of phenols is the process of breakdown of the phenols into simpler/non-toxic compounds by the action of microbes or enzymes. The two main types of biological degradation of phenols are *enzymatic degradation* and *microbial degradation.*

2.3.1 Enzymatic degradation

The enzymes, obtained from various sources are popularly being used for wastewater treatment. Among the many enzymes, *polyphenol oxidases* and *peroxidases* are widely used. According to Nagai and Suzuki [59], polyphenol oxidase is a copper-containing enzyme and the classification of polyphenol oxidase is as follows:

(a) Tyrosinase: Tyrosinase, also called catecholase, contains a binuclear copper active site that facilitates the degradation reaction of the phenols. It catalyzes the hydroxylation of mono-phenols to o-bi phenols, which is then degraded to form o-quinone by dehydrogenation reaction. The use of the tyrosinase enzyme can be coupled with various other techniques (integration into a biosensor, adsorption or immobilization) for better results

(b) Laccases: Laccases are the enzymes that catalyze the degradation reaction by reduction of oxygen to water along with oxygenation of a phenolic molecule. These enzymes are also called blue copper oxidases or blue copper proteins. They contain four neighboring copper atoms at different binding sites, where two are concerned with transfer and electron capture, and two other with the binding of oxygen. On oxidation, the phenols are converted into *o*-bi phenols which may undergo further enzymatic degradation or may undergo polymerization reaction leading to the formation of the melanin-like dark brown accumulated product. Ulfat Jan with her team [60]worked on detoxification of phenols and aromatic amines in wastewater by using polyphenol oxidases. Sources of polyphenol oxidase were gooseberry, quince leaves, tea leaves, banana fruit/ peel and potato etc. Jadhav [61] isolated and characterized the polyphenol oxidase (PPO) enzyme from banana peel. The PPO enzyme was used to treat industrial wastewater containing phenol. The parameters such as enzyme activity, optimum pH, optimum temperature, phenol degradation and phytotoxicity were examined. Complete degradation of phenols was seen within 24 hours for the smaller concentration of phenols, whereas 72 hours were required for higher phenol concentrations. The optimum temperature and pH for maximum enzyme activity were 35°C and 7, respectively. Peroxidases were found in various plants and microbes. Sources of peroxidase were radish, cabbage, tobacco etc. The different types of peroxidase enzymes were manganese peroxidase, lignin peroxidase, horse radish peroxidase, etc.

Enzyme peroxidase mechanism for phenol degradation was proposed by Chance-George. In this process, peroxidase catalyzed the one-electron oxidation of phenols. The mechanism is as follows:

$$E + H_2O_2 \longrightarrow E_i + H_2O \quad (1)$$

$$E_i + AH2 \longrightarrow E_{ii} + AH^{\cdot} \quad (2)$$

$$E_{ii} + AH_2 \longrightarrow E + AH^{\cdot} + H_2O \quad (3)$$

$$E_{ii} + H_2O_2 \longrightarrow E_{iii} + H_2O \quad (4)$$

Hydrogen peroxide (H_2O_2) oxidizes the subject enzyme **(E)** to form an active enzymatic intermediate named as **Ei** (compound I). Oxidation of phenolic aromatic compound (AH_2) takes place at the active sites of **Ei.** Free radical ($AH^{\cdot}$) produced and released in solution as a result of oxidation of aromatic compound at **Ei.** Hence **Ei** is converted to **Eii** (Compound II). The oxidation of the second molecule of aromatic compound takes place on **Eii** in the same way resulting in the production of free radical product again and the enzyme attains its original state E, hence completion of the cycle takes place. The whole peroxidase reaction is composed of three reactions illustrated by Eqs. (1–3). If hydrogen peroxide is present in excess, the reaction of Eq. (4) becomes significant because **Eiii** (compound III) is an inactive form of the enzyme. This shows that presence of excess hydrogen peroxide inhibits the enzyme activity. Conversely, the rate of reaction during the second step is limited in limited supply of hydrogen peroxide. An optimum ratio of both enzyme and hydrogen peroxide would suppress the inhibition of this enzyme [44, 62]. During the process of enzyme reduction, phenol and its derivatives are transformed into phenolic-free radicals. The phenolic-free radicals are catalytically produced during the peroxidase reaction cycle. These phenolic–free radicals either produce phenolic polymers by combining with each other or produce other radical species by undergoing radical transfer reaction with phenolic polymers or monomers present in solution [63]. The Phenolic polymer radicals may be generated by the action of peroxidases enzymes with phenolic polymers. These radical species may react with other radical species to form larger polymers or may undergo radical transfer reaction which results in the production of polymers of different sizes in solution [64]. The phenolic polymers can be precipitated out from solution and can be removed by either filtration or gravity settling [65]. Different enzymes used for polymerization of phenol with their sources are given in Table 5. Potential advantages of enzyme-based treatment [65, 66] are given below.

- Removal of phenolic compounds and similar one is carried out by selective treatment.
- Possibilities of speeding up reaction velocity minimize stay time.
- Degradation of toxic compound lethal for microbes.
- Wider operation range of concentrations of substrate.
- Operational possibility over a wide range of pH, temperature, and salinity.
- No effects of shock loading.
- The control of the process, predictable, simpler and reliable.

Table 5 Various microbial and plant enzymes and their sources involved in biodegradation of phenol [67].

Enzyme	Source	Work done	Ref.
Phenol hydroxylase	*Bacillus stearothermophilus*	This bacterium isolated from river sediment and its enzyme used for phenol degradation at level 15 mM.	[68]
Polyphenol oxidase	Mushroom	Mushroom phenol oxidase activity on organic system under optimized conditions of temperature, pH, and concentration.	[69]
Polyphenol oxidase	*Trametes trogii*	The enzyme extracted from a white-rot fungus Trametes trogii for oxidation of wide range of phenolic compounds.	[70]
Polyphenol oxidase	Transgenic tomato	Polyphenol oxidase isolated from tomato plat for the oxidation of phenolics into quinones.	[71]
Polyphenol oxidase	Potato	Immobilized enzyme isolated from potato for the removal of chlorinated phenols.	[72]
Phenol oxidase	*Lentinula edodes LE2*	Soil contaminated with pentachlorophenol was treated with *Lentinula edodes LE2and* quick removal phenol was observed upto to 99%.	[73]
Catechol 2,3-dioxygenase	*Bacillus* sp.	Thermophilic bacteria isolated from sewage effluent. its enzyme catalyze the phenol degradation by meta-cleavage pathway.	[74]
Laccase	*Rhizoctonia praticola*	Detoxification of phenolic pollutants was examined by laccase of fungal strain *Rhizoctonia praticola.*	[75]
Laccase		Characterization and purification of laccase from *Aspergillus oryzae.*	[76]
Laccase	*Pleurotus ostreatus*	Characterizationand immobilization of laccase from Pleurotus ostraetus and its use for the continuous elimination of phenolic pollutants.	[77]
Laccase	*Chalara paradoxa*	Phenol-oxidase (laccase) activity in strains of the hyphomycete Chalara paradoxa isolated from olive mill wastewater disposal ponds.	[78, 79]

Phenol oxidase	*Termitomyces albuminosus*	A novel hydrogen peroxide-dependent phenol oxidase (TAP) was isolated from the basidiomycete Termitomyces albuminosus oxidized various phenolic compounds in the presence of hydrogen peroxide.	[80]
Horseradish peroxidase	Horseradish	Removal of phenol from wastewater at broader pH range was carried out by using horseradish peroxidase enzyme isolated from horseradish roots.	[81-86]
Soybean peroxidase	Soybean	Soybean peroxidase was used for degradation and polymerization of phenols from wastewater in the presence of hydrogen peroxide. polymerization product was removed by filtration.	[65, 87, 88]
Radish peroxidase	*Raphanus sativus*	Removal of phenolic contaminants from wastewater by *Raphanus sativus peroxidase.*	[89]
Peroxidase	*Momordica charantia*	Bitter guard peroxidase efficiency for phenolic compounds removal from wastewater.	[90]
Turnip peroxidase	*Brassica napus*	Immobilized turnip peroxidase efficiency was improved in the presence of ethylene glycol for the removal of phenolics from wastewater effluents.	[91]

2.3.2 Microbial degradation

Van Schie [62] have described the utility of micro-organisms in biodegradation of Phenol. The degradation of phenol by this method is amenable due to its ease of manipulation of strains and scaling up of the process. The phenol is converted to catechol by oxygenation, which is further broken down and the compounds enter the metabolic cycle in order to produce carbon dioxide, water or other inorganic simple compounds. The degradation can be either aerobic or anaerobic condition. In aerobic degradation, the phenol is oxidized to catechol by the microbial enzyme (phenol hydrolase) in the presence of nicotinamide-adenine dinucleotide ($NADH_2$) (Fig. 9 and 10). Depending on the enzymes in the organism used, it is degraded into either 2-hydroxymuconic semi-aldehyde or cis muconic acid, that may enter the Krebs cycle and get degraded to carbon-dioxide and water. The various organisms involved in aerobic degradation are *Acinetobacter calcoaceticus, Pseudomonas, Streptomyces setonii* a *Trichosporon cutaneum* and *Candida tropicalis* etc (Table 6).

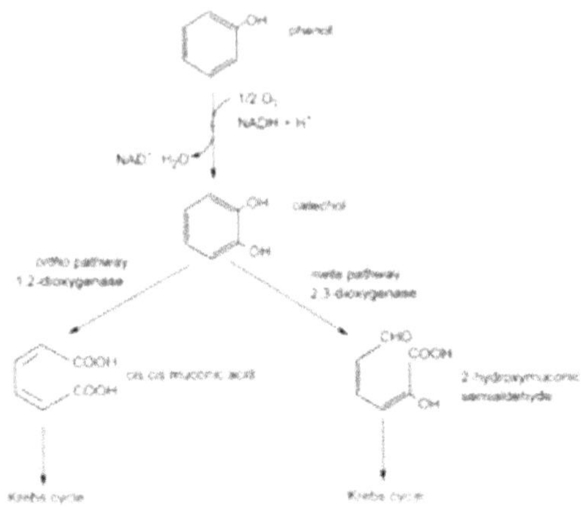

Figure 9 Schematic representation of aerobic degradation of phenol.

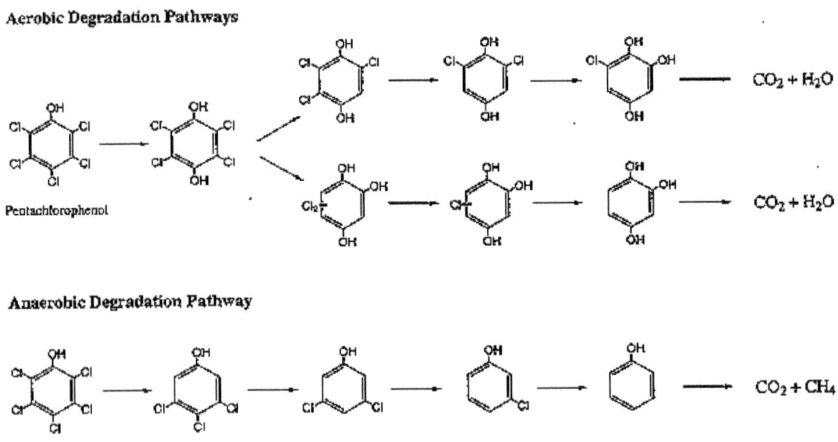

Figure 10 Aerobic and anaerobic pathway of pentachlorophenol biodegradation [92].

In anaerobic degradation, the degradation of phenols occurs in the absence of oxygen. Anaerobic degradation usually occurs under reducing conditions. The mechanism, however, varies from one organism to another. In a nitrogen reducing bacteria called *Thauera aromatic,* the phenol ring is carboxylated into 4-hydroxybenzoate which is further degraded. For anaerobic degradation, the organisms involved are *Thauera aromatica* strain K172, *Geobacter metallireducens* (iron- reducing conditions), the sulfate-reducing organism *Desulfobacterium phenolicum, Desulfotomaculumsp.* Strain Groll (in presence of bi-carbonate) and *Desulfovibrio* species. Agarry [39] worked on strains of *Pseudomonas* fluorescence at different concentrations of phenol. A comparative study of the performance of *Pseudomonas fluorescence* and other micro-organisms was carried out and complete degradation of phenol was by the action of the bacteria. The degradation time was increased from 84 to 354 hours with the increase in the concentration of phenols.

Table 6 Microorganisms used for phenol degradation under various conditions.

Microorganism	**Conditions for phenol degradation**	**Ref.**
Pseudomonas putida	Aerobic conditions	[93]
Corynebacterium sp. S. aureus, B. subtilis Proteus sp Staphylococcus sp	Resistant to 15mM of phenol	[94]
Kelibsiella Shigella Citrobacter	These bacteria have very good ability to degrade phenol in 0.2-0.9 g/L concentration range isolated from parshan lake	[95]
Pseudomonas aeruginosa Pseudomonas fluorescence	In reaction vessel with soil, these bacteria degraded the phenols efficiently	[96, 97]
Acinetobacter baumannii	Efficiently degrade phenols at high concentration range	[98]
Alcaligenes faecalis	Capable to degrade phenol at concentration range up to 600 mg/L under anoxic conditions	[99]
Candida species	Degradation of phenol on activated carbon by this sp. at phenol conc. Range up to 2 g/L	[100, 101],
Aspergillus sp	This fungal strain degraded different concentration of phenols	[102, 103]

Rhodococcus erythropolis	Usage of its catabolic gene for phenol degradation	[103, 104]
Pseudomonas stutzeri	optimal conditions of continuous bioconversion this bacterium showed the very good efficiency	[105]
Bacillus brevis	Degraded phenols at different environmental conditions	[106]

3. Conclusions

Together with the development of industrialization especially in developing countries, people have become more aware of phenol exposure effects on living organisms. Due to wide range usage of phenol in many industries, as the main chemical, the phenol is still one of the most omnipresent pollutants in the environment. The risk of phenol pollution cannot be avoided and hence the study on phenol degradation is still relevant. Among all methods available, the biological method seems to have the potential for phenol degradation in either anaerobic or aerobic condition that makes it the choice for researcher compared to other physical or chemical method. Many researchers found *Pseudomonas* sp. and *Rhodococcus* sp. to have higher potential ability to degrade phenol effectively. Toxicities of phenol and its derivatives have been associated with the ability to alter the structure of the membrane and its barrier. It leads to the imbalance of cell environment which results in the cells' death. The action of toxicity due to environmental stress causes it to be happening in microorganisms. The transition of saturation degree in membrane phospholipids cis-fatty acid into trans-fatty acid depends on the hydrophobicity of solution and its location of the group on the aromatic ring where benzene ring itself can also undergo several chemical reactions such as halogenations or substitutions. The needs to isolate and characterise microorganism, capable to tolerate laboratory and real field conditions for the efficient phenol removal will become crucial in the near future where the organism can survive under stress conditions such as high salinity, low oxygen supply, or low pH, and many other xenobiotics presences in the mixed effluent discharge in water sources. Much more studies need to be done in the future regarding to the mechanism of phenol degradation in the presence of microorganisms.

References

[1] P. Panagos, M. Van Liedekerke, Y. Yigini, L. Montanarella, Contaminated sites in Europe: review of the current situation based on data collected through a European network, J. Environ. Public Health. 2013 (2013) 1-10. https://doi.org/10.1155/2013/158764

[2] S-H. Lin, R-S. Juang, Adsorption of phenol and its derivatives from water using synthetic resins and low-cost natural adsorbents: a review, J. Environ. Manage. 90 (2009) 1336-1349. https://doi.org/10.1016/j.jenvman.2008.09.003

[3] J. Michałowicz, W. Duda, Phenols--Sources and Toxicity. Polish J. Environ. Stud. 16 (2007) 347-362.

[4] B. Bukowska. Toxicity of 2, 4-Dichlorophenoxyacetic acid molecular mechanisms, Polish J. Environ. Stud. 15 (2006) 365-374.

[5] EPA E. National Library Network. US Environmental Protection Agency. (2007) URL: http://www.epa.gov/natlibra/ols htm.

[6] ATSDR C. CERCLA priority list of hazardous substances. Agency for Toxic Substances and Disease Registry (2007).

[7] O Olujimi, O Fatoki, J Odendaal, J.Okonkwo, Endocrine disrupting chemicals (phenol and phthalates) in the South African environment: a need for more monitoring. Water SA. 36 (2010) 671-682. https://doi.org/10.4314/wsa.v36i5.62001

[8] S. Chakraborty, T. Bhattacharya, T. Patel, K. Tiwari, Biodegradation of phenol by native microorganisms isolated from coke processing wastewater, J. Environ. Biol. 31 (2010) 291-296.

[9] A.A. Gami, Phenol and its toxicity, J. Environ. Microbiol. Toxicol. 2 (2014) 173–180.

[10] A. Nuhoglu, B. Yalcin, Modelling of phenol removal in a batch reactor. Process Biochem. 40 (2005) 1233-1239. https://doi.org/10.1016/j.procbio.2004.04.003

[11] E. Díaz. Bacterial degradation of aromatic pollutants: a paradigm of metabolic versatility, Int. Microbiol, 7 (2004) 173-80.

[12] K. Jüttner, U. Galla, H. Schmieder, Electrochemical approaches to environmental problems in the process industry, Electrochim. Acta 45 (2000) 2575-2594. https://doi.org/10.1016/S0013-4686(00)00339-X

[13] M. Tomaszewska, S. Mozia, A.W. Morawski, Removal of organic matter by coagulation enhanced with adsorption on PAC, Desalination, 161 (2004) 79-87. https://doi.org/10.1016/S0011-9164(04)90042-2

[14] W. Kujawski, A. Warszawski, W. Ratajczak, T. Porebski, W. Capała, I. Ostrowska, Removal of phenol from wastewater by different separation

techniques, Desalination 163 (2004) 287-296. https://doi.org/10.1016/S0011-9164(04)90202-0

[15] Z. Lazarova, S. Boyadzhieva. Treatment of phenol-containing aqueous solutions by membrane-based solvent extraction in coupled ultrafiltration modules, Chem. Eng. J. 100 (2004) 129-138. https://doi.org/10.1016/j.cej.2004.01.028

[16] N.S. Alderman, A.L. N'Guessan, M.C. Nyman, Effective treatment of PAH contaminated Superfund site soil with the peroxy-acid process, J. Hazard Mater. 146 (2007) 652-660. https://doi.org/10.1016/j.jhazmat.2007.04.068

[17] P. Chhonkar, S. Datta, H. Joshi, H. Pathak, Impact of industrial effluents on soil health and agriculture-Indian experience: Part I-Distill a paper mill effluents, J. Sci. Ind. Res. 59 (2000) 350-361.

[18] J. Shao, Y. Cheng, C. Yang, G. Zeng, W. Liu, P. Jiao, Efficient removal of naphthalene-2-ol from aqueous solutions by solvent extraction, J. Environ. Sci. 47 (2016) 120-129. https://doi.org/10.1016/j.jes.2016.03.010

[19] M. Palma, J. Paiva, M. Zilli, A. Converti, Batch phenol removal from methyl isobutyl ketone by liquid–liquid extraction with chemical reaction, Chem. Eng. Process: Process Intensificat. 46 (2007) 764-768. https://doi.org/10.1016/j.cep.2006.10.003

[20] P. Yu, Z. Chang, Y. Ma, S. Wang, H. Cao, C. Hua, Separation of p-Nitrophenol and o-Nitrophenol with three-liquid-phase extraction system, Separat. Purificat. Tech. 70 (2009) 199-206. https://doi.org/10.1016/j.seppur.2009.09.016

[21] T. Jiao, X. Zhuang, H. He, C. Li, H. Chen, S. Zhang, Separation of phenolic compounds from coal tar via liquid–liquid extraction using amide compounds, Ind. Eng. Chem. Res. 54 (2015) 2573-2579. https://doi.org/10.1021/ie504892g

[22] N. Othman, N.F.M. Noah, L.Y. Shu, Z-Y. Ooi, N. Jusoh, M. Idroas, Easy removing of phenol from wastewater using vegetable oil-based organic solvent in emulsion liquid membrane process, Chin. J. Chem. Eng. 25 (2017) 45-52. https://doi.org/10.1016/j.cjche.2016.06.002

[23] W. Kujawski, A. Warszawski, W. Ratajczak, T. Porebski, W. Capała, I. Ostrowska, Application of pervaporation and adsorption to the phenol removal from wastewater, Separat. Purificat. Tech. 40 (2004) 123-132. https://doi.org/10.1016/j.seppur.2004.01.013

[24] M.J. González-Muñoz, S. Luque, J. Alvarez, J. Coca, Recovery of phenol from aqueous solutions using hollow fibre contactors, J. Membr. Sci. 213 (2003) 181-193. https://doi.org/10.1016/S0376-7388(02)00526-4

[25] R-S. Juang, W-C. Huang, Y-H. Hsu, Treatment of phenol in synthetic saline wastewater by solvent extraction and two-phase membrane biodegradation, J. Hazard. Mater. 164 (2009) 46-52. https://doi.org/10.1016/j.jhazmat.2008.07.116

[26] N. Rao, J.R. Singh, R. Misra, T. Nandy, Liquid–liquid extraction of phenol from simulated sebacic acid wastewater, J. Sci. Ind. Res. 68 (2009) 823-828.

[27] H. Xiuqiong, K. Huang, Y. Pinhua, C. Zhang, X. Keng, L. Pengfei, Liquid-liquid-liquid three phase extraction apparatus: operation strategy and influences on mass transfer efficiency, Chin. J. Chem. Eng. 20 (2012) 27-35. https://doi.org/10.1016/S1004-9541(12)60359-0

[28] J. Gómez, G. León, A. Hidalgo, M. Gómez, M. Murcia, G. Griñán, Application of reverse osmosis to remove aniline from wastewater, Desalination 245 (2009) 687-693. https://doi.org/10.1016/j.desal.2009.02.038

[29] R. Subha, O. Sridevi, D. Anitha, D. Sudha, Treatment methods for the removal of phenol from water-A Review. Int. Conf. Systems, Sci. Control, Commun. Eng. Tech. 01 (2015). [ICSSCCET 2015] ID. 044

[30] A. Bódalo, E. Gómez, A. Hidalgo, M. Gómez, M. Murcia, I. López, Nanofiltration membranes to reduce phenol concentration in wastewater, Desalination 245 (2009) 680-686. https://doi.org/10.1016/j.desal.2009.02.037

[31] Z. Murthy, S.K. Gupta, Thin film composite polyamide membrane parameters estimation for phenol-water system by reverse osmosis, J. Sep. Sci. Tech. 33 (1998) 2541-2557. https://doi.org/10.1080/01496399808545318

[32] L.L. Gibbs, J.F. Scamehorn, S.D. Christian, Removal of n-alcohols from aqueous streams using micellar-enhanced ultrafiltration, J. Membr. Sci. 30 (1987) 67-74. https://doi.org/10.1016/S0376-7388(00)83341-4

[33] B.R. Fillipi, J.F. Scamehorn, S.D. Christian, R.W. Taylor, A comparative economic analysis of copper removal from water by ligand-modified micellar-enhanced ultrafiltration and by conventional solvent extraction, J. Membr. Sci. 145 (1998) 27-44. https://doi.org/10.1016/S0376-7388(98)00052-0

[34] H. Adamczak, K. Materna, R. Urbański, J. Szymanowski, Ultrafiltration of micellar solutions containing phenols, J. Colloid. Interf. Sci. 218 (1999) 359-368. https://doi.org/10.1006/jcis.1999.6430

[35] G-M. Zeng, K. Xu, J-H. Huang, X. Li, Y-Y. Fang, Y-H. Qu. Micellar enhanced ultrafiltration of phenol in synthetic wastewater using polysulfone spiral membrane, J. Membr. Sci. 310 (2008) 149-160. https://doi.org/10.1016/j.memsci.2007.10.046

[36] M.F. Nazar, S.S. Shah, J. Eastoe, A.M. Khan, A. Shah, Separation and recycling of nanoparticles using cloud point extraction with non-ionic surfactant mixtures, J. colloid. Interface Sci. 363 (2011) 490-496. https://doi.org/10.1016/j.jcis.2011.07.070

[37] A. Afkhami, T. Madrakian, H. Siampour, Flame atomic absorption spectrometric determination of trace quantities of cadmium in water samples after cloud point extraction in Triton X-114 without added chelating agents, J. Hazard. Mater. 138 (2006) 269-272. https://doi.org/10.1016/j.jhazmat.2006.03.073

[38] T. Poznyak, I. Chairez, A. Poznyak, Application of a neural observer to phenols ozonation in water: Simulation and kinetic parameters identification, Water Res. 39 (2005) 2611-2620. https://doi.org/10.1016/j.watres.2005.04.061

[39] I. Chairez, A. Poznyak, T. Poznyak, Reconstruction of dynamics of aqueous phenols and their products formation in ozonation using differential neural network observers, Ind. Eng Chem. Res. 46 (2007) 5855-5866. https://doi.org/10.1021/ie0705103

[40] H. Wang, F. Zhao, S-i. Fujita, M. Arai, Hydrogenation of phenol in $scCO_2$ over carbon nanofiber supported Rh catalyst, Catal. Commun. 9 (2008) 362-368. https://doi.org/10.1016/j.catcom.2007.07.002

[41] P. Jin, R. Chang, D. Liu, K. Zhao, L. Zhang, Y. Ouyang, Phenol degradation in an electrochemical system with TiO_2/activated carbon fiber as electrode, J. Environ. Chem. Eng. 2 (2014) 1040-1047. https://doi.org/10.1016/j.jece.2014.03.023

[42] I. Udom, P.D. Myers, M.K. Ram, A. Hepp, E. Archibong, E.K. Stefanakos, Optimization of photocatalytic degradation of phenol using simple photocatalytic reactor, Am. J. Anal. Chem. 5 (2014) 743-749. https://doi.org/10.4236/ajac.2014.511083

[43] H.M. Jalali, Kinetic investigation of photo-catalytic activity of TiO_2/metal nanocomposite in phenol photo-degradation using Monte Carlo simulation, RSC Adv. 5 (2015) 36108-36116. https://doi.org/10.1039/C5RA02226F

[44] Y. Jiang, T. Waite, Degradation of trace contaminants using coupled sonochemistry and Fenton's reagent, Water Sci. Tech. 47 (2003) 85-92.

[45] Y. Xiong, C. He, H.T. Karlsson, X. Zhu, Performance of three-phase three-dimensional electrode reactor for the reduction of COD in simulated wastewater-containing phenol, Chemosphere 50 (2003) 131-136. https://doi.org/10.1016/S0045-6535(02)00609-4

[46] E. El-Ashtoukhy, Y. El-Taweel, O. Abdelwahab, E. Nassef, Treatment of petrochemical wastewater containing phenolic compounds by electrocoagulation using a fixed bed electrochemical reactor, Int. J. Electrochem. Sci. 8 (2013) 1534-1550.

[47] A.P.H. Association, A.W.W. Association, W.P.C. Federation, W.E. Federation. Standard methods for the examination of water and wastewater: Am. Public Health Association. (1915) https://www.mwa.co.th/download/file_upload/SMWW_1000-3000.pdf.

[48] S.J. Kulkarni, J.P. Kaware, Review on research for removal of phenol from wastewater. Int. J. Sci. Res. Pub. 3 (2013) 1-5.

[49] C.E. Barrera-Díaz, G.Roa-Morales, P.B. Hernández, C.M. Fernandez-Marchante, M.A. Rodrigo, Enhanced electrocoagulation: New approaches to improve the electrochemical process, J. Electrochem. Sci. Eng. 4 (2014) 285-296. https://doi.org/10.5599/jese.2014.0060

[50] T.A. Enache, A.M. Oliveira-Brett, Phenol and para-substituted phenols electrochemical oxidation pathways, Electroanal. Chem. 655 (2011) 9-16. https://doi.org/10.1016/j.jelechem.2011.02.022

[51] S. Doğan, Y. Turan, H. Ertürk, O. Arslan, Characterization and purification of polyphenol oxidase from artichoke (*Cynara scolymus* L.), J. Agr. Food. Chem. 53 (2005) 776-785. https://doi.org/10.1021/jf049053g

[52] Q. Husain, U. Jan, Detoxification of phenols and aromatic amines frompolluted wastewater by using phenol oxidases, J. Sci. Ind. Res. 59 (2000) 1-4.

[53] U. Jadhav, S. Salve, R. Dhawale, M. Padul, V. Dawkar, A. Chougale, Use of partially purified banana peel polyphenol oxidase in the degradation of various phenolic compounds and textile dye blue 2RNL, Text. Light Ind. Sci. Tech. 2 (2013) 27-35.

[54] P.M. van Schie, L.Y. Young, Biodegradation of phenol: mechanisms and applications, Bioremed. J. 4 (2000) 1-18. https://doi.org/10.1080/10588330008951128

[55] Z. Huixian, K. Taylor, Products of oxidative coupling of phenol by horseradish peroxidase, Chemosphere 94 (1994)1807-1817. https://doi.org/10.1016/0045-6535(94)90028-0

[56] L. Martirani, P. Giardina, L. Marzullo, G. Sannia, Reduction of phenol content and toxicity in olive oil mill waste waters with the ligninolytic fungus Pleurotus ostreatus, Water Res. 30 (1996) 1914-1918. https://doi.org/10.1016/0043-1354(95)00330-4

[57] H. Wright, J.A. Nicell, Characterization of soybean peroxidase for the treatment of aqueous phenols, Bioresour. Tech. 70 (1999) 69-79. https://doi.org/10.1016/S0960-8524(99)00007-3

[58] N. Duran, E. Esposito, Potential applications of oxidative enzymes and phenoloxidase-like compounds in wastewater and soil treatment: a review, Appl. catals B: Environ. 28 (2000) 83-99. https://doi.org/10.1016/S0926-3373(00)00168-5

[59] N. Pradeep, S. Anupama, K. Navya, H. Shalini, M. Idris, U. Hampannavar, Biological removal of phenol from wastewaters: a mini review, Appl. Water Sci. 5 (2015) 105-112. https://doi.org/10.1007/s13201-014-0176-8

[60] G. Gurujeyalakshmi, P. Oriel, Isolation of phenol-degrading Bacillus stearothermophilus and partial characterization of the phenol hydroxylase, Appl. Environ. Microbiol. 55 (1989) 500-502.

[61] S.G. Burton, J.R. Duncan, P.T. Kaye, P.D. Rose, Activity of mushroom polyphenol oxidase in organic medium, Biotech. Bioeng. 42 (1993) 938-944. https://doi.org/10.1002/bit.260420804

[62] A. Garzillo, M. Colao, C. Caruso, C. Caporale, D. Celletti, V. Buonocore, Laccase from the white-rot fungus Trametes trogii, Appl. Microb. Biotech. 49 (1998) 545-551. https://doi.org/10.1007/s002530051211

[63] L. Li, J.C. Steffens. Overexpression of polyphenol oxidase in transgenic tomato plants results in enhanced bacterial disease resistance, Planta 215 (2002) 239-247. https://doi.org/10.1007/s00425-002-0750-4

[64] N. Lončar, N. Božić, I. Anđelković, A. Milovanović, B. Dojnov, M . Vujčić, Removal of aqueous phenol and phenol derivatives by immobilized potato polyphenol oxidase, J. Serb. Chem. Soc. 76 (2011) 513-522. https://doi.org/10.2298/JSC100619046L

[65] B. Okeke, A. Paterson, J. Smith, I. Watson-Craik, Comparative biotransformation of pentachlorophenol in soils by solid substrate cultures of Lentinula edodes, Appl. Microb. Biotech. 48 (1997) 563-569. https://doi.org/10.1007/s002530051097

[66] S. Ali, R. Fernandez-Lafuente, D.A. Cowan, Meta-pathway degradation of phenolics by thermophilic Bacilli, Enzyme Microb. Tech. 3 (1998) 2462-2468. https://doi.org/10.1016/S0141-0229(98)00072-6

[67] J. Bollag, K. Shuttleworth, D. Anderson, Laccase-mediated detoxification of phenolic compounds, Appl. Environ. l Microb. 54 (1988) 3086-3091.

[68] P. Schneider, M.B. Caspersen, K. Mondorf, T. Halkier, L.K. Skov, P.R. Østergaard, Characterization of a Coprinus cinereus laccase, Enzyme Microb. Tech. 25 (1999) 502-508. https://doi.org/10.1016/S0141-0229(99)00085-X

[69] G. Hublik, F. Schinner. Characterization and immobilization of the laccase from Pleurotus ostreatus and its use for the continuous elimination of phenolic pollutants, Enzyme Microb. Tech. 27 (2000) 330-336. https://doi.org/10.1016/S0141-0229(00)00220-9

[70] A. Robles, R. Lucas, G.A. de Cienfuegos, A Gálvez, Phenol-oxidase (laccase) activity in strains of the hyphomycete Chalara paradoxa isolated from olive mill wastewater disposal ponds, Enzyme Microb.Tech. 26 (2000) 484-490. https://doi.org/10.1016/S0141-0229(99)00197-0

[71] M. Stanisavljević, L. Nedić, Removal of phenol from industrial wastewaters by horseradish-Cochlearia armoracia L-peroxidase. Facta universitatis-series: Work. Live, Envron. Protect. 2 (2004) 345-349.

[72] T. Johjima, M. Ohkuma, T. Kudo, Isolation and cDNA cloning of novel hydrogen peroxide-dependent phenol oxidase from the basidiomycete Termitomyces albuminosus, Appl. Microbiol. Biotech. 61 (2003) 220-225. https://doi.org/10.1016/0043-1354(95)00237-5

[73] V. Cooper, J. Nicell, Removal of phenols from a foundry wastewater using horseradish peroxidase, Water Res. 30 (1996) 954-964. https://doi.org/10.1016/0043-1354(95)00237-5

[74] Y. Wu, K.E. Taylor, N. Biswas, J.K. Bewtra, A model for the protective effect of additives on the activity of horseradish peroxidase in the removal of phenol, Enzyme Microb. Technol. 22 (1998) 315-322. https://doi.org/10.1016/S0141-0229(97)00197-X

[75] K.d.Q. Wilberg, D.G. Nunes, J. Rubio, Removal of phenol by enzymatic oxidation and flotation, Brazil J. Chem. Eng. 17 (2000) 907-914. https://doi.org/10.1590/S0104-66322000000400055

[76] Y-C Lai, S-C Lin, Application of immobilized horseradish peroxidase for the removal of p-chlorophenol from aqueous solution, Process Biochem. 17 (2005) 1167-1174. https://doi.org/10.1016/j.procbio.2004.04.009

[77] I. Alemzadeh, S. Nejati, Removal of phenols with encapsulated horseradish peroxidase in calcium alginate, Iran J. Chem. Chem. Engi. 28 (2009) 43-49.

[78] K.F. Mossallam, F.M.Sultanova, N.A. Salimova, Enzymatic removal of phenol from produced water and the effect of petroleum oil content. Thirteenth Intl Water Tech Conf, Hurghada, Egypt. (2009) 1009-1020.

[79] N. Caza, J. Bewtra, N. Biswas, K. Taylor, Removal of phenolic compounds from synthetic wastewater using soybean peroxidase, Water Res. 33 (1999) 3012-3018. https://doi.org/10.1016/S0043-1354(98)00525-9

[80] M. Ghioureliotis, J.A. Nicell, Assessment of soluble products of peroxidase-catalyzed polymerization of aqueous phenol, Enzyme Microb. Tech. 25 (1999) 185-193. https://doi.org/10.1016/S0141-0229(99)00041-1

[81] F. Naghibi, F. Pourmorad, S. Honary, M. Shamsi, Decontamination of water polluted with phenol using Raphanus sativus root, Iran. J. Pharm. Res. 2 (2010) 29-32. https://doi.org/10.1016/j.chemosphere.2006.04.049

[82] S. Akhtar, Q. Husain, Potential applications of immobilized bitter gourd (*Momordica charantia*) peroxidase in the removal of phenols from polluted water, Chemosphere, 65 (2006) 1228-1235. https://doi.org/10.1016/j.biortech.2008.04.031

[83] F. Quintanilla-Guerrero, M. Duarte-Vázquez, B. García-Almendarez, R. Tinoco, R. Vazquez-Duhalt, C. Regalado, Polyethylene glycol improves phenol removal by immobilized turnip peroxidase, Bioresourc. Tech. 99 (2008) 8605-8611.

[84] M.C. Lakshmi, V. Sridevi, A review on biodegradation of phenol from industrial effluents, J. Ind. Poll. Control. 25 (2015) 13-27.

[85] P.K. Arora, H. Bae, Bacterial degradation of chlorophenols and their derivatives, Microb Cell Fact. 13 (2014) 25- 31. https://doi.org/10.1186/1475-2859-13-25

[86] S. Agarry, B. Solomon, Kinetics of batch microbial degradation of phenols by indigenous Pseudomonas fluorescence, Int. J. Environ. Sci. Tech. 5 (2008) 223-232. https://doi.org/10.1007/BF03326016

[87] Ş. Şeker, H. Beyenal, B. Salih, A. Tanyolac, Multi-substrate growth kinetics of Pseudomonas putida for phenol removal, Appl. Microbiol. Biotech. 47 (1997) 610-614. https://doi.org/10.1007/s002530050982

[88] M. Ajaz, N. Noor, S.A. Rasool, S.A. Khan, Phenol resistant bacteria from soil: identification-characterization and genetical studies, Pak. J. Bot. 5 (2004) 5415-424.

[89] F. Kafilzadeh, M-S. Farhangdoost, Y. Tahery, Isolation and identification of phenol degrading bacteria from Lake Parishan and their growth kinetic assay, Afr. J. Biotech. 9 (2010) 6721-6726.

[90] I. Sgountzos, S. Pavlou C.Paraskeva, A. Payatakes, Growth kinetics of Pseudomonas fluorescens in sand beds during biodegradation of phenol, Biochem. Eng. J. 30 (2006) 164-173. https://doi.org/10.1016/j.bej.2006.03.005

[91] Y.Wang, J. Song, W. Zhao, X. He, J. Chen, M. Xiao, In situ degradation of phenol and promotion of plant growth in contaminated environments by a single Pseudomonas aeruginosa strain, J. Hazard. Mater. 192 (2011) 354-360. https://doi.org/10.1016/j.jhazmat.2011.05.031

[92] S. Prasad, R.S. Babu, R. Chakrapani, C. Rao, Kinetics of high concentrated phenol biodegradation by Acinetobacter baumannii, Int. J. Biotech. Biochem. 6 (2010) 609-615.

[93] S. Thomas, S. Sarfaraz, L. Mishra, L. Iyengar, Degradation of phenol and phenolic compounds by a defined denitrifying bacterial culture, World J. Microbiol. Biotech. 18 (2002) 57-63. https://doi.org/10.1023/A:1013947722911

[94] H. Ehrhardt, H. Rehm, Phenol degradation by microorganisms adsorbed on activated carbon, Appl. Microbiol. Biotech. 21 (1985) 32-36. https://doi.org/10.1007/BF00252358

[95] R.J. Varma, B.G. Gaikwad, Rapid and high biodegradation of phenols catalyzed by Candida tropicalis NCIM 3556 cells, Enzyme. Microbiol. Technol. 43 (2008) 431-435. https://doi.org/10.1016/j.enzmictec.2008.07.008

[96] C.T. dos Passos, M. Michelon, J.Burkert, S.J. Kalil, C.A.V. Burkert, Biodegradation of phenol by free and encapsulated cells of a new Aspergillus sp. isolated from a contaminated site in southern Brazil, Afr. J. Biotech. 9 (2010) 6716-6720.

[97] Y. Jiang, X. Cai, D. Wu, N. Ren, Biodegradation of phenol and m-cresol by mutated Candida tropicalis, J. Environ. Sci. 5 (2010) 621-626. https://doi.org/10.1016/S1001-0742(09)60154-6

[98] L. Zídková, J. Szőköl, L. Rucká, M. Pátek, J. Nešvera, Biodegradation of phenol using recombinant plasmid-carrying Rhodococcus erythropolis strains, Int. Biodeterior. Biodegrad. 84 (2013) 179-184. https://doi.org/10.1016/j.ibiod.2012.05.017

[99] A. Viggiani, G. Olivieri, L. Siani, A. Di Donato, A. Marzocchella, P. Salatino, An airlift biofilm reactor for the biodegradation of phenol by Pseudomonas stutzeri OX1, J. Biotech. 23 (2006) 464-477. https://doi.org/10.1016/j.jbiotec.2005.12.024

[100] V. Arutchelvan, V. Kanakasabai, R. Elangovan, S Nagarajan, V. Muralikrishnan, Kinetics of high strength phenol degradation using Bacillus brevis, J. Hazard. Mater. 129 (2006) 216-222. https://doi.org/10.1016/j.jhazmat.2005.08.040

[101] R.J. Varma, B.G. Gaikwad, Rapid and high biodegradation of phenols catalyzed by Candida tropicalis NCIM 3556 cells, Enzyme Microb. Technol. 43 (2008) 431-435. https://doi.org/10.1016/j.enzmictec.2008.07.008

[102] C.T. dos Passos, M. Michelon, J. Burkert, S.J. Kalil, C.A.V. Burkert, Biodegradation of phenol by free and encapsulated cells of a new Aspergillus sp. isolated from a contaminated site in southern Brazil, Afr. J. Biotechnol. 9 (2010) 6716-6720.

[103] Y. Jiang, X. Cai, D. Wu, N. Ren, Biodegradation of phenol and m-cresol by mutated Candida tropicalis, J. Environ. Sci. 22 (2010) 621-626. https://doi.org/10.1016/S1001-0742(09)60154-6

[104] L. Zídková, J. Szőköl, L. Rucká, M. Pátek, J. Nešvera, Biodegradation of phenol using recombinant plasmid-carrying Rhodococcus erythropolis strains, Int. Biodeter. Biodegrad. 84 (2013) 179-184. https://doi.org/10.1016/j.ibiod.2012.05.017

[105] A. Viggiani, G. Olivieri, L. Siani, A. Di Donato, A. Marzocchella, P. Salatino, P. Barbieri, E. Galli, An airlift biofilm reactor for the biodegradation of phenol by Pseudomonas stutzeri OX1, J. Biotechnol. 123 (2006) 464-477. https://doi.org/10.1016/j.jbiotec.2005.12.024

[106] V. Arutchelvan, V. Kanakasabai, R. Elangovan, S. Nagarajan, V. Muralikrishnan, Kinetics of high strength phenol degradation using Bacillus brevis, J. Hazard. Mater. 129 (2006) 216-222. https://doi.org/10.1016/j.jhazmat.2005.08.040

Chapter 7

Photocatalysis: Present, past and future

A. Manikandan [1], and E. Manikandan [2-4], S.Vadivel [5*], M.Kumaravel [5], D. Maruthamani [5], S. Hariganesh [5]

[1] Department of Chemistry, Bharath Institute of Higher Education and Research, Bharath University, Chennai – 600 073, Tamil Nadu, India

[2] Department of Physics, Thiruvalluvar University, TVUCAS Campus, Thennangur, 604408, Tamil Nadu, India

[3] UNESCO UNISA Africa Chair in Nanosciences & Nanotechnology, College of Graduate Studies, University of South Africa, Muckleneuk Ridge, PO Box 392, Pretoria, South Africa

[4] Nanosciences African Network (NANO-AFNET), iThemba LABS-National Research Foundation, 1 Old Faure Road, Somerset West, PO Box: 722, Cape Town, 7129, South Africa

[5] Department of Chemistry, PSG College of Technology, Coimbatore-641004, India

vlvelu7@gmail.com

Abstract

As one of the most attractive technologies, photocatalysis has been emerging nowadays to harvesting the solar energy for producing green fuels and a wide range of environmental applications. Due to their unique physicochemical and optical properties, a wide variety of TiO_2 based photocatalysts have emerged to drive various organic transformations and degradation reactions under light irradiation. In this chapter, we have systematically summarized the fundamentals of TiO_2 based photocatalysts, including basic mechanism of heterogeneous photocatalysis, advantages, and challenges of g- TiO_2 based photocatalysts. Through reviewing the important state-of-the-art advances on this topic, it may provide new opportunities for designing and constructing highly effective TiO_2 and various bismuth-based photocatalysts for various applications in photocatalysis and other related fields, such as solar cells and photoelectrocatalysis.

Keywords

Semiconductors, Photocatalyst, Degradation, TiO_2, Bismuth Materials

Contents

1. Introduction

Semiconductor photocatalysis has received much attention during past few decades as a promising solution for both energy and environmental problems. Since the discovery of Fujishima and Honda effect that water can be split into hydrogen and oxygen using a semiconductor (TiO_2) electrode, a lot of work has been carried out to produce hydrogen from water splitting using a variety of semiconductor photocatalysts. Nowadays, the scientific interest in heterogeneous photocatalysis has also been focused on environmental applications such as effluent treatments and air purification. Various metal oxides have been investigated and evaluated as potential photocatalysts. The most prominent examples have been TiO_2, ZnO, and SnO_2. The great interest in the development of active photocatalysts is evidenced by many research publications appeared during the last few years. The purpose of this chapter is to summarize significant developments in the field of TiO_2 and alternative bismuth-based materials for degradation of pollutants. Following a brief outline of the structural properties of TiO_2, this section is devoted to the discussion of possible modifications to improve the efficiency of the TiO_2 photocatalyst, followed by a summary of major parameters which are responsible for the degradation of organic pollutants.

1.1 Advanced oxidation processes (AOPs)

Advanced oxidation processes such as semiconductor photocatalysis, UV/H_2O_2, Photo-Fenton process and ozonation are widely used for the treatment of intractable compounds in industrial effluents. These processes involve the generation of ($^{\bullet}OH$) radicals. Generally Hydroxyl radical is considered to be a most powerful oxidizing agent with the oxidation potential of 2.8 eV and it reacts with organic compounds rapidly, the reaction leads to complete destruction of the targeted organic pollutants.

1.2 Principle of heterogeneous photocatalysis

A semiconductor can act as a photocatalyst in the photo mineralization process, due to its unique electronic structure, which is characterized by a filled valence band (VB) and an empty conduction band (CB). The photocatalytic effect of the semiconductor was first reported by Fujishima and Honda [1] for water splitting using TiO_2 semiconductor. When a semiconductor is illuminated with UV or Visible light, electrons are promoted from the VB to the CB leaving a hole as shown in Fig. 1 [2].

The effective generation of an electron-hole pair in a semiconductor particle depends on the intensity of the incident light as well as electronic characteristics of the matter. The threshold wavelength required for excitation of a semiconductor is governed by the following equation:

$$E_{bg}\,(\text{eV}) = 1240/\lambda_{bg}\,(\text{nm}) \quad (1)$$

In the absence of electron-hole scavengers, the photogenerated electron-hole pairs may recombine and dissipate the input energy as a heat within a fraction of seconds. Thus for a photocatalytic process to be efficient, the electron must be removed by an electron acceptor. So, the electron acceptors are necessary for the photocatalytic process to prevent the recombination.

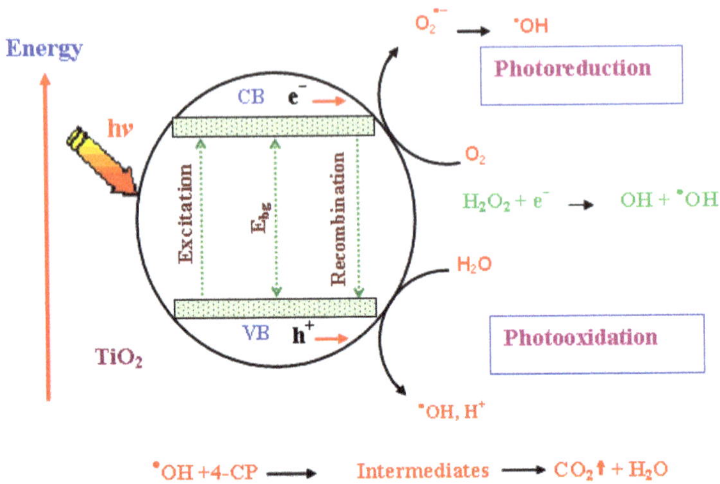

Figure 1 Electron-hole pair generation in an illuminated semiconductor particle and consecutive reactions for degradation.

Oxygen is commonly used as electron acceptor. It may react with a photogenerated electron at the surface of semiconductors. The photogenerated holes, having an affinity for electrons, are very strong oxidizing agents. After migrating to the surface of the semiconductor, they oxidize the adsorbed water molecule or hydroxide ion to form hydroxyl radicals. The photogenerated ($^{\bullet}OH$) radicals with an unpaired electron are very reactive species. Due to its very high oxidation potential, it can oxidize most of the organic pollutants into CO_2 and their mineral acids. Therefore, the overall photocatalytic process can be summarized by the following reaction.

$$\text{Organic pollutants} + O_2 \xrightarrow[\text{UV/solar light}]{\text{Semiconductor}} CO_2\uparrow + H_2O + \text{mineral acids} \tag{2}$$

Photogenerated electron and hole can also be used for reduction and oxidation before recombination. This makes the semiconductor as a combined redox catalyst in organic reactions.

1.3 Semiconductor photocatalysts for water splitting

Photocatalytic, water splitting and degradation of organic pollutants using semiconductor materials have attracted considerable interest. The development of clean and renewable energy sources is a great technological challenge in near future. One of the most attractive options is the large-scale utilization of hydrogen (H_2) as a source for electricity. However, industrial H_2 production consumes huge amounts of fossil fuels resulting in large CO_2 emission which causes atmospheric pollution. Water splitting for hydrogen production has been the main goal of semiconductor materials in photoelectrochemistry for the past few decades. H_2 is considered to be a future fuel due to its high-energy capacity per unit volume as compared to oil and petroleum [3].

The most important requirements of semiconductor photoelectrodes for water splitting process are high stability, sufficient light assimilation, favorable band gaps with respect to water oxidation potential, slow recombination rate and low cost. Among the various semiconductor materials, metal sulfides and spinel ferrites have been considered to be the most promising photocatalytic materials, due to their good photochemical stability, non-toxicity, high photostability and low band gap energy.

There are many semiconductors that are commercially available for photocatalytic application. But, only a few of them are efficiently used for the photodegradation of a wide range of organic pollutants. Table 1 summarizes the semiconducting photocatalysts commonly used with their band gap energies and corresponding absorbance values. For a semiconductor to be a photochemically active catalyst, the redox potential of the photogenerated VB hole must be positive to generate $^{\bullet}OH$ radicals and the redox potential of the photogenerated CB electron must be negative to be able to reduce adsorbed oxygen to superoxide radicals.

Table 1 Some common semiconductors, their band gap energies and corresponding threshold wavelengths.

Semiconductor	Band gap (eV)	Wavelength (nm)
SnO_2	3.9	320
TiO_2	3.2	388
ZnO	3.2	388
$BiVO_4$	2.4	516
CdS	2.4	516
In_2O_3	2.5	496
WO_3	2.8	443

Among various semiconductors, TiO_2 has become the prominent catalyst against which the photocatalytic activity of other semiconductors is measured, because of its high photocatalytic activity and stability [4]. ZnO is found to be an alternative material to TiO_2. ZnO has proved to be more efficient than TiO_2 in various applications [5,6]. CdS is not as photoactive as TiO_2 but it has been studied extensively due to its good response to wavelengths of the visible light region [7]. CdS suffer from photo-corrosion induced by self-oxidation and release of some cadmium ions into solution to cause toxicity [8,9].

$$CdS + 2H^+ \longrightarrow Cd^{2+} + S \qquad (3)$$

Other semiconductors such as WO_3, $BiVO_4$, SnO_2, In_2O_3 and $SrTiO_3$ have also been used widely for photooxidation reactions [10]. The E_g band gap of the ZnO and TiO_2 is 3.2 eV, which requires excitation wavelength in the UV region. It is found that unmodified TiO_2 has limited use in the degradation of toxic pollutants, because of its low efficiency and inability to utilize the visible light portion of the solar spectrum. Hence, the current research is focused on the tuning of physio-chemical factors of TiO_2 by various methods like doping with metals and non-metals, etc.

1.4 Applications of semiconductor photocatalysts

Qu *et al.* [11] studied the photocatalytic degradation (PCD) of rhodamine B in a TiO_2 solution using both UV and visible radiation. Photodegradation of a series of dyes (rhodamine B, sulforhodamine B, fluorescein, alizarin red and eosin) using TiO_2 catalyst under visible light illumination was studied by Wu *et al.* [12]. Wang *et al.* [13] reported the PCD of eight different commercial dyes containing various substitute groups using TiO_2 catalyst under solar irradiation. Arslan *et al.* [14] reported PCD of two different simulating textile wastewater from an integrated plant and a dyehouse mill using TiO_2 under ambient conditions.

PCD of reactive dyes (brilliant blue R, remazol black B and reactive blue) using TiO_2 and ZnO was reported by Gouvea *et al.* [15]. Galindo *et al.* [16] studied the PCD of acid orange 20 dye in the presence of TiO_2-P25 (Degussa) under UV light. CO_2 and aliphatic acids were identified as final degraded products. Decomposition of non-biodegradable methylene blue was examined using TiO_2 dispersions under UV illumination (Zhang *et al.* [17]). The PCD of eosin Y has been investigated by Poulios *et al.* [18] in aqueous solutions containing TiO_2-P25 (Degussa) and ZnO as photocatalysts. Yang *et al.* [19] reported the PCD of various dyes under visible light irradiation using TiO_2 electrodes. Wang *et al.* [20] prepared ZnO powders with various size (nanometric size: 10, 50, 200 and 1000 nm) by thermal evaporation and chemical deposition methods, and examined their photocatalytic ability under UV light for degradation of methyl orange. Ren *et al.* [21] reported the large-scale synthesis of hexagonal ZnO nanoparticles by the esterification process using zinc acetate and alcohol. Furthermore, the prepared ZnO particles exhibit high PCD of rhodamine B, indicating that the nanostructured ZnO is promising as a semiconductor photocatalyst. Mohapatra *et al.* [22] synthesized the vertically oriented TiO_2 nanotube arrays by the sono electrochemical anodization technique which was found to be an excellent photocatalyst for the decomposition of non-biodegradable various azo dyes under simulated sunlight.

Wang *et al.* [23] synthesized ZnO nanoflowers by a simple low-temperature method in the absence of surfactants. Systematic studies were carried out to investigate the factors affecting the morphology of the samples. It was found that the ZnO with flower shape morphology exhibited improved activity towards the degradation of chlorophenols under UV radiation. Pardeshi and Patil [24] synthesized ZnO crystallites by solution free mechanochemical method. Photocatalytic activity of ZnO was checked by means of oxidative PCD of resorcinol, a potent endocrine disrupting material in water using solar light. The degradation efficiency was found to be dependent on crystallite growing rate and morphology of ZnO. Lu *et al.* [25] examined ZnO photocatalytic effectiveness to successfully degrade basic blue 11 under visible light irradiation. Rao *et al.* [26] investigated the PCD of direct yellow12 by a simple batch process using ZnO under irradiation with UV light. Sun *et al.* [27] developed a novel dumb-bell shaped ZnO photocatalyst by the simple hydrothermal method. The results indicated that the prepared dumbbell-shaped ZnO microcrystal photocatalyst exhibited good photocatalytic activity and it could be considered as a promising photocatalyst for dye effluent degradation.

Tong *et al.* [28] synthesized quasi-sphere-like ZnO using an ice-water bath. This hierarchically nanostructured ZnO showed significantly improved photocatalytic activity than ZnO nanopowder. The ZnO nanostructures synthesized in the shape of particles, rods, flower-like and micro-sphere using a facile hydrothermal method has been found very useful for decolorization of acid red dye solutions under sunlight [29]. Wu *et al.* reported that CdS showed the highly efficient photocatalytic performance on reducing 4-nitroaniline [30]. Luo *et al.* synthesized CdS hollow spheres which exhibited excellent photocatalytic activity towards the degradation of rhodamine B under UV irradiation [31].

Chen *et al.* [32] synthesized CdS nanoparticles which showed excellent photocatalytic activity, cyclic stability and higher photocatalytic efficiency than that of TiO_2. Shi *et al.* [33] and Liu *et al.* [34] reported that the combination of CdS with ZnS significantly improved photocatalytic activity and minimize the usage of toxic Cd. Zhang *et al.* [35]

studied the PCD of methyl orange using $CoFe_2O_4/ZnO$ composite prepared by collosol technique. The results showed that the composite photocatalyst was able to degrade the dye substrate up to 93.9 %. In addition, this magnetic photocatalyst also showed good magnetic property and could be completely recovered using an external magnet. Zhang *et al.* used NiS supported CdS for water splitting application with a high quantum efficiency of 51.3% [36]. Hou *et al.* [37] studied the PCD of 4-chlorophenol by $ZnFe_2O_4$ nanotube array electrode. Habibi *et al.* [38] reported the PCD of an azo textile dye effluents using $ZnFe_2O_4$ -ZnO nanocomposite. Satsangi *et al.* [39] prepared a series of α-Fe_2O_3 samples with a number of (Cu and Zn) dopants and reported that 5.0 at. % Zn doped α-Fe_2O_3 film prepared by the spray pyrolysis method showed the best photoresponse as compared to other films.

1.5 Modified TiO_2 photocatalysts

TiO_2 has been a widely studied photocatalyst material for dye containing waste-water treatment, due to its ability to generate a strong oxidant hydroxyl radical, good chemical stability and non-toxicity. TiO_2 naturally occurs in three common crystalline polymorphs: anatase, rutile, and brookite. Although the anatase phase is considered as the most photocatalytically active phase, rutile is the most thermodynamically stable phase [40]. The wide energy band gap (E_g> 3.2 eV) in anatase may limit its potential application because only UV light can initiate the electron-hole generation process [41].

In general, by loading of co-catalyst to trap the photogenerated electrons or holes, the photocatalytic activity can be dramatically improved. It is a great challenge to activate TiO_2 under the sunlight, which consists of about 5% UV light and 45% visible light. Reduction of E_g makes the catalyst solar light active. Photocatalytic activity can be improved by reducing electron-hole recombination and improving the adsorption rate of the pollutant.

A number of methods have been reported for the modification of TiO_2.

I. Addition of a low percentage of metal: The metal was introduced through different ways.
    - Doping process (molecular combination of metal oxide in the lattice of TiO_2)
    - Metallization process (deposition of noble metal on the photocatalyst crystallites)
    - Impregnation method (deposition of a small amount of salt on photocatalyst surface).

    However, recently transition metals (Cr, Mn, Fe, Co, Ni, Cu, and Zn) have been used to dope on TiO_2 surface and thus, to reduce the overall catalyst production cost. The incorporation of transition metal ions into the TiO_2 lattice was found to lower the band gap (E_g) and shift the catalyst absorbance edge to the visible light region.

II. Addition of low concentrated transition metal ion (Ag^+, Zn^{2+}, Cd^{2+}, Ce^{3+}, Co^{3+}, Fe^{3+} or W^{6+}) into the solution of substrate. [42-44]

III. Addition of low percentage of non-metal: The non-metals such as S, N, C, Cl, O, P and F are introduced on TiO_2 lattice by condensation and impregnation methods.

IV. Doping by anions such as SO_4^{2-}, PO_4^{3-}.

V. Supporting the photocatalyst on porous materials of larger particle size using silica [45], alumina [46], zeolites [47] and activated carbon[48].

VI. Loading of metal halides such as AgCl, AgBr etc.,

VII. Incorporation of dyes on TiO_2 to produce a dye-sensitized catalyst for harvesting visible light [49-51].

1.6 Application of modified TiO_2 photocatalysts

As a potential answer to the World's collective environmental pollution problem, the application of hydrogen energy has attracted great attention. Extensive work has been devoted to the efficient production of hydrogen at low cost. The photocatalytic water splitting has been considered as a promising strategy for converting solar energy into hydrogen energy. The photocatalytic H_2 generation offers high energy consumption. The researchers have made enormous efforts to improve the efficiency of H_2 production by modifying TiO_2 nanostructures [52], as well as developing novel photocatalysts [53-55]. Due to the suitable energy band gap that matches to visible light absorption, chalcogenide semiconductors are regarded as good candidates for water splitting.

1.6.1 Metal loaded TiO_2

Sclafani *et al.* [56] reported the use of Ag^+ ions as redox reagent for the photocatalytic degradation of phenol using TiO_2.semiconductor. Prairie *et al.* [57] evaluated the photodegradation of salicylic acid using Pt^{4+}, Hg^{2+}, Au^{3+} and Ag^+ ions loaded TiO_2 as photocatalyst [58]. Sclafani and Herrmann [59] studied the influence of Ag and Pt-Ag bimetallic deposits on the photocatalytic activity of TiO_2 (anatase and rutile phases) in dye degradation. The degradation products and a reduction in total organic carbon on the photodecomposition of phenol was reported by Ilisz *et al.* [60] Gouvea *et al.* [61] illustrated that the toxicity of lignin and Kraft E_1 effluent due to *E. coli* can be completely removed with the use of Ag-TiO_2 photocatalyst. By using Ag-TiO_2, by photocatalyst the formation of transient toxic species for the remazol brilliant blue R dye were observed by the Chen *et al.* Chen *et al.* [62] studied the effect of metal ions (Cu^{2+}, Fe^{3+}, Zn^{2+}, Al^{3+}, and Cd^{2+}) doping in TiO_2 towards the photodegradation of sulforhodamine B

(SRB), alizarin red (AR), and malachite green (MG) dyes. UV active Ag-TiO_2 with high surface area films for efficient photodegradation of methyl orange was examined by Arabatzis *et al.* [63].

Li *et al.* [64] investigated the mechanism of photosensitization and recombination of excited electron-hole pairs by depositing Pt on TiO_2. Arabatzis *et al.* [65] studied the effect of surface modification of rough, nanocrystalline TiO_2 thin-films by gold deposition using electron beam evaporation technique towards the degradation of industrial water pollutants.

Iliev *et al.* [66] studied the degradation of xylenol orange effluent by TiO_2, modified with Pd particles. TM/TiO_2/SiO_2 photocatalysts were prepared by the photodeposition method using transition metal salts (TM = Fe^{3+}, Co^{2+}, Ni^{2+}, and Cu^{2+}) for the photodegradation of reactive brilliant red dye. The TiO_2 catalysts modified by Fe^{3+}, Co^{2+}, Ni^{2+} showed improved photocatalytic performance [67]. Xie *et al.* and Yuan *et al.* [68] developed the Nd^{3+} ion modified TiO_2 by chemical co-precipitation technique. Photocatalytic reactivity of Nd^{3+}-TiO_2 sol was evaluated by PCD of phenol under visible light irradiation.

Sahoo *et al.* [69] investigated the PCD of crystal violet in dye solutions with Ag^+ modified TiO_2 under UV and simulated solar light. It was found that Ag^+ doped TiO_2 was more efficient than undoped TiO_2. Sobana *et al.* [70] reported Ag NPs doped TiO_2 for the degradation of direct red 23 and direct blue 53. The improved photocatalytic activity of Ag-TiO_2 was due to efficient electron trapping by Ag particle. Wang *et al.* [71] synthesized Fe^3-TiO_2 nanopowders by plasma oxidative pyrolysis for use in effluent treatment. Vinu *et al.* [72] prepared the palladium substituted anatase TiO_2. The PCD of various dyes such as methylene blue, rhodamine B, and phenol was investigated under UV light. Ravichandran *et al.* [73] investigated the degradation of pentafluorobenzoic acid using Pt-doped TiO_2. Dozzi *et al.* [74] studied the effects of gold dopants on titanium dioxide towards the oxidative degradation of formic acid and acid red 1 dye under visible light irradiation.

1.6.2 Non-metal loaded TiO_2

Matos *et al.* [75] studied the synergetic effect for PCD of phenol by TiO_2/ activated carbon composite. Herrmann *et al.* [76] reported solar PCD of chlorophenols using the synergistic effect between TiO_2 activated carbon composite in the aqueous dye solution. Da Silva *et al.* and Faria *et al.* [77] reported the PCD of solophenyl green dye using TiO_2/activated carbon composite. The mechanism of degradation was attributed to synergetic effect between activated carbon and TiO_2.

Samantaray *et al.* [78] prepared the sulfate-modified TiO_2 with different crystal structures. Sulfated TiO_2 was used for the degradation of 4-nitrophenol. PCD of 4-nitrophenol depends on the surface area and the crystal form as well as the particle size of TiO_2. The photocatalytic activity of the advanced materials towards the degradation of some typical pollutants in water and air using the artificial and natural light was studied by Sakthivel *et al.* and Kisch *et al.* [79]. Tryba *et al.* [80] investigated the PCD activities of TiO_2 particles anatase phase covered by thin layer of carbon towards various dye effluents. Ohno *et al.* [81] synthesized chemically sulfur modified TiO_2 photocatalysts for the degradation of methylene blue and 2-propanol in aqueous solution under irradiation at visible light.

Li *et al.* [82] reported the photocatalytic activity of carbon-doped TiO_2 for gas-phase oxidation of benzene under solar light irradiation. The photocatalytic activity in the visible region is attributed to the presence of oxygen vacancy states between the valence band and conduction band in the TiO_2 band structure. Youji *et al.* [83] prepared TiO_2-decorated activated carbon (TiO_2/AC) composite by hydrolytic precipitation method and used for degradation of commercial dyes.

Yang *et al.* [84] reported the doping of nitrogen and carbon in TiO_2 films by sol-gel technique. Nitrogen and carbon doping into substitutional sites of TiO_2 lattice greatly reduces the band gap and shows improvement in the photocatalytic activity. Sun *et al.* [85] synthesized carbon-sulfur-co-doped TiO_2 composite by the hydrolysis of tetrabutyl titanate in a mixed aqueous solution containing thiourea and urea. The photocatalytic activity was evaluated by the PCD of 4-chloro phenols under both UV and visible light irradiation.

Chen *et al.* [86] reported the utilization of visible light more efficiently in photocatalytic reactions with carbon-doped TiO_2, nitrogen doped TiO_2 and the carbon and nitrogen co-doped TiO_2 nanoparticles. The photocatalytic activities of C-TiO_2, N-TiO_2, and C-N-TiO_2 samples with different C and N contents were evaluated by degradation of methylene blue under visible light irradiation. Xiao *et al.* [87] reported the preparation of carbon-doped TiO_2 nanoparticles by sol-gel method. The visible light photocatalytic activity of carbon-doped TiO_2 was attributed to the presence of oxygen vacancy between the valence bands and the conduction bands.

Jiang *et al.* [88] reported the visible light active N-doped TiO_2 nanomaterials by a solvothermal technique. The obtained nitrogen-doped TiO_2 were thermally stable and robust for photodegradation of methylene blue under visible light Irradiation. Hu *et al.* [89] fabricated the TiO_2 nanotube (TN) arrays by an anodic oxidation process. The experiments demonstrated that the TiO_2 Nanotube arrays displayed an excellent photocatalytic activity under sunlight irradiation towards the degradation of methylene blue.

Parida *et al.* [90] prepared the sulfated TiO_2 by incipient wetness impregnation method. It was found that sulfate modification could reduce the crystallinity size and increase the specific surface area of the catalysts. The degradation of methyl orange under solar light radiation was investigated to evaluate the photocatalytic activity of the sulfated TiO_2 catalyst. Yu *et al.* [91] reported F-doped TiO_2 nanotubes by impregnation method. Compared to pure TiO_2 nanotubes, fluoride doped TiO_2 nanotubes showed enhanced photocatalytic efficiency. Vijayabalan *et al.* [92] reported the photocatalytic activity of fluorinated TiO_2 in the degradation of a chlorotriazine dye under ambient conditions. Surface fluorination of TiO_2 enhanced the adsorption of chlorotriazine dye while improving the overall degradation rate. Ang *et al.* [93] have prepared a series of nitrogen-doped TiO_2-SiO_2 mixed oxide catalysts by a sol-gel method. The N doped TiO_2-SiO_2 catalysts exhibited superior photocatalytic activity in degradation of methylene blue under visible light compared to the pure TiO_2 catalysts. Livraghi *et al.* [94,95] reported the nitrogen-doped TiO_2, a photocatalytic material active in visible light. These results provide a characterization of the electronic states associated with N impurities in TiO_2 particles for the effective degradation of dyes under irradiation with visible light.

1.6.3 Metal halide loaded TiO_2

Hu *et al.*[96] prepared Ag/AgBr/TiO_2 composite by the precipitation approach for photocatalytic application. The results indicated that AgBr was the main photoactive species for the destruction of azo dyes and bacteria under visible light irradiation. Wang *et al.* [97] prepared the TiO_2/BaF_2 ceramic powder using the $TiCl_4$ as a precursor. Experimental results indicated that the loading of BaF_2 enhanced the photoactivity of TiO_2. Elahifard *et al.* [98,99] prepared and used Ag/AgBr/TiO_2 deposited hydroxyapatite as adsorption bioceramic photosensitive material. The prepared Ag/AgBr/TiO_2-deposited hydroxyapatite has high ability to adsorb bacteria in the dark and also possesses high photocatalytic activity under the visible light.

Li *et al.* [100] synthesized the AgI/TiO_2 photocatalyst by a feasible approach and used for dye degradation.The highly efficient visible-light-induced photocatalytic activity of the nanostructured AgI/TiO_2 was attributed to its strong absorption in the visible region and low recombination rate and the synergetic effect between the components of AgI and TiO_2 in the nanostructured AgI/TiO_2. Zang and Farnood [101] investigated the photocatalytic activity of AgBr/TiO_2 under simulated solar light irradiation for the degradation of methyl orange. Yu *et al.* [102] studied a new visible-light-driven plasmonic photocatalyst Ag/AgCl/TiO_2 nanotube arrays. The prepared metal-semiconductor nanocomposite plasmonic photocatalyst exhibited a highly PCD of methyl orange in the visible region.

1.7 Bismuth-based photocatalysts:

Bismuth-based composites attracted much interest among researchers in terms of synthesis, characterization and visible light photocatalytic activities. Many bismuth-based materials such as Bi_2S_3 and BiOX etc., classified as sulfides and oxyhalides. Most of these bismuth catalysts are active in the visible light region with a bandgap of less than 3.0 eV. The band gap of Bi_2S_3 and BiOI are less than 2.0eV, indicating their ability to absorb the visible light of longer wavelength. Therefore, the synthesis and photocatalytic activity of various bismuth-based compounds have been studied extensively [101].

1.7.1 BiOBr Semiconductors

Among the various BiOX (X= Cl, F, Br, I) BiOBr is a vital ternary compound, with a band gap of 2.64 eV. The unique tetragonal structure of BiOBr consists of the layered structure of $[Bi_2O_2]^{2+}$ sandwiched between the Br atoms. It has PbFCl type crystal structure is and D4h space group, which brings about an effective separation of the photoinduced electrons, and the holes for enhanced photocatalytic activity. However, the photocatalytic performance of BiOBr has been quite limited due to its high recombination rate which originated from its electron-hole pairs. According to previous reports, synergistic modifications in BiOBr brought through coupling with BiOCl and $ZnFe_2O_4$ or doping with noble metals such as Ag, Au, Pt have improved the photocatalytic activity [102].

1.7.2 Bi_2S_3 semiconductors

In recent years, there has been increasing interest among researchers in metal chalcogenides semiconductors owing to their unique properties and potential applications in various fields. Among these, Bi_2S_3 has been the most studied semiconductor which mostly exists in the orthorhombic phase. In particular, Bi_2S_3 nanomaterials have potential applications in various fields including, lithium-ion batteries, photovoltaic cells, photocatalysts and thermoelectric cooling devices, etc. Bi_2S_3 nanomaterials in the form of nanowires, nanoribbons, and nanorods have been synthesized by various techniques such as hydrothermal, solvothermal, physical vapor deposition (PVD), microwave synthesis and wet chemical synthesis methods. However, the participation of thiourea, thioacetamide and sodium sulfide as sulfur source induces various health and environmental risks. Such issues can be overcome by using various biomolecules assisted synthesis methods. The photocatalytic properties of different types of Bi_2S_3

nanostructures have been reported earlier. For example, though Bi_2S_3 is a good visible light active photocatalytic material but its electron-hole recombination rate is much faster that limits its practical application. Therefore, many efforts have been taken to improve the photocatalytic efficiency of Bi_2S_3 by combining with other semiconductors such as ZnO, BiOCl, CuS, CdS and other carbon materials [102].

Conclusions

Recently, many efforts have been put into the development of novel semiconductor materials for the degradation of organic pollutants and water purification by the photocatalytic process. The photocatalytic process has gained much attention from the researchers as the basic idea behind this process is quite simple. In this chapter, we have discussed in detail about various semiconductor photocatalysts that are already available and to provide an overview of recent developments that have emerged in this field to develop novel materials as a photocatalyst under visible light. We also encourage further research on this hot topic as there is a lot of space to be explored in this fascinating field of the photocatalyst.

References

[1] Fujishima, K. Honda, Electrochemical photolysis of water at a semiconductor electrode, Nature 238 (1972) 37-38. https://doi.org/10.1038/238037a0

[2] D. Chen, A.K. Ray, Photodegradation kinetics of 4-nitrophenol in TiO_2 suspension, Water Res. 32 (1998) 3223-3234. https://doi.org/10.1016/S0043-1354(98)00118-3

[3] A.A. Tahir, K.G.U. Wijayantha, Photoelectrochemical water splitting at nanostructured $ZnFe_2O_4$ electrodes, J. Photochem. Photobiol. A: Chem. 216 (2010) 119-125. https://doi.org/10.1016/j.jphotochem.2010.07.032

[4] L.B. Reutergadh, M. Iangphasuk, Photocatalytic decolourization of reactive azo dye: a comparison between TiO_2 and CdS photocatalysis, Chemosphere 35 (1997) 585-596. https://doi.org/10.1016/S0045-6535(97)00122-7

[5] Poulios, M. Kositzi, A. Kouras, Photocatalytic decomposition of triclopyr over aqueous semiconductor suspensions, J. Photochem. Photobiol. A: Chem. 115 (1998) 175-183. https://doi.org/10.1016/S1010-6030(98)00259-7

[6] Poulios, I. Tsachpinis, Photodegradation of the textile dye Reactive Black 5 in the presence of semiconducting oxides, J. Chem. Technol. Biotechnol.74 (1999) 349-

357. https://doi.org/10.1002/(SICI)1097-4660(199904)74:4<349::AID-JCTB5>3.0.CO;2-7

[7] A.P. Davis, C.P. Huang, The removal of substituted phenols by a photocatalytic oxidation process with cadmium sulfide, Water Res.24 (1990)543-550. https://doi.org/10.1016/0043-1354(90)90185-9

[8] N. Kakuta, J.M. White, A. Campion, A.J. Bard, M.A. Fox, S.E. Webber, Surface analysis of semiconductor-incorporated polymer systems. 1. Nafion and cadmium sulfide-Nafion, J. Phys. Chem.89 (1985)48-52. https://doi.org/10.1021/j100247a014

[9] M.A. Fox, M.T. Dulay, Heterogeneous photocatalysis, Chem. Rev. 93 (1993) 341-357. https://doi.org/10.1021/cr00017a016

[10] R.P.S. Suri, J. Liu, D.W. Hand, J.C. Crittenden, D.L. Perram, M.E. Mullins, Heterogeneous photocatalytic oxidation of hazardous organic contaminants in water, Water Environ. Res.65 (1993)665-673. https://doi.org/10.2175/WER.65.5.9

[11] P. Qu, J. Zhao, T. Shen, H. Hidaka, TiO_2-assisted photodegradation of dyes: A study of two competitive primary processes in the degradation of RB in an aqueous TiO_2 colloidal solution, J. Mol. Catal. A: Chem. 129 (1998) 257-268. https://doi.org/10.1016/S1381-1169(97)00185-4

[12] T. Wu, G. Liu, J. Zhao, Evidence for H_2O_2 generation during the TiO_2-assisted photodegradation of dyes in aqueous dispersions under visible light illumination, J. Phys. Chem. B 103 (1999) 4862-4867. https://doi.org/10.1021/jp9846678

[13] Y. Wang, Solar photocatalytic degradation of eight commercial dyes in TiO_2 suspension, Wat. Res. 34 (2000) 990-994. https://doi.org/10.1016/S0043-1354(99)00210-9

[14] Arslan, A. Balcioglu, D.W. Bahnemann, Heterogeneous photocatalytic treatment of simulated dyehouse effluents using novel TiO_2-photocatalysts, Appl. Catal. B 26 (2000) 193-206. https://doi.org/10.1016/S0926-3373(00)00117-X

[15] C.A.K. Gouvea, F. Wypych, S.G. Moraes, N. Duran, N. Nagata, P.P. Zamora, Semiconductor-assisted photocatalytic degradation of reactive dyes in aqueous solution, Chemosphere 40 (2000) 433-440. https://doi.org/10.1016/S0045-6535(99)00313-6

[16] Galindo, P. Jacques, A. Kalt, Photooxidation of the phenylazonaphthol AO20 on TIO_2: kinetic and mechanistic investigations, Chemosphere 45 (2001) 997-1005. https://doi.org/10.1016/S0045-6535(01)00118-7

[17] T. Zhang, T. Oyama, A. Aoshima, H. Hidaka, J. Zhao, N. Serpone, Photooxidative N-demethylation of methylene blue in aqueous TiO_2 dispersions under UV irradiation, J. Photochem. Photobiol. A 140 (2001) 163-172. https://doi.org/10.1016/S1010-6030(01)00398-7

[18] Poulios, E. Micropoulou, R. Panou, E. Kostopoulou, Photooxidation of eosin Y in the presence of semiconducting oxides, Appl. Catal. B: Environ. 41 (2003) 345-355. https://doi.org/10.1016/S0926-3373(02)00160-1

[19] J. Yang, C. Chen, J. Hongwei, M. Wanhong, J. Zhao, Mechanism of TiO_2-assisted photocatalytic degradation of dyes under visible irradiation: photoelectro catalytic study by TiO_2-film electrodes, J. Phys. Chem. B 109 (2005) 21900-21907. https://doi.org/10.1021/jp0540914

[20] X. Wang, S. Meng, X. Zhang, H. Wang, W. Zhong, Q. Du, Multi-type carbon doping of TiO_2 photocatalyst, Chem. Phys. Lett. 444 (2007) 292-296. https://doi.org/10.1016/j.cplett.2007.07.026

[21] X. Ren, D. Han, D. Chen, F. Tang, Large-scale synthesis of hexagonal cone-shaped ZnO nanoparticles with a simple route and their application to photocatalytic degradation, Mater. Res. Bull.42 (2007) 807-813. https://doi.org/10.1016/j.materresbull.2006.08.030

[22] S.K. Mohapatra, N. Kondamudi, S. Banerjee, M. Misra, Functionalization of self-organized TiO_2 nanotubes with pd nanoparticles for photocatalytic decomposition of dyes under solar light illumination, Langmuir 24 (2008) 11276-11281. https://doi.org/10.1021/la801253f

[23] Y. Wang, X. Li, N. Wang, X. Quan, Y. Chen, Controllable synthesis of ZnO nano flowers and their morphology-dependent photocatalytic aactivities, Sep. Purif. Technol. 62 (2008) 727-732. https://doi.org/10.1016/j.seppur.2008.03.035

[24] S.K. Pardeshi, A.B. Patil, Effect of morphology and crystallite size on solar photocatalytic activity of zinc oxide synthesized by solution free mechanochemical method, J. Mol. Catal. A: Chem. 308 (2009) 32-40. https://doi.org/10.1016/j.molcata.2009.03.023

[25] Lu, Y. Wu, F. Mai, W. Chung, C. Wu, W. Lin, C. Chen, Degradation efficiencies and mechanisms of the ZnO-mediated photocatalytic degradation of Basic Blue 11

under visible light irradiation, J. Mol. Catal. A: Chem. 310 (2009) 159-165. https://doi.org/10.1016/j.molcata.2009.06.011

[26] A.N. Rao, B. Sivasankar, V. Sadasivam, Kinetic studies on the photocatalytic degradation of Direct Yellow 12 in the presence of ZnO catalyst, J. Mol. Catal. A: Chem. 306 (2009) 77-81. https://doi.org/10.1016/j.molcata.2009.02.028

[27] J.H. Sun, S.Y. Dong, Y.K. Wang, S.P. Sun, Preparation and photocatalytic property of a novel dumbbell-shaped ZnO microcrystal photocatalyst, J. Hazard. Mater. 172 (2009) 1520-1526. https://doi.org/10.1016/j.jhazmat.2009.08.022

[28] Y. Tong, J. Cheng, Y. Liub, G.G. Siu, Enhanced photocatalytic performance of ZnO hierarchical nanostructures synthesized via a two-temperature aqueous solution route, Scr. Mater. 60 (2009) 1093-1096. https://doi.org/10.1016/j.scriptamat.2008.12.060

[29] M.S. Mohajerania, A. Laka, A. Simchia, Effect of morphology on the solar photocatalytic behavior of ZnO nanostructures, J. Alloys Compd. 485 (2009) 616-620. https://doi.org/10.1016/j.jallcom.2009.06.054

[30] W.M. Wu, G.D. Liu, Q.H. Xie, S.J. Liang, H.R. Zheng, R.S. Yuan, W.Y. Su, L. Wu, A simple and highly efficient route for the preparation of p-phenylenediamine by reducing 4-nitroaniline over commercial CdS visible light-driven photocatalyst in water, Green Chem. 14 (2012) 1705-1709. https://doi.org/10.1039/c2gc35231a

[31] M. Luo, Y. Liu, J.C. Hu, H. Liu, J.L. Li, One-Pot Synthesis of CdS and Ni-Doped CdS hollow spheres with enhanced photocatalytic activity and durability, ACS Appl. Mater. Interf. 4 (2012) 1813-1821. https://doi.org/10.1021/am3000903

[32] F. Chen, Y. Cao, D. Jia, X. Niu, Facile synthesis of CdS nanoparticles photocatalyst with high performance, Ceram. Int. 39 (2013) 1511-1517. https://doi.org/10.1016/j.ceramint.2012.07.098

[33] J. Shi, H. Cui, Z. Liang, X. Lu, Y. Tong, C. Su, The roles of defect states in photoelectric and photocatalytic processes for ZnxCd1−xS, Energy Environ. Sci. 4 (2011) 466-470. https://doi.org/10.1039/C0EE00309C

[34] M. Liu, L. Wang, G. Lu, X. Yao, L. Guo, Twins in Cd1−xZnxS solid solution: Highly efficient photocatalyst for hydrogen generation from water Energy Environ. Sci. 4 (2011) 1372-1378. https://doi.org/10.1039/c0ee00604a

[35] G. Zhang, W. Xu, Z. Li, W. Hu, Y. Wang, Preparation and characterization of multi-functional CoFe2O4–ZnO nanocomposites. Magn. Magn. Mater. 321 (2009) 1424-1427. https://doi.org/10.1016/j.jmmm.2009.02.057

[36] W. Zhang, Y.B. Wang, Z. Wang, Z.Y. Zhong, R. Xu, Highly efficient and noble metal-free NiS/CdS photocatalysts for H_2 evolution from lactic acid sacrificial solution under visible light, Chem. Commun. 46 (2010) 7631-7633. https://doi.org/10.1039/c0cc01562h

[37] Y. Hou, X. Li, Q. Zhao, X. Quan, G. Chen, Electrochemically assisted photocatalytic degradation of 4-chlorophenol by $ZnFe_2O_4$–modified TiO_2 nanotube array electrode under visible light irradiation, Environ. Sci. Technol. 44 (2010) 5098-5103. https://doi.org/10.1021/es100004u

[38] M.H. Habibi, A.H. Habibi, Nanostructure composite $ZnFe_2O_4$–$FeFe_2O_4$–ZnO immobilized on glass: Photocatalytic activity for degradation of an azo textile dye F3B, J. Ind. Eng. Chem. 20 (2014) 68-73. https://doi.org/10.1016/j.jiec.2013.04.025

[39] V.R. Satsangi, S. Kumari, A.P. Singh, R. Shrivastav, S. Dass, Nanostructured hematite for photoelectrochemical generation of hydrogen, Int. J. Hydrogen Energy 33 (2008) 312-318. https://doi.org/10.1016/j.ijhydene.2007.07.034

[40] Chatterjee, S. Dasgupta, Visible light induced photocatalytic degradation of organic pollutants, J. Photochem. Photobiol. C6 (2005) 186-205. https://doi.org/10.1016/j.jphotochemrev.2005.09.001

[41] K. Lv, H. Zuo, J. Sun, K. Deng, S. Liu, X. Li, D. Wang, (Bi, C and N) codoped TiO_2 nanoparticles, J. Hazard. Mater. 161 (2009) 396-401. https://doi.org/10.1016/j.jhazmat.2008.03.111

[42] W. Choi, A. Termin, M.R. Hoffmann, The role of metal ion dopants in quantum-sized TiO_2: correlation between photoreactivity and charge carrier recombination dynamics, J. Phys. Chem. 98 (1994) 13669-13679. https://doi.org/10.1021/j100102a038

[43] V. Brezova, A. Blazkova, L. Karpinsky, J. Groskova, B. Havlinova, V. Jorik, M. Ceppan, Phenol decomposition using Mn^+/TiO_2 photocatalysts supported by the sol-gel technique on glass fibres, J. Photochem. Photobiol. A 109 (1997) 177-183. https://doi.org/10.1016/S1010-6030(97)00121-4

[44] Dvoranova, V. Brezova, M. Mazur, M.A. Malati, Investigations of metal-doped titanium dioxide photocatalysts, Appl. Catal. B 37 (2002) 91-105. https://doi.org/10.1016/S0926-3373(01)00335-6

[45] G.P. Lepore, L. Persaud, C.H. Langford, Supporting titanium dioxide photocatalysts on silica gel and hydrophobically modified silica gel, J. Photochem. Photobiol. A98 (1996)103-111. https://doi.org/10.1016/1010-6030(95)04242-3

[46] C. Minero, F. Catozzo, E. Pelizzetti, Role of adsorption in photocatalyzed reactions of organic molecules in aqueous titania suspensions, Langmuir 8 (1992) 481-486. https://doi.org/10.1021/la00038a029

[47] S. Sampath, H. Uchida, H. Yoneyama, photocatalytic degradation of gaseous pyridine over zeolite-supported titanium dioxide, J. Catal. 149 (1994) 189-194. https://doi.org/10.1006/jcat.1994.1284

[48] J. Arana, J.M. D. Rodriguez, E.T. Rendon, C.G. Cabo, O.G. Diaz, J.A.H. Melinan, J.P. Pena, G. Colon, J.A. Navio, TiO_2 activation by using activated carbon as a support: part I. Surface characterisation and decantability study, Appl. Catal. B 44 (2003) 161-172. https://doi.org/10.1016/S0926-3373(03)00107-3

[49] D. Chatterjee, S. Dasgupta, N.N. Rao, Visible light assisted photodegradation of halocarbons on the dye modified TiO_2 surface using visible light, Sol. Energy Mater. Sol. Cells 90 (2006) 1013-1020. https://doi.org/10.1016/j.solmat.2005.05.016

[50] C.F. Wang, C.T. Yu, B.H. Lin, J.H. Lee, Synthesis and characterization of TiO_2/BaF_2/ceramic radio-sensitive photocatalyst, J. Photochem. Photobiol A: Chem. 182 (2006) 93-98. https://doi.org/10.1016/j.jphotochem.2006.01.020

[51] L. Ge, M. Xu, Influences of the Pd doping on the visible light photocatalytic activities of $InVO_4$–TiO_2 thin films, Mater. Sci. Eng. B 131 (2006) 222-229. https://doi.org/10.1016/j.mseb.2006.04.021

[52] X. Zong, H.J. Yan, G.P. Wu, G.J. Ma, F.Y. Wen, L. Wang, Enhancement of photocatalytic H_2 evolution on CdS by loading MoS_2 as cocatalyst under visible light irradiation, J. Am. Chem. Soc. 130 (2008) 7176-7177. https://doi.org/10.1021/ja8007825

[53] H.G. Kim, D.W. Hwang, J.S. Lee, An undoped, single-phase oxide photocatalyst working under visible light, J. Am. Chem. Soc. 126 (2004) 8912-8913. https://doi.org/10.1021/ja049676a

[54] K. Maeda, T. Takata, M. Hara, N. Saito, Y. Inoue, H. Kobayashi, GaN:ZnO solid solution as a photocatalyst for visible-light-driven overall water splitting, J. Am. Chem. Soc. 127 (2005) 8286-8287. https://doi.org/10.1021/ja0518777

[55] S.H. Shen, L. Zhao, L.J. Guo, Cetyltrimethylammoniumbromide (CTAB)-assisted hydrothermal synthesis of $ZnIn_2S_4$ as an efficient visible-light-driven photocatalyst for hydrogen production, Int. J. Hydrogen Energy 33 (2008) 4501-4510. https://doi.org/10.1016/j.ijhydene.2008.05.043

[56] Sclafani, L. Palminsano, E. Davi, Photocatalytic degradaton of phenol in aqueous polycrystalline TiO_2 dispersions: the influence of Fe^{3+}, Fe^{2+} and Ag^{+} on the reaction rate, J. Photochem. Photobiol A: Chem. 56 (1991) 113-123. https://doi.org/10.1016/1010-6030(91)80011-6

[57] M.R. Prairie, L.R. Evans, B.M. Stange, S.L. Martinez, An investigation of titanium dioxide photocatalysis for the treatment of water contaminated with metals and organic chemicals, Environ. Sci. Technol. 27 (1993) 1776. https://doi.org/10.1021/es00046a003

[58] Lassaletta, A.R.G. Elipe, A. Justo, A. Fernandez, F.J. Ager, M.A. Respaldiza, J.C. Soares, M.F. Da Silva, Thermal and photochemical methods for the preparation of thin films of cermet materials, J. Mater. Sci. 31 (1996) 2325-2332. https://doi.org/10.1007/BF01152941

[59] Sclafani, J.M. Herrmann, Influence of metallic silver and of platinum-silver bimetallic deposits on the photocatalytic activity of titania (anatase and rutile) in organic and aqueous media, J. Photochem. Photobiol A: Chem. 113 (1998) 181-188. https://doi.org/10.1016/S1010-6030(97)00319-5

[60] Ilisz, A. Dombi, Investigation of the photodecomposition of phenol in near-UV-irradiated aqueous TiO_2 suspensions. II. Effect of charge-trapping species on product distribution, Appl. Catal. A: Gen. 180 (1999) 35. https://doi.org/10.1016/S0926-860X(98)00375-5

[61] C.A.K. Gouvea, F. Wypych, S.G. Moraes, N. Duran, P.P. Zamora, Semiconductor-assisted photodegradation of lignin, dye, and kraft effluent by Ag-doped ZnO, Chemosphere 40 (2000) 427-432. https://doi.org/10.1016/S0045-6535(99)00312-4

[62] C. Chen, X. Li, W. Ma, J. Zhao, Effect of Transition Metal Ions on the TiO_2-assisted photodegradation of dyes under visible irradiation: A probe for the interfacial electron transfer process and reaction mechanism, J. Phys. Chem. B 106 (2002) 318-324. https://doi.org/10.1021/jp0119025

[63] M. Arabatzis, T. Stergiopoulos, D. Andreeva, S. Kitova, S.G. Neophytides, P. Falaras, Characterization and photocatalytic activity of Au/TiO_2 thin films for azo-

dye degradation, J. Catal. 220 (2003) 127-135. https://doi.org/10.1016/S0021-9517(03)00241-0

[64] B. Li, X. Z. Li, The enhancement of photodegradation efficiency using Pt–TiO_2 catalyst, Chemosphere 48 (2002) 1103-1111. https://doi.org/10.1016/S0045-6535(02)00201-1

[65] I.M. Arabatzis, T. Stergiopoulos, M.C. Bernard, D. Labou, S.G. Neophytides, P. Falaras, Silver-modified titanium dioxide thin films for efficient photodegradation of methyl orange, Appl. Catal. B: Environ. 42 (2003) 187-201. https://doi.org/10.1016/S0926-3373(02)00233-3

[66] V. Iliev, D. Tomova, L. Bilyarska, L. Petrov, Photooxidation of xylenol orange in the presence of palladium-modified TiO_2 catalysts, Catal. Commun. 5 (2004) 759-763. https://doi.org/10.1016/j.catcom.2004.09.005

[67] Chun, T. Yuchao, T. Hongxiao, Characterization and photocatalytic activity of transition-metal-supported surface bond-conjugated TiO_2/SiO_2, Catal. Today 90 (2004) 325-330. https://doi.org/10.1016/j.cattod.2004.04.042

[68] Y. Xie, C. Yuan, Photocatalysis of neodymium ion modified TiO_2 sol under visible light irradiation, Appl. Surf. Sci. 221 (2004) 17-24. https://doi.org/10.1016/S0169-4332(03)00945-0

[69] C. Sahoo, A.K. Gupta, P. Anjali, Photocatalytic degradation of Crystal Violet (C.I. Basic Violet 3) on silver ion doped TiO_2, Dyes Pigments 66 (2005) 189-196. https://doi.org/10.1016/j.dyepig.2004.09.003

[70] N. Sobana, M. Muruganadham, M. Swaminathan, Nano-Ag particles doped TiO_2 for efficient photodegradation of direct azo dyes, J. Mol. Catal. A: Chem. 258 (2006) 124-132. https://doi.org/10.1016/j.molcata.2006.05.013

[71] Wang, G. Zhang, Z. Zhang, X. Zhang, G. Zhao, F. Wen, Z. Pan, Y. Li, P. Zhang, P. Kang, Investigation on photocatalytic degradation of ethyl violet dyestuff using visible light in the presence of ordinary rutile TiO_2 catalyst doped with upconversion luminescence agent, Water Res. 40 (2006) 2143-2150. https://doi.org/10.1016/j.watres.2006.04.009

[72] R. Vinu, G. Madras, Synthesis and photoactivity of Pd substituted nano-TiO_2, J. Mol. Catal. A: Chem. 291 (2008) 5-11. https://doi.org/10.1016/j.molcata.2008.07.005

[73] L. Ravichandran, K. Selvam, B. Krishnakumar, M. Swaminathan, Photovalorisation of pentafluorobenzoic acid with platinum doped TiO_2, J. Hazard. Mater. 167 (2009) 763-769. https://doi.org/10.1016/j.jhazmat.2009.01.048

[74] M.V. Dozzi, L. Prati, P. Canton, E. Selli, Effects of gold nanoparticles deposition on the photocatalytic activity of titanium dioxide under visible light, Phys. Chem. Chem. Phys. 11 (2009) 7171-7180. https://doi.org/10.1039/b907317e

[75] Matos, J. Laine, J.M. Herrmann, Synergy effect in the photocatalytic degradation of phenol on a suspended mixture of titania and activated carbon, Appl. Catal B 18 (1998) 281-291. https://doi.org/10.1016/S0926-3373(98)00051-4

[76] J.M. Herrmann, J. Matos, J. Disdier, C. Guillard, J. Laine, S. Malato, Blanco, Solar photocatalytic degradation of 4-chlorophenol using the synergistic effect between titania and activated carbon in aqueous suspension, Catal. Today 54 (1999) 255-265. https://doi.org/10.1016/S0920-5861(99)00187-X

[77] C.G. Da Silva, J.L. Faria, Photochemical and photocatalytic degradation of an azo dye in aqueous solution by UV irradiation, J. Photochem. Photobiol. A: Chem. 155 (2003) 133-143. https://doi.org/10.1016/S1010-6030(02)00374-X

[78] S.K. Samantaray, P. Mohapatra, K. Parida, Physico-chemical characterisation and photocatalytic activity of nanosized SO_4^{2-}/TiO_2 towards degradation of 4-nitrophenol, J. Mol. Catal. A: Chem. 198 (2003) 277-287. https://doi.org/10.1016/S1381-1169(02)00693-3

[79] S. Sakthivel, H. Kisch, Daylight photocatalysis by carbon-modified titanium dioxide, Angew. Chem. Int. Ed. 42 (2003) 4908-4911. https://doi.org/10.1002/anie.200351577

[80] B. Tryba, T. Tsumura, M. Janus, A.W. Morawski, M. Inagaki, Carbon-coated anatase: adsorption and decomposition of phenol in water, Appl. Catal. B 50 (2004) 177-183. https://doi.org/10.1016/j.apcatb.2004.01.003

[81] T. Ohno, M. Akiyoshi, T. Umebayashi, K. Asai, T. Mitsui, M. Matsumura, Preparation of S-doped TiO_2 photocatalysts and their photocatalytic activities under visible light, Appl. Catal. A: Gen. 265 (2004) 115-121. https://doi.org/10.1016/j.apcata.2004.01.007

[82] Y. Li, D. S. Hwang, N.H. Lee, S.J. Kim, Synthesis and characterization of carbon-doped titania as an artificial solar light sensitive photocatalyst, Chem. Phys. Lett. 404 (2005) 25-29. https://doi.org/10.1016/j.cplett.2005.01.062

[83] Youji, L. Xiaodong, L. Junwen, J. Yin, Photocatalytic degradation of methyl orange by TiO_2-coated activated carbon and kinetic study, Water Res. 40 (2006) 1119-1126. https://doi.org/10.1016/j.watres.2005.12.042

[84] J. Yang, H. Bai, X. Tan, J. Lian, IR and XPS investigation of visible-light photocatalysis—Nitrogen–carbon-doped TiO_2 film, Appl. Surf. Sci. 253 (2006) 1988-1994. https://doi.org/10.1016/j.apsusc.2006.03.078

[85] H. Sun, Y. Bai, Y. Cheng, W. Jin, N. Xu, Preparation and characterization of visible-light-driven carbon−sulfur-codoped TiO_2 photocatalysts, Ind. Eng. Chem. Res. 45 (2006) 4971-4976. https://doi.org/10.1021/ie060350f

[86] D. Chen, Z. Jiang, J. Geng, Q. Wang, D. Yang, Carbon and nitrogen co-doped TiO_2 with enhanced visible-light photocatalytic activity, Ind. Eng. Chem. Res. 46 (2007)2741-2746. https://doi.org/10.1021/ie061491k

[87] Q. Xiao, J. Zhang, C. Xiao, Z. Si, X. Tan, Solar photocatalytic degradation of methylene blue in carbon-doped TiO_2 nanoparticles suspension, Sol. Energy 82 (2008) 706-713. https://doi.org/10.1016/j.solener.2008.02.006

[88] Z. Jiang, F. Yang, N. Luo, B.T.T. Chu, D. Sun, H. Shi, T. Xiao, P.P. Edwards, Solvothermal synthesis of N-doped TiO_2 nanotubes for visible-light-responsive photocatalysis, Chem. Commun. 47 (2008) 6372-6374. https://doi.org/10.1039/b815430a

[89] X. Hu, T. Zhang, Z. Jin, J. Zhang, W. Xu, J. Yan, J. Zhang, L. Zhang, Y. Wu, Fabrication of carbon-modified TiO_2 nanotube arrays and their photocatalytic activity, Mater. Lett. 62 (2008) 4579-4581. https://doi.org/10.1016/j.matlet.2008.08.051

[90] K.M. Parida, N. Sahu, N.R. Biswal, B. Naik, A.C. Pradhan, Preparation, characterization, and photocatalytic activity of sulfate-modified titania for degradation of methyl orange under visible light, J. Colloid Interface Sci. 318 (2008) 231-237. https://doi.org/10.1016/j.jcis.2007.10.028

[91] Y. Yu, H.H. Wu, B.L. Zhu, S.R. Wang, W.P. Huang, S.H. Wu, S.M. Zhang, Preparation, characterization and photocatalytic activities of F-doped TiO_2 nanotubes, Catal. Lett. 121 (2008) 165-171. https://doi.org/10.1007/s10562-007-9316-1

[92] Vijayabalan, K. Selvam, R. Velmurugan, M. Swaminathan, Photocatalytic activity of surface fluorinated TiO_2-P25 in the degradation of Reactive Orange 4, J. Hazard. Mater. 172 (2009) 914-921. https://doi.org/10.1016/j.jhazmat.2009.07.082

[93] T.P. Ang, C.S. Toh, Y.F. Han, Synthesis, characterization, and activity of visible-light-driven nitrogen-doped TiO_2–SiO_2 mixed oxide photocatalysts, J. Phys. Chem. C 113 (2009) 10560-10567. https://doi.org/10.1021/jp9000658

[94] Woan, G. Pyrgiotakis, W. Sigmund, Photocatalytic Carbon-Nanotube–TiO_2 Composites, Adv. Mater. 21 (2009) 2233-2239. https://doi.org/10.1002/adma.200802738

[95] S. Livraghi, M.C. Paganini, E. Giamello, A. Selloni, C.D. Valentin, G. Pacchioni, Origin of photoactivity of nitrogen-doped titanium dioxide under visible light, J. Am. Chem. Soc. 128 (2006) 15666-15671. https://doi.org/10.1021/ja064164c

[96] C. Hu, Y. Lan, J. Qu, X. Hu, A. Wang, Ag/AgBr/TiO_2 visible light photocatalyst for destruction of azo dyes and bacteria, J. Phys. Chem. B 110 (2006) 4066-4072. https://doi.org/10.1021/jp0564400

[97] X.H. Wang, J.G. Li, H. Kamiyama, Y. Moriyoshi, T. Ishigaki, Wavelength-sensitive photocatalytic degradation of methyl orange in aqueous suspension over iron(III)-doped TiO_2 nanopowders under UV and visible light irradiation, J. Phys. Chem. B 110 (2006) 6804-6809. https://doi.org/10.1021/jp060082z

[98] M.R. Elahifard, S. Rahimnejad, S. Haghighi, M.R. Gholami, Apatite-coated Ag/AgBr/TiO_2 visible-light photocatalyst for destruction of bacteria, J. Am. Chem. Soc. 129 (2007) 9552-9553. https://doi.org/10.1021/ja072492m

[99] C.T. Yu, C.F. Wang, W.Z. Wang, Decomposition of organic resin by radio-sensitive photocatalyst, J. Photochem. Photobiol. A 186 (2007) 369-375. https://doi.org/10.1016/j.jphotochem.2006.09.007

[100] Y. Li, H. Zhang, Z. Guo, J. Han, X. Zhao, Q. Zhao, S.J. Kim, Highly efficient visible-light-induced photocatalytic activity of nanostructured AgI/TiO_2 photocatalyst, Langmuir 24 (2008) 8351-8357. https://doi.org/10.1021/la801046u

[101] Y. Zang, R. Farnood, Photocatalytic activity of AgBr/TiO_2 in water under simulated sunlight irradiation, Appl. Catal. 79 (2008) 334-340. https://doi.org/10.1016/j.apcatb.2007.10.019

[102] J. Yu, G. Dai, B. Huang, Fabrication and characterization of visible-light-driven plasmonic photocatalyst Ag/AgCl/TiO_2 nanotube arrays, J. Phys. Chem. C 113 (2009) 16394-16401. https://doi.org/10.1021/jp905247j

Chapter 8

Photocatalytic Decomposition of Organic Dyes

Dolly Rana[1], Ashish Soni[1], Gaurav Sharma[1], Amit Kumar[1], Deepika Jamwal[1,2*], Akash Katoch[3*]

[1] School of Chemistry, Faculty of Basic Sciences, Shoolini University, Solan, HP 173212, India

[2]Department of Chemistry, Panjab University, Chandigarh, 160014, India

[3]Institute Instrumentation Centre, Indian Institute of Technology Roorkee, Roorkee 247667, India

jamwaldeepika@gmail.com[1,2], akashfic@iitr.ac.in[3], katochakash@gmail.com[3]

Abstract

In ecological and environmental protection, photocatalytic degradation of toxic organic dyes is a challenging task. Oxide-semiconductor is a special category of materials with a nanostructure that has obtained more attention in recent times owing to their fascinating properties and an extensive choice of applications in photocatalysis. This chapter includes a critical review of oxide-semiconductor nanomaterials towards photocatalytic wastewater purification. The structural aspects, nanostructure formation process and the various parameters affecting catalytic activity, photocatalytic applications of ZnO, TiO_2, Fe_3O_4, and CeO_2 based catalysts for efficient photocatalytic degradation of organic dyes are reviewed. It also includes a brief discussion on the mechanism of photocatalysis. In this regard, various oxide-semiconductor and their composite nanostructures for degradation of various organic dyes are sketched.

Keywords

Organic Dyes, Photocatalysis, Oxide-Semiconductor, Nanomaterials

Contents

1. Introduction

Industry development is extensively connected with the disposal of a large number of various toxic pollutants, which are harmful to the environment, hazardous to human health, and difficult to degrade by natural means. Some organic compounds are major contaminators in industrial wastewater [1,2]. Discharges of chemical process industries such as pulp and paper, pharmaceutical agrochemical, dyestuff, food processing, petrochemical, tanning, steel, fiber wood, preservatives of foodstuffs, coal gasification, polymeric resin production, oil refining, coking plants etc. are the main sources of these organic pollutants [3-5]. Environmental problems caused by hardly-degradable and harmful organic pollutants have risen to be a grave menace to human growth and well-being. Organic pollutants once released into the aquatic ecosystem, they can cause various environmental problems, such as increasing biochemical oxygen, clogging sewage treatment plants and adversely affecting aquatic biota [6,7]. Therefore, before discharging the wastewater into the aquatic environment an effective and economical technique needs to be developed to lessen the concentration of organic pollutants.

Dyes are one group of organic pollutants used widely in textile, paper, dye houses and printing industries. The dye industry discharges one of the most problematic wastewaters to be considered not only for their high chemical and biological oxygen demands, content in toxic compounds and suspended solids but also for color, which is the first contaminant to be recognized by the human eye. Dyes may considerably influence photosynthetic activity in aquatic life due to reduced light penetration and may also be toxic to some aquatic life [8,9]. Dyeing of textiles requires high amounts of water and some harmful chemical substances too. Over 10,000 types of dyes are used for dyeing or printing of fabrics. These harmful chemicals in the form of dyes can be acquired by human beings through the skin, bronchially or through digestion. Textile materials can cause allergic reactions, or even can be carcinogenic and mutagenic [10]. Dyes commonly have complex aromatic molecular structures and synthetic origin, which make them more stable non-biodegradable, so these harmful dyes that affect human and environmental health should be taken into consideration [11]. Presently, various technologies such as coagulation and adsorption slightly concentrate or separate these pollutants from water, but cannot completely remove or demolish them into biodegradable or less toxic organic compounds. Photocatalysis has endorsed a lot of changes over the past two decades [12]. For the photodegradation of organic pollutant and generation of energy, important advancements are being made in the synthesis of nanostructured and novel materials [13]. The improvement of a simple recyclable photocatalyst and also the recovery of the deactivated photocatalyst required for reducing the total cost, and lowering overall usage of the photocatalytic material [14]. Organic contaminants have become of more interest, due to the inability of traditional water-treatment technologies to completely decompose these contaminants in aqueous media. The presence of organic contaminants in groundwater, drinking water, sewage effluents, and sludge poses a serious threat to human and aquatic organisms [15,16]. Conventional analysis methods only transfer pollutants from one medium to another but the semiconductor photocatalyst converts contaminants to harmless products such as CO_2 and H_2O [17,18]. In addition, the reaction conditions are mild, the reaction time is moderate and can be used to gaseous, aqueous and solid phase treatments. So, to degrade organic pollutant the use of semiconductor nanomaterials-based photocatalysts perceived as one of the most hopeful areas of research and application. These materials include treatment of environmental pollution, pharmaceutical industries, and medicine (destruction of cancer and viruses) [19]. Accordingly, oxide semiconductor photocatalyst has the advantage of not only minimizing running costs but also producing the desired product in the most profitable and adequate way. Many semiconductor materials such as CdS [20], ZnS [21], Ag_3PO_4 [22] TiO_2 [23], $ZnWO_4$ [24], CeO_2 [25] and Fe_2O_3 [26] etc. can be used as photocatalysts for the degradation of organic pollutants. Consequently, to establish an advanced and stable system with high efficiency, particularly sunlight utilization, it still remains an urgent challenge. The significant characteristics of the photocatalytic system are the desired band gap, high surface area, suitable morphology reusability, and stability. Among the various photocatalyst, metal oxides have wide applications as photocatalyst and the oxides of vanadium, chromium, zinc, titanium cerium and tin having the aforementioned characteristics follow similar primary photocatalytic processes such as light absorption, which induces a charge separation process with the formation of positive holes which are capable of oxidizing organic substrates [27]. In this chapter, we have mainly focused on the contribution and aspects of oxide semiconductor nanoparticles for photodegradation of dyes.

2. Features of oxide semiconductors for photocatalysis

The dominant benefits of semiconductor photocatalysis are that they offer a satisfying substitute for the energy-intensive treatment methods and have the capacity to use pollution-free and renewable solar energy. As visible light constitute a large fraction of the solar energy, one of the big dispute of photocatalyst study is to dream up new catalyst that can display high activity under illumination by visible light. Generally, the catalyst can be photoactivated by a photon with irradiation energy equal to or higher than its band gap energy (E_g). An electron moves from the valence band to the conduction band at the photocatalytic surface which accelerates the generation of holes (h^+) in the valence band (see equation i). Then an O_2 molecule is reduced to $^{\cdot}O^{2-}$ by the combination with an electron (e^-), which further turns into hydroxyl radical ($^{\cdot}OH$) (described in Eqs. 2, 3, 4 and 5). In the intervening time, the interaction of holes (h^+) with hydroxyl (OH^-) may produce a hydroxyl radical ($^{\cdot}OH$) (Eqs. 6 and 7). The resulting $^{\cdot}OH$ radical acts as a strong oxidizing agent and converts organic substances into carbon dioxide (CO_2), water (H_2O) and other less hazardous molecules than the original substances (Eq. 8) [28,29].

$$\text{Photocatalyst} + h\nu \rightarrow h^+ + e^- \quad (1)$$

$$e^- + O_2 \rightarrow {}^{\cdot}O_2^- \quad (2)$$

$${}^{\cdot}O_2^- + H^+ \rightarrow {}^{\cdot}OOH \quad (3)$$

$$2{}^{\cdot}OOH \rightarrow O_2 + H_2O_2 \quad (4)$$

$$H_2O_2 + h\nu \rightarrow 2{}^{\cdot}OH \quad (5)$$

$$h^+ + OH^- \rightarrow {}^{\cdot}OH \quad (6)$$

$$h^+ + H_2O \rightarrow {}^{\cdot}OH + H^+ \quad (7)$$

$$\text{Pollutant} + {}^{\cdot}OH \rightarrow CO_2 + H_2O + \text{degradation product} \quad (8)$$

The photo induced electron/hole in oxide semiconductors precedes catalytic oxidation or reduction of the species on the surface, though the electron/hole recombination determines the extent of photocatalytic activity. Slower the rate of electron-hole recombination, higher is the photodegradation of the organic pollutants. The band gap of oxide semiconductors effectively fulfills requirement essential for excellent photocatalyst. The charge transport developed due to quantum confinement effect in oxide-semiconductor particles drive at the nanoscale is promisingly contributes towards their photocatalytic activity. Moreover, other aspects such as ease of fabrication, non-toxic nature, stability under harsh environmental conditions, higher reproducibility are encouraging features that make the reliable candidate for photocatalysis applications.

3. Zinc oxide nanomaterials as photocatalyst

Currently, due to the excellent properties of the II-VI semiconductor zinc oxide (ZnO) is considered as one of the most flexible and functional materials which can be used for various high-technological applications. Many approaches have been developed to fabricate ZnO nanoparticles with different shapes and sizes including hydrothermal method [30], template synthesis [31], solvothermal method [32], chemical vapor deposition [33] and biosynthesis method [34] etc. Because of the non-toxicity, easy fabrication, high electrochemical, photocatalytic and adsorptive properties, ZnO is treated as one of the effective adsorbents for the removal of various toxic and hazardous chemicals and dyes. In addition, their composites have also been discussed. ZnO nanoparticles, nanorods, and nanoflowers synthesized through the hydrothermal process have been investigated for the photodegradation of methyl orange dye (Fig. 1).

Upon irradiation, the color of methyl orange solution was rapidly degraded with the addition of ZnO nanorods. Methyl orange decomposition efficiency of ZnO nanorods and nanoflowers after 4 cycles was 91.5 and 90% respectively [35]. In another study, ZnO/SnO_2 hetero-nanofibres of about 250 nm in diameter and several micrometers in length synthesized via electrospinning method. They have been thoroughly investigated for recycling ability. The photodegradation abilities of ZnO/SnO_2 nanostructures for various dyes (methylene blue, congo red, eosin red and methyl orange) were studied and were found useful for fast photodegradation of dyes (Fig. 2).

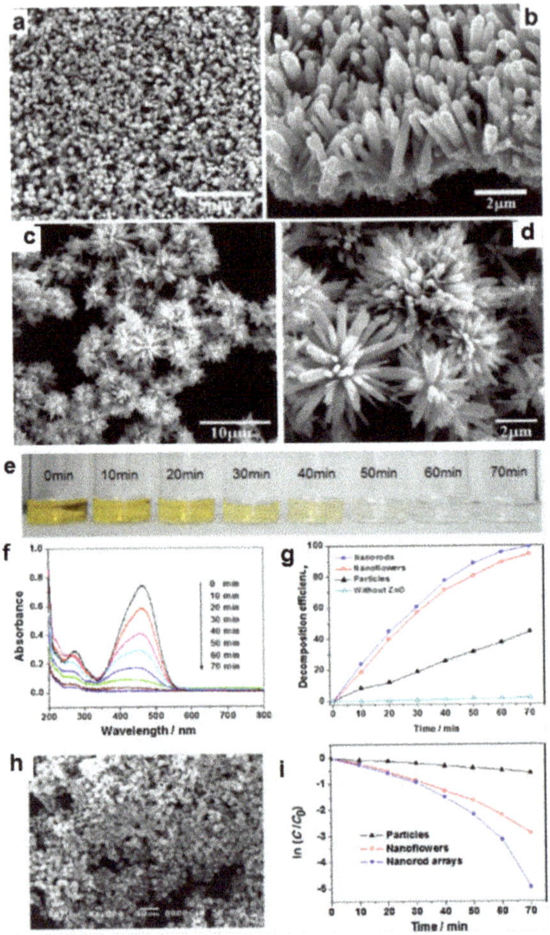

Figure 1(a and b) SEM images of the ZnO nanorod grown on the Cu substrate, (c and d) collected at the bottom of the autoclave. (e) Photographs of methyl orange solutions upon photodegradation catalyzed by ZnO nanorods (f) Spectral changes of methyl orange solutions upon photodegradation catalyzed by ZnO nanorod. (g) Comparison of decomposition efficiency of different ZnO samples for the degradation of methyl orange as well as that only induced by UV irradiation in the absence of any ZnO samples. (h)SEM image of the commercial ZnO particles. (i) Photodegradation kinetics of MO in the presence of ZnO micro/nanostructures [35].

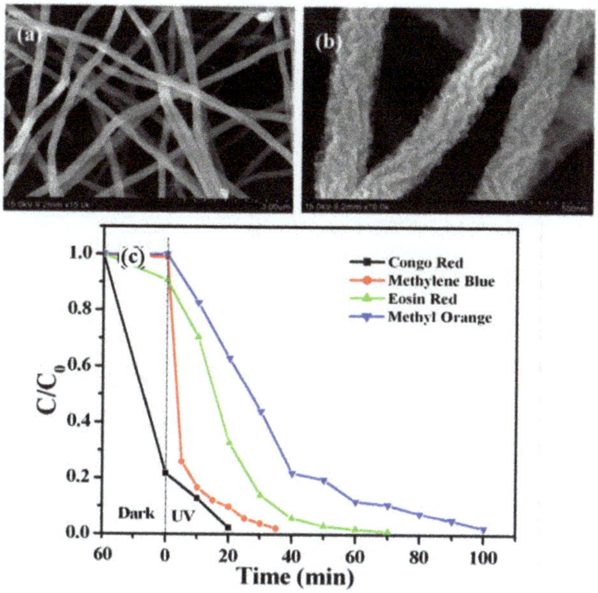

Figure 2(a-b) SEM images of ZnO/SnO_2. (c) Photocatalysis degradation rate of the four dyes [36].

The enhanced photocatalytic ability of ZnO/SnO_2 hetero-nanofibres were attributed to the effective separation of the photogenerated electron/hole and high adsorption capacity. Photocatalytic degradation towards methylene blue, congo red, eosin red and methyl orange were 97.3, 96.8, 98.9 and 97.7% respectively. Based on this investigation, ZnO/SnO_2 hetero-nanofibers possess functional potential applications for wastewater purification with excellent recycling ability [36]. A green synthesis route was used to synthesize pure ZnO and ZnO/starch quantum dots with size range of 5-10 nm. Further, the solar photocatalytic activity for rhodamine B (RhB) dye of both the ZnO quantum dots was examined at the end of 30, 60, 90 and 120 minutes. In this, ZnO/starch quantum dots show more decomposition of RhB than pure ZnO quantum dots under the solar light environment (Fig. 3) [37].

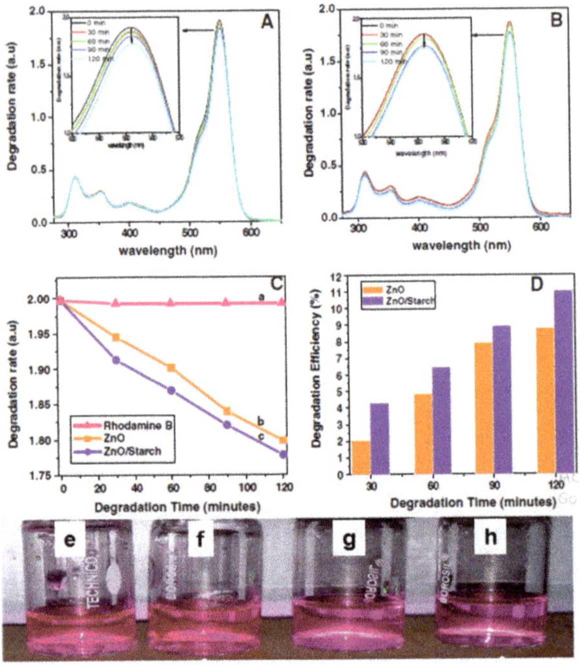

Figure 3 (a) Solar photodegradation spectrum of RhB dye in pure ZnO QDs, (b) ZnO/starch QDs, (c) photodegradation rate of QDs with control (RhB dye) and (d) photodegradation efficiency of QDs. (e) solar photodegradation of RhB dye before irradiation, (f) RhBdye after irradiation (120 min),(g) pure ZnO QDs with RhB after irradiation (120 min) and (h) ZnO/starch QDs with RhB after irradiation (120 min) [37].

ZnO@ZnS core-shell nanoparticles were synthesized via co-precipitation method. The photocatalytic activity of these core-shell nanoparticles was investigated for degradation of the congo red dye. Congo red dye degradation experiments indicate that the ZnO@ZnScore-shell nanoparticles degraded 70% dye after illumination for the nanoparticles [38]. Thus, ZnO nanomaterials exhibit wonderful photocatalytic activity and also show same effectiveness in the photodegradation of other environmentally hazardous organic pollutants under the natural solar light. A summary of the ZnO for the degradation of various dyes is summarized in Table 1.

Table 1 Summary of organic dyes removal by ZnO based nanomaterials.

Photocatalyst	Dye	Photocatalytic Degradation Time (min)	Efficiency (%)	Ref.
ZnO CNTs	RhB	60	49	[39]
ZnO	MO, MB	180	98.3	[40]
ZnO	RhB, MO	60	92	[41]
ZnO	DR 81, VB, Rh 6G, FG, EBT, AO	90	59.06, 56.66, 71.40, 78.60, 62.50, 61.60	[42]
ZnO@ZnS Core-shell	CR	120	70	[38]
Zn_4O	MO	120	44.5	[43]
ZnO	MO	120	83.99	[44]
ZnO/CuO	MB, MO	120	90	[45]
ZnO	MO	90	90.5	[35]
ZnO	MO	150	94.3	[46]
ZnO	MO	160	97	[47]
ZnO	MB	90	90.8	[48]
ZnO	MG	120	93.09	[49]
ZnO	MR	180	92.45	[50]
ZnO	RhB	45	40	[51]
ZnO NWs	MR	80	90	[52]
ZnO	MO	60	100	[53]
ZnO	RhB	120	11	[37]
ZnO	MB	125	70-75	[54]
ZnO/SnO_2 NFs	MB, CR, ER, MO	20	95.5	[36]
ZnO	MO	210	92	[55]

CR=Congo red, ER=Eosin red, RhB=Rhodamine B, MO=Methyl orange, DR81=Direct Red 81, VB=Victoria blue, Rh6G=Rhodamine 6G, FG=Fast green, EBT=Eriochrome black T, AO=Acridine orange, MB=Methylene blue, MR=Methyl red, MG=Malachite green. CNTs=Carbon nanotubes, NWs=Nanowires, NFs=Nanofibres

4. Titanium oxide nanomaterials as photocatalyst

Among the nano photocatalysis used in the treatment of environmental wastewater, titanium oxide nanostructures have been extensively studied [56]. Since the discovery of the phenomenon of the photocatalytic splitting of water on a TiO_2 electrode under UV light, enormous attempts committed to titania research have led to most promising applications in the fields of photovoltaics, photo electrochromic's, ceramics, sensors, and photocatalysis. As most advantageous semiconductor photocatalysts, TiO_2-based materials are accordingly expected to play a major role to check serious environmental and pollution challenges and ease the energy crisis through the use of renewable solar energy [57-59]. For the synthesis of titanium oxide nanoparticles, a number of techniques are available and these include hydrothermal [60], solvothermal [61], sol-gel [62], simple chemical method [63] and chemical vapor deposition [64] etc. and have been successfully utilized for the photocatalytic degradation of dyes.

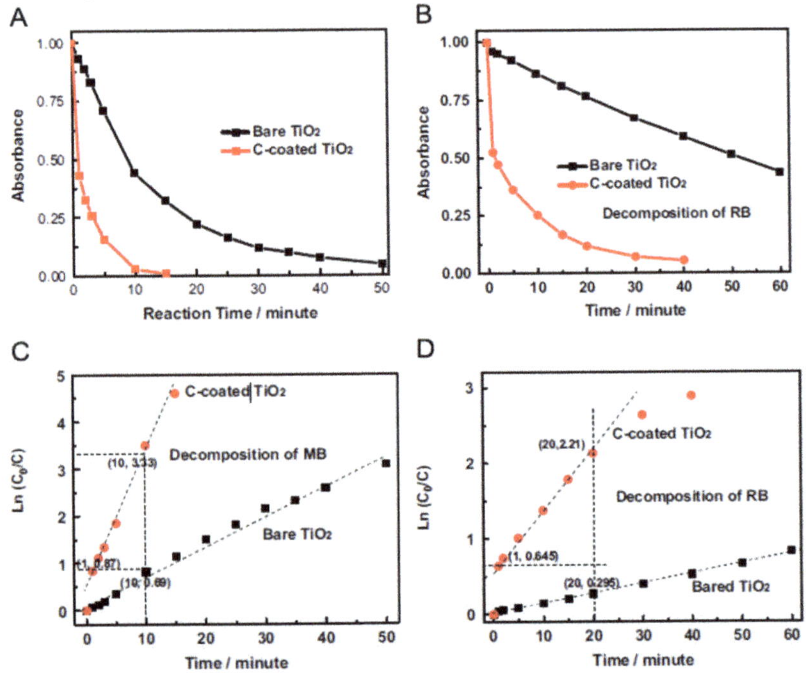

Figure 4 (a) The maximum optical absorbance change of methylene blue and (b) rhodamine B solution as a function of solar irradiation time with pristine and C-coated TiO_2 nanocrystals. (c) $ln(C_0/C)$ of methylene blue and (d) rhodamine B as a function of solar irradiation time with pristine and C-coated TiO_2 nanocrystals [65].

In another study, the carbon-coated TiO_2 nanocrystals synthesized were formed about 10 times lithium rate improvement and 4 times the photocatalytic performance compared to Li-based photocatalysts. To reveal the photocatalytic activity of TiO_2 and C-coated TiO_2, photocatalytic decomposition of methylene blue (MB) and rhodamine B (RhB) solution was conducted. After the start of the reaction, the UV-Visible absorption spectra of MB and RB were monitored. The result showed that C-coated TiO_2 nanocrystals have an about four-time better photocatalytic activity than the pristine TiO_2 nanocrystals in decomposing of MB, and for RB it took around 40 min for C-coated TiO_2 nanocrystals to decolorize the RB, while a large amount of RB still remained after 60 min irradiation with pristine TiO_2 nanocrystals (Fig. 4).

The adding of a fine layer of amorphous carbon on crystalline materials may be used as a different approach for improving the photocatalytic performance of TiO_2 and other materials as well [65]. In another work, the 1D CdS nanowires@TiO_2 nanoparticles with very high photocatalytic activity for organic dye degradation was reported [66]. The 1D CdS nanowires @TiO_2 nanoparticles were reported with improved photocatalytic activity in comparison to the 1D CdS nanowires, for degradation of organic dyes (methyl orange, methylene blue, and rhodamine B) under visible light radiation due to the longer lifetime of photogenerated electron-hole pairs of ID CdS nanowires@TiO_2 nanoparticles, in contrast to the original 1D CdS nanowires (Scheme 1).

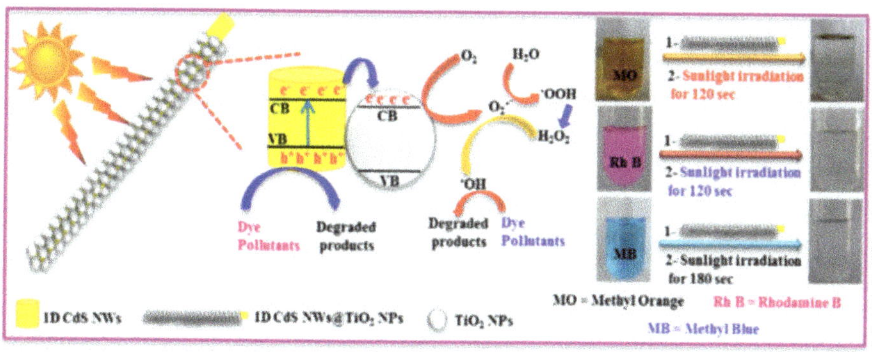

Scheme 1 Illustration of the proposed reaction mechanism for degradation of dye pollutants over the 1D CdS NWs@TiO_2 NPs photocatalyst [66].

Thus, the formation of a heterostructure between ID CdS nanowires and TiO_2 nanoparticles improved the charge separation efficiency. In another study, bifunctional TiO_2/Ag_3PO_4/graphene composites have been successfully synthesized by a combination of ion-exchange and hydrothermal methods. The improved photocatalytic activity of composites is attributed to the effective separation of photoexcited electron-hole pairs and fast charge transfer between components in the composites. Removal of the majority of Rhodamine B and methyl orange under visible light irradiation was obtained in 6 and 12 min, achieving a photocatalytic efficiency of around 60 and 100% respectively [67]. Overall, this discussion brings to attention the advancements of titanium oxide nanoparticles in their use for water-treatment processes. Further, since TiO_2 NPs can degrade organic pollutants and at the same time can oxidize heavy metal species [64]. A summary of the TiO_2 based photocatalyst for the degradation of various dyes is summarized in Table 2.

Table 2 Summary of organic dyes removal by TiO_2 based nanomaterials.

Photocatalyst	Dye	Photocatalytic degradation time (min)	Efficiency (%)	Ref.
Ag-TiO_2 NCs	RhB	48 hours	15.8	[68]
TiO_2/ZrO_2NCs	MO	105	59	[69]
CdS NWs@TiO_2	MO, MB, RB	2, 2, 3	99.7	[66]
TiO_2 NPs	MO, MB , Rh B	2,	98.2	[65]
TiO_2 NPs	RhB	40	93	[70]
TiO_2/Ag_3PO_4/graphene composites	RhB, MB	6, 12	60, 100	[67]
TiO_2 NPs	RB5	5	13.2-15.3	[71]
Fe-doped TiO_2	MB	5 h	92	[72]
TiO_2 NPs	MO	100	90	[73]
TiO_2/s-PS aerogels	MB	180	30	[74]
TiO_2 NPs	MO, MB	60	80	[75]
$CuFe_2O_4$-TiO_2-Ag NCs	RhB, MB	60	90	[76]
TiO_2NFs	MB	120	70	[77]
TiO_2NRs	MO	240	80	[78]
TiO_2 NPs	RB	150	30.12	[79]
TiO_2 NPs	MO	300	90	[80]

RhB=Rhodamine B, MO=Methyl orange, MB=Methylene blue, RB5=reactive black 5, RB=Remazol brown. NCs=Nano composites, NWs=Nanowires, NFs=Nanofibers, NRs=Nanorods, NPs=Nanoparticles

5. Iron oxide nanomaterials as photocatalyst

Iron oxide, as an ordinary compound is broadly distributed in nature and can be synthesized on large scale. Over the past few decades, magnetic iron oxide nanoparticles because of their significance in basic research were extensively synthesized with various morphologies and structures [81]. In addition, due to a wide range of applications including pigments, magnetic fluids, catalysis, magnetic resonance imaging (MRI), biosensors, targeted drug delivery, data storage and the environmental remediation, magnetic iron oxides are of immense interest to researchers[82,83]. The different crystalline iron oxides have unique magnetic, catalytic, biochemical and other properties that make them suitable for specific technical and biomedical applications [84,85,86].

One of the specific morphology, three-dimensional (3D) cauliflower-like α-Fe_2O_3 based nanostructure was prepared by one-step toluene-water biphasic interfacial reaction route. The prepared microstructures were found to exhibit exceptional water treatment performance with high removal capacities for organic dyes. The 3D cauliflower-like α-Fe_2O_3 nanostructures with high surface area and porous structure showed better removal capacities for organic dye in wastewater and structurally enhanced visible-light photocatalytic activity in the presence of H_2O_2 when compared with the commercial α-Fe_2O_3 powders (Fig. 5) [87].

Figure 5 (a and b) FESEM images of the microstructures prepared at different temperatures, (c) UV–visible absorption spectra of congo red solutions in the presence of cauliflower-likeα-Fe_2O_3 at different time intervals and (d) adsorption rates of the congo red on newly prepared cauliflower-like α-Fe_2O_3 [87].

The higher surface area provided by the porous structure as well as 3D features probably the reason for the promoted catalytic activity of the 3D cauliflower-like α-Fe_2O_3 nanostructures. On the other hand, the combination of graphene oxide (GO) and Fe_2O_3 has been found effective for degradation of organic contaminants. The GO-Fe_2O_3 hybrid material found eminently effective for the degradation of Rhodamine B (RhB) in aqueous solution with H_2O_2 under wide pH range of 2.09-10.09 and showed high photocatalytic activity after seven reaction cycles. The GO- Fe_2O_3 catalyst was also found efficient for the degradation of 4-nitrophenol (4-NP) [88]. On the other and photocatalytic degradation performance for congo red (CR), eosin red (ER) and methylene blue (MB) dyes was studied under visible light for α-Fe_2O_3. The order of degradation rate for the dyes was as MB (80%), ER (84%) and CR (98%) [89]. A summary of the iron oxide-based photocatalyst for the degradation of various dyes is summarized in Table 3.

Table 3 Summary of organic dyes removal by iron oxide-based nanomaterials.

Photocatalyst	**Dye**	**Photocatalytic degradation time (min)**	**Efficiency (%)**	**Ref.**
Fe_3O_4@SiO_2@TiO_2 NPs	RhB	120	92.03	[90]
Fe_3O_4–Ag_3PO_4	MB	30	79.3	[91]
Fe_3O_4@His@Ag	MO, MB	8, 4	85	[92]
Fe_2O_3-$BiVO_4$ NPs	RhB	80	100	[93]
α -Fe_2O_3 NPs	RhB	120	40	[87]
Fe_2O_3	Orange II	90	75	[94]
Graphene oxide-Fe_2O_3	RhB	80	60	[88]
α -Fe_2O_3	CR	120	55.5	[95]
α -Fe_2O_3	CR, ER, MB,	25, 50, 45	98, 84, 80	[89]
g-C_3N_4–Fe_3O_4	RhB	6 cycles	90	[96]
Diatomite-Fe_2O_3	RhB	5 cycles	90	[97]
γ-Fe_2O_3/α-Fe_2O_3/Carbon aerogel	RhB	3 cycles	98	[98]
α-Fe_2O_3	RhB	160	98	[99]
Graphene-Fe_2O_4	MB	120	98	[100]
Fe_2O_3–TiO_2	MB	100	81	[101]
α -Fe_2O_3 NPs	MO	210	80	[102]
Fe_3O_4	RhB	360	82	[103]
Fe_2O_3 pillared clay	MB	150	41	[104]
α-Fe_2O_3 NPs	RhB	140	79	[105]
α -Fe_2O_3/$NiTiO_3$NFs	RhB	180	90.4	[106]
α -Fe_2O_3 NPs	MB, ER, CR	50	57, 63, 93	[107]

CR=Congo red, ER=Eosin red, RhB=Rhodamine B, MO=Methyl orange, MB=Methylene blue, NPs=Nanoparticles, NFs=Nanofibres

Various major issues have to be addressed before recommending photocatalysts for large-scale industrial applications. Apart from offering easy separation of the photocatalyst from the reaction system, the magnetic iron oxide semiconductor system has obtained an exclusive place in the progress of heterogeneous photocatalysis.

6. Cerium oxide nanomaterials as photocatalyst

Cerium (CeO_2), as one of the most valuable rare earth metal oxides, has redox properties, metal-support interactions, and high oxygen storage capacity [108]. These uncommon properties make a CeO_2 an auspicious contender for using as catalysts, adsorbents [109], luminescence, polishing powder, UV blocking, and shielding materials [110,111], etc. Different nanostructures of CeO_2 and its based composites such as nanoparticles, microspheres [112], nanofibers, nanosheets, nanotubes, thin films [113] have been received much attention for their many exclusive and superior advantages, such as the small size and the high surface-to-volume ratio [114].

Undoped CeO_2 and Fe-doped CeO_2 nanoparticles were synthesized by a homogeneous/impregnation method. The superior photocatalytic performance of the Fe-doped CeO_2 films, compared with undoped CeO_2 films, was ascribed mainly to the decrease in band gap energy and an increase in the specific surface area of the material. The presence of Fe as found from XPS analysis was supposed to act as an electron acceptor and/or hole donor (Fig. 6), facilitating longer lived charge carrier separation in Fe-doped CeO_2 film as confirmed by photoluminescence spectroscopy.

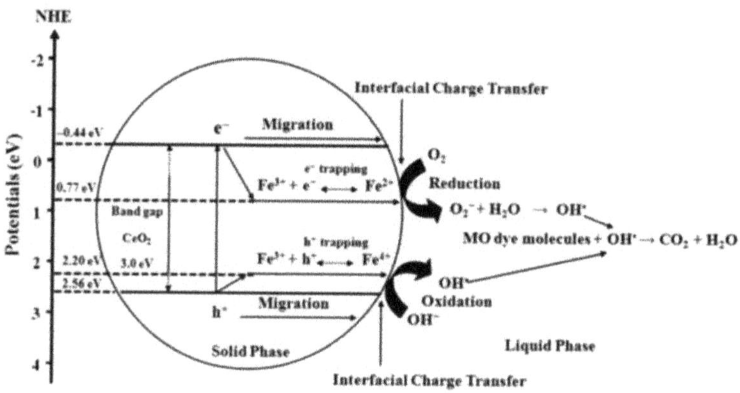

Figure 6 Proposed mechanism for the photoexcited electron-hole separation and transport processes at the Fe-doped CeO_2 interface under visible light irradiation [115].

The 1.50 mol% Fe-doped CeO_2 film was found to be optimal iron doping concentration for methyl orange degradation by D. Channei et al. [115]. Zhang et al. successfully synthesized porous CeO_2 nanofibers. The absorption performance of nanofibers showed that the CeO_2 nanofibers can remove the methyl orange in very short time i.e. the methyl orange removal was 50% in 5 min (Fig. 7). The accelerated adsorption is favorable for the water treatment [116].

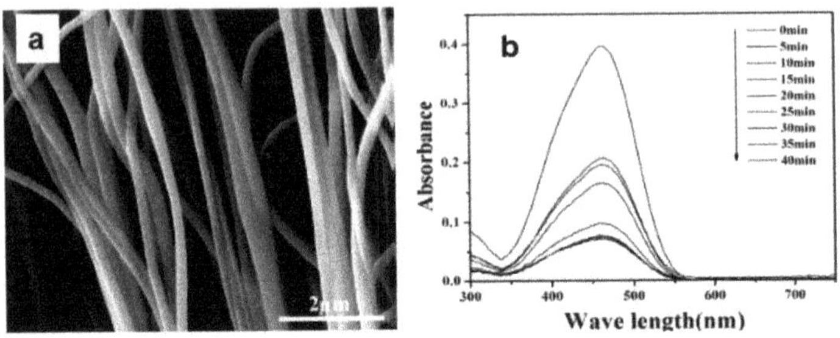

Figure 7 (a) SEM images of CeO_2 nanofibers and (b) absorption spectra of methyl orange solutions [116].

Photocatalytic degradation behavior of orange 16 reactive dye was successfully studied using cerium oxide nanoparticles. For CeO_2 nanoparticles, no color of reactive orange dye was observed after 120 min [117]. Photodegradation process by CeO_2 has attracted increasing attention during the past decades. However, its application is limited due to some main practical challenges such as low adsorption capacity for hydrophobic contaminants, post-recovery of the CeO_2 particles after water treatment and the uniform dispersion in aqueous suspension. The recently developed counter measures for improving the performance of CeO_2-based photocatalytic degradation of organic pollutants with respect to the above limitations have been reported [118-120]. A summary of the CeO_2 based photocatalysts for the degradation of various dyes is summarized in Table 4.

Table 4 Summary of organic dyes removal by CeO_2 based nanomaterials.

Photocatalyst	Dye	Photocatalytic Degradation time (min)	Efficiency (%)	Ref.
CeO_2-ZnO	MB	300	30	[121]
CeO_2/TiO_2 NCs	Bph-dye	3 h	72	[122]
CeO_2 NPs	Orange 16	2h	100	[117]
$CuBi_2O_4/CeO_2$NCs	CR	100	83.05	[118]
$CuBi_2O_4/CeO_2$	CR	100	83.05	[119]
Fe^{3+} Doped Mesoporous CeO_2	RhB	12h	85	[120]
CeO_2 Micro/Nanostructure	RhB	90	74.9	[121]
CeO_2	MB	10	90	[122]
CeO_2	MB	10	78	[123]
CeO_2NPs	Rh 6G	18 cycles	80	[124]
CeO_2NPs	CR	240	90	[125]
CeO_2	RhB	100	50	[126]
CeO_2 SiO_2 NPs	MB	180	85	[127]
ZnS decorated CeO_2	RhB	120	92	[128]
Co-doped CeO_2 NPs	MB	420	98	[129]
CeO_2nanostructure	MB	50	61.6	[130]
Fe–doped CeO_2	MO	120	10	[115]
CeO_2 NFs	MO	5	50	[116]

7. Conclusions

Recent efforts in photocatalytic degradation of organic dyes using different types of oxide-semiconductor nanomaterials have been encapsulated in the present chapter. It is concluded that above-discussed materials are very good photocatalysts and the addition of dopants to some of the pure nanomaterials can consider embellishing catalyst performance. Various dopants do not have the same effect on the interaction with electrons and holes due to the different position of the dopants in the host lattice. The search for new photocatalysts having desired characteristics to induce the oxidation of organic pollutants under visible light irradiation is encouraging. The metal oxides should be ecologically affable and preparation via inexpensive routes should be the main attention. These are the major concerns from the synthetic and industrial points of view which should be the primary focus of the researchers involved.

References

[1] A.A. Adeyemo, I.O. Adeoye, O.S. Bello, Metal-organic frameworks as adsorbents for dye adsorption: Overview, prospects and future challenges, Toxicol. Environ. Chem. 94 (2012) 1846-1863. https://doi.org/10.1080/02772248.2012.744023

[2] G. Liu, X. Li, J. Zhao, H. Hidaka, N. Serpone, Photooxidation pathway of sulforhodamine-B. Dependence on the adsorption mode on TiO_2 exposed to visible light radiation, Environ. Sci. Technol. 34 (2000) 3982-3990. https://doi.org/10.1021/es001064c

[3] N.S. Arul, D. Mangalaraj, T.W. Kim, P.C. Chen, N. Ponpandian, P. Meena, Y. Masuda, Synthesis of CeO_2 nanorods with improved photocatalytic activity: Comparison between precipitation and hydrothermal process, J. Mater. Sci. Mater. Electron. 24 (2013) 1644-1650. https://doi.org/10.1007/s10854-012-0989-x

[4] Kumar, M. Naushad, A. Rana, Inamuddin, Preeti, G. Sharma, A.A. Ghfar, F.J. Stadler, M.R. Khan, ZnSe-WO3 nano-hetero-assembly stacked on Gum ghatti for photo-degradative removal of Bisphenol A: Symbiose of adsorption and photocatalysis, Int. J. Biol. Macromol. 104 (2017) 1172–1184. https://doi.org/10.1016/j.ijbiomac.2017.06.116

[5] G. Sharma, S. Bhogal, M. Naushad, Inamuddin, A. Kumar, F.J. Stadler, Microwave assisted fabrication of La/Cu/Zr/carbon dots trimetallic nanocomposites with their adsorptional vs photocatalytic efficiency for remediation of persistent organic pollutants, J. Photochem. Photobiol. A Chem. 347 (2017) 235–243. https://doi.org/10.1016/j.jphotochem.2017.07.001

[6] Y. Li, X. Dong, J. Li. Synthesis and characterization of super paramagnetic composite photocatalyst Titania/silica/nickel ferrite, Particuology 9 (2011) 475-479. https://doi.org/10.1016/j.partic.2011.04.005

[7] Y. Wang, J. Zhang, X. Liu, S. Gao, B. Huang, Y. Dai, Y. Xu. Synthesis and characterization of activated carbon-coated $SiO_2/TiO_{2-x}C_x$ nanoporous composites with high adsorption capability and visible light photocatalytic activity, Mater. Chem. Phys. 135 (2012) 579-586. https://doi.org/10.1016/j.matchemphys.2012.05.029

[8] M.A. Nasalevich, M van der Veen, F Kapteijn, J.Gascon, Metal–organic frameworks as heterogeneous photocatalysts: Advantages and challenges, Cryst. Eng. Comm. 16 (2014) 4919-4926. https://doi.org/10.1039/C4CE00032C

[9] S. Sarkar, R. Das, H. Choi, C. Bhattacharjee, Involvement of process parameters and various modes of application of TiO_2 nanoparticles in heterogeneous photocatalysis of pharmaceutical wastes–a short review, RSC Adv. 4 (2014) 57250-57266. https://doi.org/10.1039/C4RA09582K

[10] Some Aromatic Amines, Organic Dyes, and Related Exposures, Monographs on the evaluation of the carcinogenic risk of chemicals to humans, World health organization, International agency for research on cancer (IARC), Lyon, France 99 (2008) Retrieved from: https://monographs.iarc.fr/ENG/Monographs/vol99/mono99.pdf .

[11] S.H.S. Chan, T.Y. Wu, J.C. Juan, C.Y. Teh, Recent developments of metal oxide semiconductors as photocatalysts in advanced oxidation processes (AOPs) for treatment of dye waste-water, J. Chem. Technol. Biotechnol. 86 (2011) 1130–1158. https://doi.org/10.1002/jctb.2636

[12] M. Dai, H.X. Li, J.P. Lang, New approaches to the degradation of organic dyes, and nitro-andchloroaromatics using coordination polymers as photocatalysts, Cryst. Eng. Comm. 17 (2015) 4741-4753. https://doi.org/10.1039/C5CE00619H

[13] T.S. Natarajan, M. Thomas, K. Natarajan, H.C. Bajaj, R.J. Tayade, Study on UV-LED/TiO_2 process for degradation of Rhodamine B dye, Chem. Eng. J. 169 (2011) 126-134. https://doi.org/10.1016/j.cej.2011.02.066

[14] I. Dror, D. Baram, B. Berkowitz. Use of nanosized catalysts for transformation of chloro-organic pollutants, Environ. Sci. Technol. 39 (2005) 1283-1290. https://doi.org/10.1021/es0490222

[15] M. Dai, W.Y. Yan, Z.G. Ren, H.F. Wang, W.J. Gong, F.L. Li, J.P. Lang, Zinc (II) coordination polymers of tetrakis (4-pyridyl) cyclobutane and various dicarboxylates: Syntheses, structures and luminescent properties, Cryst. Eng. Comm. 14 (2012) 6230-6240. https://doi.org/10.1039/c2ce25482d

[16] D. Liu, H.F. Wang, B.F. Abrahams, J.P. Lang, Single-crystal-to-single-crystal transformation of a two-dimensional coordination polymer through highly selective [2+2] photodimerization of a conjugated dialkene, Chem. Commun. 50 (2014) 3173-3175. https://doi.org/10.1039/c3cc49749f

[17] Y.T. Liang, B.K. Vijayan, K.A. Gray, M.C. Hersam, Minimizing graphene defects enhances titania nanocomposite-based photocatalytic reduction of CO_2 for improved solar fuel production, Nano. Lett. 11 (2011) 2865-2870. https://doi.org/10.1021/nl2012906

[18] M. Farrokhi, S.C. Hosseini, J.K. Yang, M. Shirzad-Siboni, Application of ZnO-Fe_3O_4nanocomposite on the removal of azo dye from aqueous solutions: kinetics and equilibrium studies,Water, Air, Soil Pollut. 225 (2014) 2113. https://doi.org/10.1007/s11270-014-2113-8

[19] B.Q. Song, C. Qin, Y.T. Zhang, X.S. Wu, K.Z. Shao, Z.M. Su, The copper (I) metal azolate framework showing unusual coordination mode for the 1, 2, 4-triazole derivative and photocatalytic activity, Dalton Trans. 44 (2015) 3954-3958. https://doi.org/10.1039/C4DT03933E

[20] J. Yang, J. Yu, J. Fan, D. Sun, W. Tang, X. Yang, Biotemplated preparation of CdS nanoparticles/bacterial cellulose hybrid nanofibers for photocatalysis application, J. Hazard Mater. 189 (2011) 377-383. https://doi.org/10.1016/j.jhazmat.2011.02.048

[21] X. Xu, L. Hu, N. Gao, S. Liu, S. Wageh, A.A. Al-Ghamdi, X. Fang. Controlled growth from ZnS nanoparticles to ZnS-CdS nanoparticle hybrids with enhanced Photoactivity, Adv. Funct. Mater. 25 (2015) 445-454. https://doi.org/10.1002/adfm.201403065

[22] F. Pang, X. Liu, M. He, J. Ge, Ag_3PO_4 colloidal nanocrystal clusters with controllable shape and superior photocatalytic activity, Nano. Res. 8 (2015) 106-116. https://doi.org/10.1007/s12274-014-0580-2

[23] V. Vaiano, O. Sacco, D. Sannino, P. Ciambelli, Photocatalytic removal of spiramycin from wastewater under visible light with N-doped TiO_2photocatalysts, Chem. Eng J. 261 (2015) 3-8. https://doi.org/10.1016/j.cej.2014.02.071

[24] J. Lu, M. Liu, S. Zhou, X. Zhou, Y. Yang, Electrospinning fabrication of $ZnWO_4$ nanofibers and photocatalytic performance for organic dyes, Dyes Pigm. 136 (2017) 1-7. https://doi.org/10.1016/j.dyepig.2016.08.008

[25] T.V.M. Sreekanth, G.R. Dillip, Y.R. Lee, Picrasmaquassioides mediated cerium oxide nanostructures and their post-annealing treatment on the microstructural, morphological and enhanced catalytic performance, Ceram Int. 42 (2016) 6610-6618. https://doi.org/10.1016/j.ceramint.2015.12.171

[26] M.H. Khedr, K.A. Halim, N.K. Soliman, Synthesis and photocatalytic activity of nano-sized iron oxides, Mater. Lett. 63 (2009) 598-601. https://doi.org/10.1016/j.matlet.2008.11.050

[27] Y.S. Zhao, C. Sun, J.Q. Sun, R. Zhou, Kinetic modeling and efficiency of sulfate radical-based oxidation to remove p-nitroaniline from wastewater by persulfate/Fe

$_3O_4$ nanoparticles process, Sep. Purif. Technol. 142 (2015) 182-188. https://doi.org/10.1016/j.seppur.2014.12.035

[28] S. Kitagawa, R. Kitaura, S.I. Noro, Functional porous coordination polymers, Angew. Chem. Int. Ed. 43 (2004) 2334-2375. https://doi.org/10.1002/anie.200300610

[29] Moulton, M.J. Zaworotko, From molecules to crystal engineering: supramolecular isomerism and polymorphism in network solids, Chem. Rev. 101 (2001) 1629-1658. https://doi.org/10.1021/cr9900432

[30] N.T. Khoa, S.W. Kim, D. Van Thuan, D.H. Yoo, E.J. Kim, S.H. Hahn, Hydrothermally controlled ZnO nanosheet self-assembled hollow spheres/hierarchical aggregates and their photocatalytic activities, Cryst. Eng. Comm. 16 (2014) 1344-1350. https://doi.org/10.1039/C3CE41763H

[31] L. Wang, Z. Lou, T. Fei, T. Zhang, Templating synthesis of ZnO hollow nanospheres loaded with Au nanoparticles and their enhanced gas sensing properties, J. Mater. Chem. 22 (2012) 4767-4771. https://doi.org/10.1039/c2jm15342d

[32] T. Thilagavathi, D. Geetha, Nano ZnO structures synthesized in presence of anionic and cationic surfactant under hydrothermal process, Appl. Nanosci. 4 (2014)127-132. https://doi.org/10.1007/s13204-012-0183-8

[33] Karunakaran, S. SakthiRaadha, P. Gomathisankar, P. Vinayagamoorthy, The enhanced photocatalytic and bactericidal activities of carbon microsphere-assisted solvothermally synthesized cocoon-shaped Sn^{4+} doped ZnO nanoparticles, Dalton Trans. 42 (2013) 13855-13865. https://doi.org/10.1039/c3dt51058a

[34] S. Gao, H. Zhang, X. Wang, R. Deng, D. Sun, G. Zheng, ZnO-based hollow microspheres: Biopolymer-assisted assemblies from ZnO_nanorods, J. Phys. Chem. B 110 (2006) 15847-15852. https://doi.org/10.1021/jp062850s

[35] S. Ma, R. Li, C. Lv, W. Xu, X. Gou, Facile synthesis of ZnO nanorod arrays and hierarchical nanostructures for photocatalysis and gas sensor applications, J. Hazard Mater. 192 (2011) 730-740. https://doi.org/10.1016/j.jhazmat.2011.05.082

[36] X. Chen, F. Zhang, Q. Wang, X. Han, X. Li, J. Liu, F. Qu, The synthesis of ZnO/SnO_2 porous nanofibers for dye adsorption and degradation, Dalton Trans. 44 (2015) 3034-3042. https://doi.org/10.1039/C4DT03382E

[37] K. Vidhya, M. Saravanan, G. Bhoopathi, V.P. Devarajan. S. Subanya, Structural and optical characterization of pure and starch-capped ZnO quantum dots and their

photocatalytic activity, Appl. Nanosci. 5 (2015) 235-243. https://doi.org/10.1007/s13204-014-0312-7

[38] A. Sadollahkhani, O. Nur, M. Willander, I. Kazeminezhad, V. Khranovskyy, M.O. Eriksson, R. Yakimova, P. O. Holtz, Adetailed optical investigation of ZnO@ZnScore-shell nanoparticles and their photocatalytic activity at different pH values, Ceram. Int. 41 (2015) 7174-7184. https://doi.org/10.1016/j.ceramint.2015.02.040

[39] M. Ahmad, E. Ahmed, Z.L. Hong, J.F. Xu, N.R. Khalid, A. Elhissi, W. Ahmed, A facile one-step approach to synthesizing ZnO/graphene composites for enhanced degradation of methylene blue under visible light, Appl. Surf. Sci. 274 (2013) 273-281. https://doi.org/10.1016/j.apsusc.2013.03.035

[40] R. Saravanan, E. Sacari, F. Gracia, M.M. Khan, E. Mosquera, V.K. Gupta, Conducting PANI stimulated ZnO system for visible light photocatalytic degradation of coloured dyes, J. Mol. Liq. 221 (2016) 1029-1033. https://doi.org/10.1016/j.molliq.2016.06.074

[41] J. Liu, Z.Y. Hu, Y. Peng, H.W. Huang, Y. Li, M. Wu, B.L. Su, 2D ZnO mesoporous single-crystal nanosheets with exposed {0001} polar facets for the depollution of cationic dye molecules by highly selective adsorption and photocatalytic decomposition, Appl. Catal. B 181 (2016) 138-145. https://doi.org/10.1016/j.apcatb.2015.07.054

[42] S. Chaudhary, Y. Kaur, A. Umar, G.R. Chaudhary, 1-Butyl-3-methylimidazolium tetrafluoroborate functionalized ZnO nanoparticles for removal of toxic organic dyes, J. Mol. Liq. 220 (2016) 1013-1021. https://doi.org/10.1016/j.molliq.2016.05.011

[43] M.C. Das, H. Xu, Z. Wang, G. Srinivas, W. Zhou, Y.F. Yue, B. Chen, A Zn_4 O-containing doubly interpenetrated porous metal–organic framework for photocatalytic decomposition of methyl orange, Chem. Commun. 47 (2011) 11715-11717. https://doi.org/10.1039/c1cc12802g

[44] T. Karnan, S.A.S. Selvakumar, Biosynthesis of ZnO nanoparticles using rambutan (Nepheliumlappaceuml.) peel extract and their photocatalytic activity on methyl orange dye, J. Mol. Struct. 1125 (2016) 358-365. https://doi.org/10.1016/j.molstruc.2016.07.029

[45] R. Saravanan, S. Karthikeyan, V.K. Gupta, G. Sekaran, V. Narayanan, A. Stephen, Enhanced photocatalytic activity of ZnO/CuO nanocomposite for the degradation

of textile dye on visible light illumination, Mater Sci. Eng. ***C*** 33 (2013) 91-98. https://doi.org/10.1016/j.msec.2012.08.011

[46] V.C. Bhusal, Photolytic degradation of organic dyes using photocatayst nanoparticles for waste water treatments, Int. J. Sci. Res. 5 (2016).

[47] M.E. Hassan, J. Chen, G. Liu, D. Zhu, J. Cai, Enhanced photocatalytic degradation of methyl orange dye under the daylight irradiation over CN-TiO_2 modified with OMS-2, Mat. 7 (2014) 8024-8036. https://doi.org/10.3390/ma7128024

[48] A.N. El-Shazly, M.M. Rashad, E.A. Abdel-Aal, I.A. Ibrahim, M.F. El-Shahat, A.E. Shalan, Nanostructured ZnO photocatalysts prepared via surfactant assisted Co-precipitation method achieving enhanced photocatalytic activity for the degradation of methylene blue dyes, J. Chem. Eng. 4 (2016) 3177-3184. https://doi.org/10.1016/j.jece.2016.06.018

[49] L. Saikia, D. Bhuyan, M. Saikia, B. Malakar, D.K. Dutta, P. Sengupta ,Photocatalytic performance of ZnO nanomaterials for self-sensitized degradation of malachite green dye under solar light, Appl. Catal. A: General 490 (2015) 42-49. https://doi.org/10.1016/j.apcata.2014.10.053

[50] L. Fu, Z. Fu, Plectranthusamboinicus leaf extract–assisted biosynthesis of ZnO nanoparticles and their photocatalytic activity, Ceram. Int. 41(2015) 2492-2496. https://doi.org/10.1016/j.ceramint.2014.10.069

[51] Y. Miao, H. Zhang, S. Yuan, Z. Jiao X. Zhu, Preparation of flower-like ZnO architectures assembled with nanosheets for enhanced photocatalytic activity, J. Colloid Interface Sci. 462 (2016) 9-18. https://doi.org/10.1016/j.jcis.2015.09.064

[52] A. Sugunan, V.K. Guduru, A. Uheida, M.S. Toprak, M. Muhammed, Radially oriented ZnO nanowires on flexible poly-l-lactide nanofibers for continuous-flow photocatalytic water purification, J. Am. Ceram. Soc. 93 (2010) 3740-3744. https://doi.org/10.1111/j.1551-2916.2010.03986.x

[53] K.M. Lee, C.W. Lai, K.S. Ngai, J.C. Juan, Recent developments of zinc oxide based photocatalyst in water treatment technology: a review, ***Water*** Res. 88 (2016) 428-448. https://doi.org/10.1016/j.watres.2015.09.045

[54] M.M. Hossain, B.C. Ku, J.R. Hahn, Synthesis of an efficient white-light photocatalyst composite of graphene and ZnO nanoparticles: application to methylene blue dye decomposition, Appl. Surf. Sci. 354 (2015) 55-65. https://doi.org/10.1016/j.apsusc.2015.01.191

[55] R. Kumar, G. Kumar, A. Umar, ZnO nano-mushrooms for photocatalytic degradation of methyl orange, Mater. Lett. 97 (2013) 100-103. https://doi.org/10.1016/j.matlet.2013.01.044

[56] C.H. Ao, S.C. Lee, Enhancement effect of TiO_2 immobilized on activated carbon filter for the photodegradation of pollutants at typical indoor air level, Appl. Catal. B: Environ. 44 (2003) 191-205. https://doi.org/10.1016/S0926-3373(03)00054-7

[57] C.P. Athanasekou, S. Morales-Torres, V. Likodimos, G.E. Romanos, L.M Pastrana-Martinez, P. Falaras, A.M. Silva, Prototype composite membranes of partially reduced graphene oxide/TiO_2 for photocatalytic ultrafiltration water treatment under visible light, Appl. Catal. B: Environ. 158 (2014) 361-372. https://doi.org/10.1016/j.apcatb.2014.04.012

[58] M.F.J. Dijkstra, A. Michorius, H. Buwalda, H.J. Panneman, J.G.M. Winkelman, A.A.C.M. Beenackers, Comparison of the efficiency of immobilized and suspended systems in photocatalytic degradation. Catal. Today 66 (2001) 487-494. https://doi.org/10.1016/S0920-5861(01)00257-7

[59] P. Fu, Y. Luan, X. Dai, Preparation of activated carbon fibers supported TiO_2 photocatalyst and evaluation of its photocatalytic reactivity, J. Mol. Catal: Chem. 221 (2004) 81-88. https://doi.org/10.1016/j.molcata.2004.06.018

[60] F. Dong, S. Guo, H. Wang, X. Li, Z. Wu, Enhancement of the visible light photocatalytic activity of C-doped TiO_2nanomaterials prepared by a green synthetic approach, J. Phys. Chem. C 115 (2011) 13285-13292. https://doi.org/10.1021/jp111916q

[61] X. Wu, S. Yin, Q. Dong, C. Guo, H. Li, T. Kimura, T. Sato, Synthesis of high visible light active carbon doped TiO_2 photocatalyst by a facile calcination assisted solvothermal method, Appl. Catal. B: Environ. 142 (2013) 450-457. https://doi.org/10.1016/j.apcatb.2013.05.052

[62] R.C. Suciu, E. Indre, T.D. Silipas, S. Dreve, M.C. Rosu, V. Popescu, TiO_2 thin films prepared by sol-gel method, J. Phys. 182 (2009) 1-4.

[63] J.C. Zhao, C.C. Chen, W.H. Ma, Photocatalytic degradation of organic pollutants under visible light irradiation, Top. Catal. 35 (2005) 269-278. https://doi.org/10.1007/s11244-005-3834-0

[64] G. Williams, B. Seger, P. V. Kamat, TiO_2-graphene nanocomposites. UV-assisted photocatalytic reduction of graphene oxide, ACS Nano. 2 (2008) 1487-1491. https://doi.org/10.1021/nn800251f

[65] T. Xia, W. Zhang, Z. Wang, Y. Zhang, X. Song, J. Murowchick, V. Battaglia, G. Liu, X. Chen, Amorphous carbon-coated TiO_2 nanocrystals for improved lithium-ion battery and photocatalytic performance, Nano Energy 6 (2014) 109-118. https://doi.org/10.1016/j.nanoen.2014.03.012

[66] A. Arabzadeh, A. Salimi, One dimensional CdS nanowire@TiO_2 nanoparticles core-shell as high performance photocatalyst for fast degradation of dye pollutants under visible and sunlight irradiation, J. Colloid Interface Sci. 479 (2016) 43-54. https://doi.org/10.1016/j.jcis.2016.06.036

[67] X. Yang, J. Qin, Y. Jiang, R. Li, Y. Li, H. Tang, Bifunctional TiO_2/Ag_3PO_4/graphene composites with superior visible light photocatalytic performance and synergistic inactivation of bacteria, RSC Adv. 4 (2014) 18627-18636. https://doi.org/10.1039/C4RA01559B

[68] Q. Xiang, J. Yu, B. Cheng, H.C. Ong, Microwave-hydrothermal preparation and visible-light photoactivity of plasmonic photocatalyst Ag-TiO_2 nanocomposite hollow spheres, Chem. Asian J. 5 (2010) 1466-1474. https://doi.org/10.1002/asia.200900695

[69] Zhang, F. Zeng, Structural, photochemical and photocatalytic properties of zirconium oxide doped TiO_2 nanocrystallites, Appl. Surf. Sci. 257 (2010) 867-871. https://doi.org/10.1016/j.apsusc.2010.07.083

[70] Q. Zhang, D.Q. Lima, I. Lee, F. Zaera, M. Chi, Y. Yin, A highly active titanium dioxide based visible-light photocatalyst with nonmetal doping and plasmonic metal decoration, Angew. Chem. Int. Ed. 50 (2011) 7088-7092. https://doi.org/10.1002/anie.201101969

[71] M.R. Eskandarian, M. Fazli, M.H. Rasoulifard, H. Choi, Decomposition of organic chemicals by zeolite-TiO_2 nanocomposite supported onto low density polyethylene film under UV-LED powered by solar radiation, Appl. Catal. B: Environ. 183 (2016) 407-416. https://doi.org/10.1016/j.apcatb.2015.11.004

[72] S. Sood, A. Umar, S.K. Mehta, S.K. Kansal, Highly effective Fe-doped TiO_2 nanoparticles photocatalysts for visible-light driven photocatalytic degradation of toxic organic compounds, J .Colloid Interface Sci. 450 (2015) 213-223. https://doi.org/10.1016/j.jcis.2015.03.018

[73] K.R. Reddy, M. Hassan, V.G. Gomes, Hybrid nanostructures based on titanium dioxide for enhanced photocatalysis, Appl. Catal. A: General. 489 (2015) 1-16. https://doi.org/10.1016/j.apcata.2014.10.001

[74] V. Vaiano, O. Sacco, D. Sannino, P. Ciambelli, S. Longo, V. Venditto, G. Guerra, N-doped TiO2/s-PS aerogels for photocatalytic degradation of organic dyes in wastewater under visible light irradiation, J. Chem. Technol. Biotechnol. 89 (2014) 1175-1181. https://doi.org/10.1002/jctb.4372

[75] I. Barton, V. Matejec, J. Matousek, Photocatalytic activity of nanostructured TiO_2 coating on glass slides and optical fibers for methylene blue or methyl orange decomposition under different light excitation, J. Photochem. Photobiol. A. Chem. 317 (2016) 72-80. https://doi.org/10.1016/j.jphotochem.2015.11.009

[76] S. Masoumi, G. Nabiyouni, D. Ghanbari, Photo-degradation of azo dyes: Photo catalyst and magnetic investigation of $CuFe_2O_4$–TiO_2 nanoparticles and nanocomposites, J. Mater. Sci. Mater. Engg. 27 (2016) 9962-9975. https://doi.org/10.1007/s10854-016-5067-3

[77] P.S. Sauda, B. Pant, M. Park, S.H. Chae, S.J. Park, M.E. Newehy, S.S.A. Deyab, H.Y. Kim, Preparation and photocatalytic activity of fly ash incorporated TiO_2 nanofibres for effective removal of organic pollutants, Ceram. Int. 41 (2015) 1771-1777. https://doi.org/10.1016/j.ceramint.2014.09.123

[78] S.A. Bakar, G. Byzynski, C. Ribeiro, Synergistic effect on the photocatalytic activity of N-doped TiO_2 nanorods synthesised by novel route with exposed (110) facet, J. Alloys Compd. 666 (2016) 38-49. https://doi.org/10.1016/j.jallcom.2016.01.112

[79] K. Santhi, P. Manikandan, C. Rani, S. Karuppuchamy, Synthesis of nanocrystalline titanium dioxide for photodegradation treatment of remazol brown dye, Appl. Nanosci. 5 (2015) 373-378. https://doi.org/10.1007/s13204-014-0327-0

[80] T. Luttrell, S. Halpegamage, J. Tao, A. Kramer, E. Sutter, M. Batzill, Why is anatase a better photocatalyst than rutile? Model studies on epitaxial TiO_2 films Sci. Rep. 4 (2014) 40-43.

[81] M. Yamaura, R.L. Camilo, L.C. Sampaio, M.A. Macedo, M. Nakamura, H.E. Toma, Preparation and characterization of (3-aminopropyl) triethoxysilane-coated magnetite nanoparticles, J. Magn. Magn. Mater. 279 (2004) 210–217. https://doi.org/10.1016/j.jmmm.2004.01.094

[82] I. Shakir, M. Shahid, D.J. Kang, MoO_3 and Cu 0.33 MoO_3 nanorods for unprecedented UV/Visible light photocatalysis, Chem. Commun. 46 (2010) 4324-4326. https://doi.org/10.1039/c000003e

[83] S.G. Zhang, Y. Zhang, Y. Wang, S.M. Liu, Y.Q. Deng, Sonochemical formation of iron oxide nanoparticles in ionic liquids for magnetic liquid marble, Phys. Chem. Chem. Phys. 14 (2012) 5132-5138. https://doi.org/10.1039/c2cp23675c

[84] W. Li, Y. Deng, Z. Wu, X. Qian, J. Yang, Y. Wang, D. Gu, F. Zhang, B. Tu, D. Zhao, Hydrothermal etching assisted crystallization: a facile route to functional yolk-shell titanate microspheres with ultrathin nanosheets-assembled double shells, J. Am. Chem. Soc. 133 (2011) 15830-15833. https://doi.org/10.1021/ja2055287

[85] H. Wang, L. Sun, Y. Li, X. Fei, M. Sun, C. Zhang, Y. Li, Q. Yang, Layer-by-layer assembled Fe_3O_4@C@CdTe core/shell microspheres as separable luminescent probe for sensitive sensing of Cu^{2+} ions, Langmuir 27 (2011) 11609-11615. https://doi.org/10.1021/la202295b

[86] Y. Li, Y. Hu, H. Jiang, X. Hou, C. Li, Phase-segregation induced growth of core–shell α-Fe_2O_3/SnO_2heterostructures for lithium-ion battery, Cryst. Eng. Comm. 15 (2013) 6715-6721. https://doi.org/10.1039/c3ce40737c

[87] X.L. Cheng, J.S. Jiang, C.Y. Jin, C.C. Lin, Y. Zeng, Q.H. Zhang, Cauliflower-like α-Fe_2O_3microstructures: Toluene–water interface-assisted synthesis, characterization, and applications in wastewater treatment and visible-light photocatalysis, Chem. Eng. J. 236 (2014) 139-148. https://doi.org/10.1016/j.cej.2013.09.089

[88] S. Guo, G. Zhang, Y. Guo, C.Y. Jimmy, Graphene oxide-Fe_2O_3 hybrid material as highly efficient heterogeneous catalyst for degradation of organic contaminants, Carbon 60 (2013) 437-444. https://doi.org/10.1016/j.carbon.2013.04.058

[89] Y. Jiao, Y. Liu, F. Qu, A. Umar, X. Wu, Visible-light-driven photocatalytic properties of simply synthesized α-Iron (III) oxide nanourchins, J. Colloid Interface Sci. 451 (2015) 93-100. https://doi.org/10.1016/j.jcis.2015.03.055

[90] Chen, F. Yan, Q. Chen, Y. Wang, L. Han, Z. Chen, S. Fang, Fabrication of Fe_3O_4@ SiO_2@ TiO_2 nanoparticles supported by graphene oxide sheets for the repeated adsorption and photocatalytic degradation of rhodamine B under UV irradiation, Dalton Trans. 43 (2014) 13537-13544. https://doi.org/10.1039/C4DT01702A

[91] Li, L. Mao, Magnetically separable Fe_3O_4–Ag_3PO_4 sub-micrometre composite: Facile synthesis, high visible light-driven photocatalytic efficiency and good recyclability, RSC Adv. 2 (2012) 5108-5111. https://doi.org/10.1039/c2ra20504a

[92] M. Amir, U. Kurtan, A. Baykal, Rapid color degradation of organic dyes by Fe_3O_4@His@Ag recyclable magnetic nanocatalyst, J. Ind. Eng. Chem. Res. 27 (2015) 347-353. https://doi.org/10.1016/j.jiec.2015.01.013

[93] P. Cai, S. M. Zhou, D.K. Ma, S.N. Liu, W. Chen, S.M. Huang, Fe_2O_3-modified porous $BiVO_4$ nanoplates with enhanced photocatalytic activity, Nano. Lett. 7 (2015) 183-193. https://doi.org/10.1007/s40820-015-0033-9

[94] J. Feng, X. Hu, P.L. Yue, H.Y. Zhu, G.Q. Lu, Degradation of azo-dye orange II by a photoassisted Fenton reaction using a novel composite of iron oxide and silicate nanoparticles as a catalyst, Ind. Eng. Chem. Res. 42 (2003) 2058-2066. https://doi.org/10.1021/ie0207010

[95] Z. Jia, J. Liu, Q. Wang, S. Li, Q. Qi, R. Zhu, Synthesis of 3D hierarchical porous iron oxides for adsorption of Congo red from dye wastewater J. Alloys Compd. 622 (2015) 587-595. https://doi.org/10.1016/j.jallcom.2014.10.125

[96] S. Kumar, B. Kumar, A. Baruah, V. Shanker, Synthesis of magnetically separable and recyclable g-C_3N_4-Fe_3O_4 hybrid nanocomposites with enhanced photocatalytic performance under visible-light irradiation, J. Phys. Chem. C 117 (2013) 26135-26143. https://doi.org/10.1021/jp409651g

[97] Liang, S. Zhou, Y. Chen, F. Zhou, C. Yan, Diatomite coated with Fe_2O_3 as an efficient heterogeneous catalyst for degradation of organic pollutant, J. Taiwan Inst. Chem. Eng. 49 (2015) 105-112. https://doi.org/10.1016/j.jtice.2014.11.002

[98] Y.F. Lin, C.Y. Chang, Magnetic mesoporous iron oxide/carbon aerogel photocatalysts with adsorption ability for organic dye removal, RSC Adv. 4 (2014) 28628-28631. https://doi.org/10.1039/c4ra03436h

[99] M.A. Mahadik, S.S. Shinde, V.S. Mohite, S.S. Kumbhar, A.V. Moholkar, K.Y. Rajpure, C.H. Bhosale, Visible light catalysis of rhodamine B using nanostructured Fe_2O_3, TiO_2 and TiO_2/Fe_2O_3 thin films, J. Photochem. Photobiol. B 133 (2014) 90-98. https://doi.org/10.1016/j.jphotobiol.2014.01.017

[100] S. Nazim, T. Kousar, M. Shahid, M.A. Khan, G. Nasar, M. Sher, M.F. Warsi, New graphene-$Co_xZn_{(1-x)}Fe_2O_4$ nano-heterostructures: Magnetically separable visible light photocatalytic materials, Ceram. Int. 42 (2016) 7647-7654. https://doi.org/10.1016/j.ceramint.2016.01.177

[101] A. Abbasi, D. Ghanbari, M.S. Niasari, M. Hamadanian, Photo-degradation of methylene blue: Photocatalyst and magnetic investigation of Fe_2O_3–TiO_2

nanoparticles and nanocomposites, J. Mater. Sci. Mater. Electron. 27 (2016) 4800-4809. https://doi.org/10.1007/s10854-016-4361-4

[102] P. Sharma, R. Kumar, S. Chauhan, D. Singh, M.S. Chauhan Facile growth and characterization of α-Fe2O_3 nanoparticles for photocatalytic degradation of methyl orange, J. Nanosci. Nanotechnol. 14 (2014) 6153-6157. https://doi.org/10.1166/jnn.2014.8734

[103] X.C. Song, X. Wang, Y.F. Zheng, R. Ma, H.Y. Yin, Synthesis and electrocatalytic activities of Co_3O_4 nanocubes, J. Nanopart. Res. 13 (2011) 1319-1324. https://doi.org/10.1007/s11051-010-0127-8

[104] A.A. Tireli, F.C.F. Marcos, L.F. Oliveira, I. Rosario Guimaraes, M.C. Guerreiro, J.P. Silva Influence of magnetic field on the adsorption of organic compound by clays modified with iron. Appl. Clay Sci. 97 (2014), 1-7. https://doi.org/10.1016/j.clay.2014.05.014

[105] A. Umar, M.S. Akhtar, G.N. Dar, S. Baskoutas, Low-temperature synthesis of α-Fe_2O_3 hexagonal nanoparticles for environmental remediation and smart sensor applications, Talanta 116 (2013) 1060-1066. https://doi.org/10.1016/j.talanta.2013.08.026

[106] Y. Zhang, J. Gu, M. Murugananthan, Y. Zhang, Development of novel Fe_2O_3/$NiTiO_3$ heterojunction nanofibers material with enhanced visible-light photocatalytic performance, J. Alloys Compd. 630 (2015) 110-116. https://doi.org/10.1016/j.jallcom.2014.12.193

[107] X. Zheng, Y. Jiao, F. Chai, F. Qu, A. Umar, X. Wu, Template-free growth of well-crystalline α-Fe_2O_3 nanopeanuts with enhanced visible-light driven photocatalytic properties, J .Colloid Interface Sci. 457 (2015) 345-352. https://doi.org/10.1016/j.jcis.2015.07.023

[108] M.J. Godinho, R.F. Gonçalves, L.S. Santos, J.A. Varela, E. Longo, E.R. Leite, Room temperature co-precipitation of nanocrystalline CeO_2 and Ce 0.8 Gd 0.2 O 1.9−δ powder, Mater. Lett. 61 (2007) 1904-1907. https://doi.org/10.1016/j.matlet.2006.07.152

[109] C. Sun, J. Sun, G. Xiao, H. Zhang, X. Qiu, H. Li, L. Chen, Mesoscale organization of nearly monodisperse flowerlike ceria microspheres, J. Phys. Chem. B. 110 (2006) 13445-13452. https://doi.org/10.1021/jp062179r

[110] S. Maensiri, C. Masingboon, P. Laokul, W. Jareonboon, V. Promarak, P.L. Anderson, S. Seraph in, Egg white synthesis and photoluminescence of plate like

clusters of CeO_2 nanoparticles, Cryst. Growth Des. 7 (2007) 950-955. https://doi.org/10.1021/cg0608864

[111] Z. Liu, Q. Liu, Y. Huang, Y. Ma, S. Yin, X. Zhang, Y. Chen, Organic photovoltaic devices based on a novel acceptor material: graphene, Adv. Mater. 20 (2008) 3924-3930. https://doi.org/10.1002/adma.200800366

[112] J. Wang, D.N. Tafen, J.P. Lewis, Z. Hong, A. Manivannan, M. Zhi, N. Wu, Origin of photocatalytic activity of nitrogen-doped TiO_2 nanobelts, J. Am. Chem. Soc. 131 (2009) 12290-12297. https://doi.org/10.1021/ja903781h

[113] K.S. Lin, S. Chowdhury, Synthesis, characterization, and application of 1-D cerium oxide nanomaterials: a review, Int. J. Mol. Sci. 11 (2010) 3226-3251. https://doi.org/10.3390/ijms11093226

[114] T. Xu, L. Zhang, H. Cheng, Y. Zhu, Significantly enhanced photocatalytic performance of ZnO via graphene hybridization and the mechanism study, Appl. Catal. B Environ. 101 (2011) 382-387. https://doi.org/10.1016/j.apcatb.2010.10.007

[115] D. Channei, B. Inceesungvorn, N. Wetchakun, S. Ukritnukun, A. Nattestad, J. Chen, S. Phanichphant, Photocatalytic degradation of methyl orange by CeO_2 and Fe–doped CeO_2 films under visible light irradiation, Sci. Rep. 4 (2014) 5757. https://doi.org/10.1038/srep05757

[116] Y. Zhang, R. Shi, P. Yang, X. Song, Y. Zhu, Q. Ma, Fabrication of electronspun porous CeO_2 nanofibers with large surface area for pollutants removal, Ceram. Int. 42 (2016) 14028-14035. https://doi.org/10.1016/j.ceramint.2016.06.009

[117] Balavi, S. Samadanian-Isfahani, M. Mehrabani-Zeinabad, M. Edrissi, Preparation and optimization of CeO_2 nanoparticles and its application in photocatalytic degradation of reactive orange 16 dye, Powder Technol. 249 (2013) 549-555. https://doi.org/10.1016/j.powtec.2013.09.021

[118] Kaspar, P. Fornasiero, N. Hickey, Automotive catalytic converters: current status and some perspectives, Catal. Today 77 (2003) 419-449. https://doi.org/10.1016/S0920-5861(02)00384-X

[119] H. Ma, M. Wang, R. Yang, W. Wang, J. Zhao, Z. Shen, S. Yao, Radiation degradation of congo red in aqueous solution, Chemosphere 68 (2007) 1098-1104. https://doi.org/10.1016/j.chemosphere.2007.01.067

[120] Y. Wang, Y. Wang, Y. Meng, H. Ding, Y. Shan, X. Zhao, X. Tang, A highly efficient visible-light-activated photocatalyst based on bismuth-and sulfur-

codoped TiO_2, J. Phys. Chem. C 112 (2008) 6620-6626. https://doi.org/10.1021/jp7110007

[121] G.A. Apostolescu, C. Cernatescu, C. Cobzaru, R.E. Tataru-Farmus, N. Apostolescu, Studies on the photocatalytic degradation of organic dyes using CeO_2-ZnO mixed oxides, Environ. Eng. Manag. J. 14 (2015) 415-420.

[122] S. Ameen, M.S. Akhtar, H.K. Seo, H.S. Shin, Solution-processed CeO_2/TiO_2 nanocomposite as potent visible light photocatalyst for the degradation of bromophenol dye, J. Chem. Eng. 247 (2014) 193-198. https://doi.org/10.1016/j.cej.2014.02.104

[123] S.S. Lee, W. Song, M. Cho, H.L. Puppala, P. Nguyen, H. Zhu, V.L. Colvin, Antioxidant properties of cerium oxide nanocrystals as a function of nanocrystal diameter and surface coating, ACS Nano. 7 (2013) 9693-9703. https://doi.org/10.1021/nn4026806

[124] H. Li, G. Wang, F. Zhang, Y. Cai, Y. Wang, I. Djerdj, Surfactant-assisted synthesis of CeO_2 nanoparticles and their application in wastewater treatment, RSC Adv. 2 (2012) 12413-12423. https://doi.org/10.1039/c2ra21590j

[125] A.D. Liyanage, S.D. Perera, K. Tan, Y. Chabal, K.J. Balkus Jr, Synthesis, characterization, and photocatalytic activity of Y-Doped CeO_2 nanorods, ACS Catal. 4 (2014) 577-584. https://doi.org/10.1021/cs400889y

[126] R.M. Mohamed, E.S. Aazam, Synthesis and characterization of CeO_2-SiO_2 nanoparticles by microwave-assisted irradiation method for photocatalytic oxidation of methylene blue dye, Int. J. Photoenergy 2012 (2012) 928760 (1-9).

[127] Qi, K. Zhao, G. Li, Y. Gao, H. Zhao, R. Yu, Z. Tang, Multi-shelled CeO_2 hollow microspheres as superior photocatalysts for water oxidation, Nanoscale 6 (2014) 4072-4077. https://doi.org/10.1039/C3NR06822F

[128] Saranya, K.S. Ranjith, P. Saravanan, D. Mangalaraj, R.T.R. Kumar, Cobalt-doped cerium oxide nanoparticles: Enhanced photocatalytic activity under UV and visible light irradiation, Mater. Sci. Semicond. Process. 26 (2014) 218-224. https://doi.org/10.1016/j.mssp.2014.03.054

[129] T.V.M. Sreekanth, G.R. Dillip, Y.R. Lee Picrasmaquassioides mediated cerium oxide nanostructures and their post-annealing treatment on the microstructural, morphological and enhanced catalytic performance, Ceram. Int. 42 (2016) 6610-6618. https://doi.org/10.1016/j.ceramint.2015.12.171

[130] N. Zhang, S. Liu, X. Fu, Y.J. Xu, A simple strategy for fabrication of plum-pudding type Pd@CeO_2 semiconductor nanocomposite as a visible-light-driven photocatalyst for selective oxidation, J. Phys. Chem. C 115 (2011) 22901-22909. https://doi.org/10.1021/jp205821b

Chapter 9

Photocatalytic Degradation of Organic Pollutants by Zinc Oxide Composite

Deepika Jamwal[1,2], Jae Young Park[3], Amit Kumar[1], Gaurav Sharma[1], Dolly Rana[1*], Akash Katoch[4*]

[1]School of Chemistry, Faculty of Basic Sciences, Shoolini University, Solan, HP 173212, India

[2]Depatment of Chemistry, Panjab University, Chandigarh, 160014, India

[3]Surface R&D Group, Korea Institute of Industrial Technology (KITECH), 156, Gaetbeol-ro, Yeonsu-gu, Incheon 21999, South Korea

[4]Institute Instrumentation Centre, Indian Institute of Technology Roorkee, Roorkee 247667, India

*ranadolly079@gmail.com[1], akashfic@iitr.ac.in[4], katochakash@gmail.com[4]

Abstract

Photocatalysis has been proved to be a promising approach for the degradation of organic pollutants. Oxide semiconductors have shown their recognition in the area of catalysis. Particularly, oxide semiconductor based composites have shown improved photodegradation of organic pollutants by modification of their electronic and structural properties than their single counterparts. ZnO has an advantage of being non-toxic, cost-effective and environmentally friendly to become an efficient photocatalyst that can be utilized for preparing different kinds of composites. This chapter summarizes the recent progress in the fabrication and photocatalytic efficiency of various types of ZnO composites for degradation of organic pollutants.

Keywords

Catalysis, Semiconductor, Composites, Organic Pollutants, Oxides

Contents

1. Introduction

We are exposed to various organic pollutants (surfactants, pesticides, dyes, phenols, chloro compounds, nitrogen-containing compounds etc.) instigating from different industries (leather, paper, pharmaceutical and steel plant) and agricultural wastes (usage of pesticides, herbicides). These pollutants contaminate water resources and cause a hazardous effect on the human health, aquatic life as well as environment [1-5]. For the removal of various organic pollutants, various methods have been so far adopted such as physical, chemical and biological methods [6, 7]. Amongst these, photocatalytic degradation and physical adsorption are considered to be efficient because of their high efficiency, economic suitability, and easy fabrication. The processes have attracted much attention for degradation of organic pollutants [8-11].

In the past two decades, several semiconductor-based photocatalysts; ZnO, TiO_2, CuO, SnO_2, WO_3, Fe_2O_3 and Bi_2O_3 etc. have widely been preferred for degradation of organic pollutants due to their superior photocatalytic properties [12-18]. Among these materials, ZnO exhibits promising properties such as cost-effectiveness, non-toxic nature, biologically and chemically inert, photo and thermal stability as well as high quantum yield [19]. Pure ZnO based photocatalysts have a band gap (3.2 eV) at room temperature and high electron mobility (200-300 cm^{-2} $V^{-1}s^{-1}$) that promotes electron transfer efficiency [20]. To broaden the absorption range i.e. to improve the photoactivity of the catalyst, significant development and modification of metal-oxide-semiconductors are highly desirable.

The extending of bandgap corresponds to the visible region can be achieved by improving various aspects such as structural modifications, the addition of foreign element, introducing metal doping, carbon-based composites and achieving quantum confinement effect by means of controlling size up to nano level [21]. Particularly, the superior charge transport ability of semiconductor-based materials due to their better electronic structure having adequate band gap separation between valence and conduction bands compared to other class of materials is effective for photocatalytic activity. The electrons present in the valence band jumped to the conduction band and leave a hole behind when the energy of an incident photon is equal to or exceeds the band gap energy of metal-oxide semiconductors. The photogenerated electron/hole induces reduction and oxidation respectively, at the surface. The ability of metal-oxide-semiconductors mainly depends upon their ability to generate electron/hole pair under illumination of incident light photons and their recombination. However, faster recombination of electrons and holes hampers the photocatalytic efficiency by trapping the photo-induced charge carriers [22]. The most efficient way to improve the separation of charge carrier is to couple ZnO with other semiconductor-based oxide or sulphide [23,24] as well as with high mesoporous materials (zeolite, clay, alumina, silica) [25-28] and carbon-based materials (graphene and its derivatives, CNT) [29-31], Doping with metal or non-metal [32,33], fabrication of composite with polymeric materials [34], dye sensitization [35] and metal oxide with hierarchical architectures are another possible method [36]. In this chapter, we aimed to cover the different kinds of ZnO composites investigated to degrade organic pollutants in the last five years.

2. Photocatalytic ZnO/graphene composites

It has been found that the combination of semiconductor material and carbon-based materials may enhance the photocatalytic efficiency of photocatalysts because the existence of semiconductor particles on the graphene surface avoids the aggregation of graphene sheets and preserves the surface area at a high level. It provides unique pore structure, electronic properties, adsorptive capacity, and acidity. These materials include activated carbon [37], CNT [31] and graphene and its derivatives [29,30].

Graphene, a 2D material has excellent electrical, thermal, optical and mechanical properties due to its unique structure. It consists of sp^2-bonded carbon atoms that are connected in hexagonal shapes with each carbon atom covalently bonded to three other carbon atoms. Graphene not only provides reaction active sites due to the higher specific surface area but also possesses very high charge carrier mobility that can be effective to enhance the charge separation in order to improve the photocatalytic process [38]. The reduced graphene oxide also improves charge separation by initializing hetero junction induced electric files at the interface with oxide semiconductors. The difference in Fermi level between reduced graphene oxide and oxide semiconductor promptly transfer an electron to the surface of graphene and stimulates faster reduction reaction, thereby contributes towards photocatalytic efficiency by enhancing separation competence of photogenerated electron-hole pairs [39]. The π-π stacking between aromatic rings of graphene and incoming moiety also significantly accelerates their adsorption on the oxide semiconductors [40]. This section covers the recent progress and work done on organic pollutant degradation using various forms of ZnO composites prepared using graphene and graphene oxides.

2.1 ZnO/graphene composites

A number of ZnO/graphene composites used for photocatalytic degradation of azo dyes. The ZnO/graphene composites prepared via chemical precipitation method showed 2.4 times enhancement in their photocatalytic degradation efficiency towards methylene blue (MB) compared to the bare ZnO [41]. The ZnO/graphene composites processed through simple chemical precipitation method followed by the heat treatment led to the formation of multilayer graphene accompany by the loss of 2D sheet-like structure with reduced surface area, thereby lowers photocatalytic activity of synthesized composite. The graphene loaded in ZnO nanofibers were prepared by electrospinning with 0.5% of graphene showed excellent photocatalytic activity towards degradation of MB under UV irradiation. The better electron transport due to the superior electrical conductivity of graphene [42] and strong interaction between ZnO and defect sites of graphene were attributed to the improved photocatalytic activity of ZnO/graphene composites [41]. The higher adsorption of MB on graphene is through π-π conjugation between aromatic regions of graphene and their oxidation by the charge transfer from ZnO to graphene under irradiation of light. Another study proposed graphene-ZnO composite with the quasi-shell-core structure for photocatalytic degradation of rhodamine B (RhB) [43]. According to this study, the electric field established between core ZnO and shell graphene layer has been responsible to decrease charge transfer resistance to the photogenerated electrons and for improved photocatalytic dye degradation capability. The studies concluded that the optimal ratio of ZnO and graphene is necessary to achieve higher photocatalytic degradation efficiency. The graphdiyne-ZnO nanohybrids used for degradation of two azo dyes (MB and RhB) showed better results than bare ZnO [44]. The electron acceptor nature of the graphdiyne nanosheets was similar to that of graphene which was confirmed the photoluminescent analysis. In another study, photocatalytic degradation of reactive black 5 (RB5) under visible light was accomplished using solvent exfoliated graphene /ZnO photocatalysts [45] with remarkable RB5 degradation of 97.0% with a rate constant of 0.0199 min^{-1} under a low-power 15 W energy saving light bulb for 3 h. The optimum ZnO content was 69.0 wt%, which demonstrated a 10-fold photocatalytic degradation enhancement after graphene hybridization.

An organic pollutant, 4-nitrophenol was degraded using ZnO/graphene composites and the reduction of nitrobenzene to aniline with a yield of 97.8 % was achieved using graphene-ZnO-Au nanocomposites [46, 47]. The photo performance of the catalysts depends on the amount of graphene present in the composite. The contribution of graphene and Au towards the photocatalytic activity of ZnO was also studied where graphene acted as an inhibitor towards the electron/hole recombination, due to excellent electron transport properties, and the Au-assisted inefficient charge separation.

2.2 ZnO/reduced graphene oxide composites

Hierarchical ZnO and reduced graphene oxide synthesized via hydrothermal method have been used as photocatalyst for the photodegradation of azure B dye [48], and effect of reduced graphene oxide content on photodegradation of azure B dye was also investigated. It was suggested that the optimum content of reduced graphene oxide effectively improved the photocatalytic activity of the catalyst. In another case, degradation of RhB from water was investigated using similar photocatalyst [49]. The composite showed an excellent cycling performance for RhB removal up to 99% recovery over several cycles via simulated sunlight irradiation. The composite structure was further modified by coating ZnO nanospheres with reduced graphene oxide, which was established by electrostatic attraction between positively charged ZnO nanospheres and negatively charged reduced graphene oxide [49]. About 10% reduced graphene oxide exhibits a 5-fold enhancement of photoactivity compared to bare ZnO nanospheres which are attributed to the decreased recombination of the electron/hole pairs and the improved absorption capacity of ultraviolet light. The photocatalytic degradation efficiency was closely related to the fate and transfer of photogenerated electrons. The highly enhanced photoactivity could be attributed to the fact that sufficient interfacial contact between ZnO nanospheres and RGO nanosheets greatly improves the fate and transfer of photogenerated electron-hole pairs from ZnO.

In another study, Ag-Ag_2O-ZnO/graphene oxide nanocomposites were used as a photocatalyst to degrade acid blue 74 dye under visible light irradiation in synthetic wastewater containing the dye [50]. The results indicated that Ag-Ag_2O-ZnO/ graphene oxide nanocomposite has a higher photocatalytic activity (90% removal) compared to Ag-Ag_2O-ZnO (85% removal) and ZnO (75% removal). The high visible light photocatalytic performance of the Ag-Ag-Ag_2O-ZnO/ graphene oxide nanocomposite was due to: (i) the surface plasmon resonance of Ag, (ii) minimized electron-hole pair recombination by enhanced interfacial electron transport due to the *p-n* heterojunction between Ag_2O and ZnO (iii) the outstanding electron mobility of graphene which enhances the transfer of photogenerated electrons and holes from ZnO to the graphene, thus, minimizing the rate of recombination of the electron-hole pairs.

Table 1 Photocatalytic ZnO/graphene and ZnO/reduced graphene oxide-based composites used for the photodegradation of organic pollutants.

Photo-catalyst	Organic pollutants	Photocatalytic efficiency w.r.t time & light source	Content	Concentration of photocatalyst & organic pollutant	Synthesis method	Ref.
ZnO/ GCs	MB	100%, 100 min UV-visible light irradiation (xenonarc lamp 300 W)	600 mg $Zn(O_2CCH_3)_2$	50 mg/100 mL & (1×10^{-5} M)	Precipitation method	[41]
ZnO/ GNFs	MB	80%, 4 h UV light irradiation	0.5 wt.% G	0.1 & 0.1 wt.%	Electro-spinning method	[42]
ZnO/ GQSCCs	RhB	100%, 20 min UV-visible light irradiation (xenon arc lamp 300 W)	10 mg G	0.1 g/100 mL & 10 mg/L	Wet-chemical method	[43]
ZnO/ GY Cs	MB&RhB	68 & 55% UV light irradiation	0.5 % GY	5 mg/L &1×10^{-3} M	Hydrothermal method	[44]
ZnO/ SEGCs	RB-5	97.0%,3 h visible light irradiation (energy saving light bulb 15 W)	69.0 wt% ZnO	30 mg & 20 mg/L (50 ml)	Chemical deposition-calcination method	[45]
ZnO/ GCs	NB	95% UV light irradiation	weight ratio, RGO:$Zn(O_2CCH_3)_2$: 1.4 %	100 mg & 6×10^{-5} M/100 mL	Hydrothermal method	[46]
Au/ZnO /GCs	NB	97.8 %, 140 min UV light	--	4 mg/mL & 5 mM (50 mL) pH = 10	Hydrothermal method	[47]

		irradiation (UV-vis lamp 500 W)				
ZnOHC/ RGOCs	AB	~99%,20 min UV light irradiation (mercury vapour lamp 125W)	10 % RGO	0.1 g/L & 1.7×10^{-5} M	Hydrothermal method	[48]
ZnONS/ RGOCs	RhB	100%, 2 h ~99%, 20 min UV light irradiation (xenon arc lamp 300 W)	10% RGO	20 mg/ 80 mL & 10 ppm	Hydrothermal reduction	[49]
Ag-Ag_2O/ ZnO/ GCs	AB-74	90%, 210 min. visible light irradiation (ozone free xenon lamp150 W)	--	0.1 g/100 mL & 20 mg/L	Co-precipitation method	[50]
ZnO/ GCs	MB	100%, 90 min visible light irradiation (xenon arc lamp 150 W)	--	30 mg & 20 mg/L (30 mL)	Solvothermal method	[51]
ZnO/ RGOCs	RhB	100%, 60 min visible light irradiation (metal halogen lamp)	--	4.0 mg/L & 6 ppm (100 mL)	Photoreduction method	[52]
ZnO/ RGOCs	CV	~95%, 80 min Visible light irradiation (xenon arc lamp)	--	--	Chemical method	[53]
ZnOQD/ /GCs	MO	80%, 30 min UV-visible light irradiation (high-pressure mercury lamp 100 W) & (metal halide lamp 175 W) pH = 6.5	5 % G	50 mg/100 mL & 1.5×10^{-4} M	Chemical method	[54]

ZnO/ 3D-GCs	MO	92%, 2 h UV light irradiation	--	120 mg/20 mL & 5 mM (20 mL)	Chemical vapor deposition & hydrothermal method	[55]
ZnO/ RuO_2/ RGOCs	MB	100%, 60 min simulated sunlight irradiation	--	100 mg/100 mL & 10 mg/L	Hydrothermal method & precipitation method	[56]
ZnO/ TiO_2/ RGOCs	MB	92%, 120 min visible light irradiation (300 W xenon lamp)	0.5wt% RGO	0.1 g/L & 0.3 mg/L (100 mL)	Solvothermal method	[57]
Ag/ZnO /RGOCs	RhB	90%, 10 min UV-visible light irradiation (UV lamp 40 W) & (xenon lamp 350 W)	--	20 mg & 10^{-5} M (100 mL)	Solution-based synthetic method	[58]
Ag/ZnO /GCs	MB, XT, RhB&MO	visible light irradiation (metal halogen lamp 400 W)	--	30 mg/ 30 mL & 20 mg/L	Solvothermal method	[59]

Acid blue-74=AB-74,Azure-B=AB, Crystal Violet=CR, Rhodamine-B=RhB, Methyl orange=MO,MB=Methylene blue, NB=Nitrobenzene, Reactive black-5=RB-5,Xanthene=XT, Cs=Composites or nanocomposites,G=Graphene, RGO=Reduced graphene oxide, NFs= Nanofibres, Quasi-Shell-Core=QSC, Graphdiyne=GY, Solvent exfoliated graphene=SEG, Quantum dots=QD, Nanohybrid=NHs, Hierarchical=HC, Nanospheres=NS.

3. Photocatalytic ZnO/carbon nanotube based semiconductor composites

Carbon nanotubes (CNT), another class in the carbon family has shown their recognition and promising aspects in the photocatalysis application because of their good catalytic activity on verge of their large surface area, chemical stability, high mechanical strength and extraordinary electronic properties [60]. The mesoporous character of CNT provides channels or supports for diffusion of reactive species for the photocatalytic activity [61]. In recent years, a lot of publications have been reported on CNT-ZnO composites for photocatalytic degradation of various organic pollutants. ZnO/CNTs hierarchical microsphere composites synthesized via a facile chemical deposition route were used for photodegradation of methylene blue [62]. ZnO/CNTs hierarchical microsphere composites with 1.1 wt% CNTs showed the optimized photocatalytic activity and decrease for higher concentrations of CNT. The two major reasons were proposed for better photocatalysis; (1) light harvesting competition between CNTs and ZnO appears with an increase of CNTs amount (2) the excessive CNTs can act as a kind of recombination center instead of providing an electron pathway.

In another study, 1-D ZnO nanowire/multi-walled CNT composites synthesized by one-step hydrothermal method were used for photocatalytic degradation performances on methylene blue and rhodamine 6 Gunder UV irradiation. These materials showed 3 times higher photocatalytic efficiency than that of pure ZnO nanowires. Furthermore, when ZnO nanoparticles decorated on multi-walled CNT were used for photocatalytic degradation of RhB and methyl orange (MO), showed better photocatalytic activity than bare ZnO [63, 64]. The enhanced photocatalytic activity was attributed to the fast transfer of photogenerated electrons from ZnO to CNTs, leading to low recombination rate of photoinduced electron-hole pairs.

Recently, few publications have appeared on metal-doped ZnO/ multi-walled carbon nanotubes (MWCNTs) nanocomposites. Ahmad et al., [63] successfully synthesized Cu-doped ZnO nanoparticles embedded on MWCNTs via non-toxic sol method which exhibited extended light absorption in visible light region and possessed better charge separation capability, respectively as compared to Cu-doped ZnO, pure ZnO, and ZnO/CNTs composite. The photocatalytic activity was tested by degradation of MO dye under visible light irradiation. In another study, multi-walled CNT/ metal-doped ZnO nanohybrid materials (metal: Mn, Mg, and Co) were developed, where, the MWNT/Mg-doped ZnO hybrids showed the high photocatalytic ability for MO, in which the degraded rate for MO reached 100% in 60 min. It was concluded that the enhancement in photocatalytic activity was because of the superb electrical property of MWNTs and the formation of many defects in the ZnO crystals by Mg-doping. The synergic effect between MWCNT and ZnO hinders the recombination of electron-hole pairs, whereas the doped Mg replaces Zn site or join interstitial site in the ZnO lattice and generates intermediate energy gap between valence band and conduction band of ZnO that likely to promote the absorption of photons by generating the large number of electron-hole pairs, thereby improved overall photoactivity [65].

Table 2 ZnO/carbon nanotub based composites used for the photodegradation of organic pollutants.

Photo-catalyst	Organic pollutants	Photocatalytic efficiency w.r.t time & light source	Content	Concentration of photocatalyst & organic pollutant	Synthesis method	Ref.
ZnO/CNTs HMSCs	MB	92.3%, 60 min UV-visible light irradiation (mercury lamp 500 W) & (xenon lamp 500 W)	1.1 wt% CNTs	0.17g/L & 30 mg/L	Chemical deposition method	[62]
ZnO/MCNTs	Rh B	~61%, 30 min sunlight irradiation (ultrasound natural and low power)	20 wt.% MCNTs	30 mg/ 30 mL & 20 mg/L	Sol method	[63]
ZnO/MCNTs	MO	94.12%, 30 min UV light irradiation (high pressure mercury lamp 150 W)	--	0.5% wt & 10 mg/L (70 mL)	--	[64]
Mn, Mg, Co/ZnO/MCNTs NH	MO	100 %, 60 min UV light irradiation (highpressure osram ultra-vitalux lamp 100 W)	--	0.25 g/L & 20 mg/L	Co-deposition method	[65]
ZnONW/MCNTs Cs	MB& Rh6G	100%, 30 min UV light irradiation	10 wt.% MCNTs	10 mg &10 mg/L (50 mL)	Hydrothermal method	[66]
ZnO/MCNTsCs	RhB	80%, 60 min UV light irradiation (high-pressure mercury lamp 300 W)	--	50 mg & 1.0×10^{-5} mol/L (50 mL)	Ammonia method	[67]
Cu/ZnO/MCNTs	MO	120 min visible light irradiation (metal halogen lamp 400 W)	Cu: 3 wt.%, CNTs: 10 wt.%	30 mg & 20 ppm (30 mL)	Sol method	[68]

Rhodamine-B=RhB, Rhodamine-6G=Rh6GMethyl orange=MO, MB=Methylene blue, Cs=Composites or nanocomposites, Carbon nanotubes=CNTs, Multiwalled carbon nanotubes=MCNTs, Nanohybrid=NH, Hierarchical microsphere=HMS, Nanowires=NW.

4. Photocatalytic ZnO-silica, alumina, zeolite and clay based composites

A number of attempts have also been made to resist the recombination of electron-hole pair counts which is a negative process and responsible for the lower photocatalytic efficiency. Mesoporous materials (zeolite, clay, alumina and silica etc.) and their modified forms used as support for semiconductors and coupling of two or more semiconductors are the well-known strategies that have been used to decrease the recombination process [25-28]. These mesoporous materials are eco-friendly in nature, thermally stable and cost-effective. For example, zeolites with uniform nanoscale pore size, superior cation exchange capacity, high adsorption capacity and extensive surface area have been applied in the designing of effective photocatalytic systems [26,69,70]. Moreover, they have the capacity to extending self life and reusability of the photocatalysts. Due to high adsorption capacity, they carry molecules of organic pollutants near the catalyst surface, where hydroxyl and superoxide radicals generated, and for this reason, photodegradation efficiency tends to increase [71].

Moreover, porous nature is significant in terms of obtaining less agglomerated semiconductor material, thereby helping to improve the surface area or active sites required to enhance photocatalytic degradation of organic pollutants. On the other hand, zeolites tend to produce a strong electric field, effectively distribute photogenerated electrons in the zeolitic network are supportive for reducing electron-hole pair recombination [72].

ZnO/mordenite zeolite nanoparticles were prepared via precipitation process using ultrasonic irradiation and hydrothermal methods [73]. Mordenite, a zeolite, is a mineral with an ideal composition (Na_8) $[Al_8Si_{40}O_{96}].nH_2O$. Due to high thermal and catalytic activity, mordenite actively performs photocatalytic oxidation of AB92 under UV irradiation. The photocatalytic efficiencies of ZnO/mordenite with ZnO, ZnO/activated carbon and ZnO/Al_2O_3 nanocomposites were compared and results revealed that the insertion of ZnO on different supports results in the decrease of the particle size, increase in the surface area and reduction in electron-hole recombination rate in comparison to pure ZnO. Photocatalytic activity of catalysts was investigated by degradation of AB92 in water. Mass ratios 0.25:1 of ZnO and mordenite showed the highest photocatalytic activity [73].

The photodegradation of trichloroethylene (chlorinated hydrocarbons) was performed using heterogeneous nano-ZnO/laponite composites. A direct comparison of the photocatalytic degradation rates of trichloroethylene between TiO_2 powder and ZnO/laponite composites balls suggested ZnO/laponite composites as a valid alternative to suspended TiO_2 powder for mineralizing trichloroethylene [74]. In another study, metaldehyde (molluscicide) was removed through photocatalytic reactions using nano-sized ZnO/laponite composite. Statistical results showed that linear effects of time, quadratic/linear effects of nano-ZnO/laponite composites mass and the interaction of pH and nano-ZnO/laponite composites mass proved to be the most significant variables for degrading metaldehyde [75].

The degradation of humic acid was investigated using nano-sized ZnO/laponite composite. The photocatalytic degradation of humic acid was found to depend on pH and catalyst concentration. Experimental results showed that the photocatalytic efficiency was higher in an acidic medium which was ascribed to the protonation of functional groups at low pH and the creation of the less negatively charged humic acid reduces the intermolecular repulsion between the humic acid and the catalyst surfaces. Humic acid decomposition increased as the dose of catalyst increased up to an optimum value of 20 mg/L. Moreover, the effect of H_2O_2 was also studied through homogeneous and heterogeneous photocatalysis, in which heterogeneous photocatalysis showed better removal because of the combined effects of both the catalyst and added H_2O_2 [76].

A remarkable performance regarding photodecolorization of methylene blue (97.6%), malachite green (89.5%), congo red (85.6%), and MO (80.1%) by an electrogenerated ZrO_2 and ZnO supported aluminosilicate protonated Y zeolite prepared by the facile synthesis has been reported for the first time. In addition, the catalyst remained stable even after five cycling runs and the leaching test showed negligible leaching effects [77].

Rectorite is a kind of layered silicate, the structure and characteristics are much similar to those of montmorillonite. It is referred to as an interstratified clay mineral which can adsorb organic substance both on its external surfaces and within its interlaminar spaces by interaction or substitution. For these reasons, many efforts have been focused on the use of rectorite-based materials to adsorb or catalyze organic pollutants. The nanometer ZnO/rectorite composites prepared by hydrothermal intercalation method have been subjected to examine their photocatalytic activity for photocatalytic decolorization of methylene blue under simulated sunlight irradiation. The suggested catalyst is stable low cost and reusable which requires sunlight only and does not need the addition of H_2O_2 [78].

The photocatalytic activity of ZnO/CuO supported on clinoptilolite nanoparticles regarding degradation of benzophenone has been reported. Experimental results illustrated that both supporting and coupling of two semiconductors played important role in enhancing their photocatalytic activity with respect to monocomponent catalysts and unsupported one. Also, the types of semiconductors in the coupled system affect the photocatalytic activity; therefore the supported ZnO-CuO had better activity than the ZnS-CuS/clinoptilolite nanoparticles [79].

Table 3 Silica, alumina, zeolite, and clay based ZnO composites used for the photodegradation of organic pollutants.

Photo-catalyst	Organic pollutants	Photocatalytic efficiency w.r.t time & light source	Content	Concentration of photocatalyst & organic pollutant	Synthesis method	Ref.
ZnO/ MDT	AB-92	88%, 2 h UV light irradiation (pyrex reactor equipped UV lamp)	Mass ratio, 0.25:1 ZnO :MDT	100 ppm & 20 ppm	Precipitation method&hy drothermal method	[73]
ZnO/ LPTCs	TCE	100%, 60 min UV light irradiation (photocatalytic reactor, PVC plastic, 4 UV-C lamps 8 W)	--	20 g/L &10 mg/L	--	[74]
ZnO/ LPTCs	MA	UV light irradiation (photocatalytic reactor, PVC plastic, 4 UV-C lamps 8 W)	--	28 g & 278 $\mu g/dm^3$	---	[75]
ZnO/ ZrO/ ASPY	MB	97.6%, 1 h visible light irradiation	1 wt.% ZrO_2–1 wt.% ZnO/ ASPY	0.12 g & 10 mg/L (200 mL)	Electro-chemical method	[77]
ZnO/ RRTCs	MB	99%, 2 h simulated solar light	weight ratio, ZnO: RTT 1:3	0.9 g/L & 15 mg/L	Hydrotherm al intercalation method	[78]
ZnO/ CuO/ CPTCs	BP	UV light irradiation (medium-pressure mercury-vapor lamp 35 W)	weight % ratio, ZnO:C uO 0.80:3. 18	0.12 g/L & 30 mg/L	Calcination method	[79]

ZnO/ CPT	PH	69%, 3 h UV light irradation (photocatalyst reactor, medium pressure mercury lamp 75 W)	---	0.25 g/L & 20 ppm	Calcination method	[80]
ZnO/ CPT	4-NP	UV light irradiation (photocatalyst reactor, medium pressure mercury lamp 75 W)		0.25 g/L & 10 ppm (20 mL)	Calcination method	[81]
Pd/ ZnO/ KLT	MB	98% , 60 min UV light irradiation (High-pressure mercury lamp 30W)	--	100 mg & 5 mg/L (250 mL)	---	[82]
ZnO/ MMT	MB	97% ,150 min (photocatalyst reactor, UV-B lamp 20 W)	5 wt.% MMT	5 g/L & 10 ppm	Sol-gel intercalation method	[83]

Acid blue-92=AB-92, Trichloroethylene=TCE, Metaldehyde=MA, Benzophenone=BP, Phenylhydrazine=PH, 4-Nitrophenol=4-NP, Montmorillonite=MMT, Kaolinite=KLT, Clinoptilolite=CPT, Rectorite=RTT, Aluminosilicate protonated Y zeolite=ASPY, Laponite=LPT, Mordenite=MDT, Cs=Composites or nanocomposites.

5. Photocatalytic ZnO-luminescent agent based composites

In order to overcome the limited availability of ultraviolet light which is only 4% of solar light, researchers have doped upconversion luminescence agents such as Er^{3+}:$Y_3Al_5O_{12}$, Er^{3+}:$YAlO_3$, $NaYF_4$:Yb, $NaYF_4$:Tm, YF_3:Yb^{3+} and YF_3:Tm^{3+}etc. with semiconductors to enhance their photocatalytic efficiency [84-88]. These materials have the capability to transforming visible light and near-infrared (NIR) light to the ultraviolet light. Under solar light irradiation, the upconversion luminescence process can take place readily for the continuous irradiation under desired wavelengths. In that case, the luminescence agents can create full use of solar energy and supply the rich ultraviolet light for satisfying the actual requirement of semiconductor photocatalyst [88].

Comparative studies of photocatalytic activities were performed between Er^{3+}:$Y_3Al_5O_{12}$/TiO_2-ZnO and Er^{3+}:$Y_3Al_5O_{12}$/TiO_2 under solar light irradiation. Results revealed that Er^{3+}:$Y_3Al_5O_{12}$/TiO_2-ZnO have superior photocatalytic activities, the degree of degradation of azo fuchsine dyes was 96.65% within 60 min after irradiation [89]. *β*-

$NaYF_4:Yb^{3+}$ and Tm^{3+}/ZnO composites were prepared via high-temperature thermolysis method and their NIR photocatalytic behavior was studied for the degradation of RhB. β-$NaYF_4:Yb^{3+}$, Tm^{3+} materials a luminescence agents have potential to upconversion of NIR-to-UV which provided UV light to the ZnO semiconductor. Comparative studies of energy transfer in the composites and the mixtures of β-$NaYF_4:Yb^{3+}$, β-$NaYF_4:Tm^{3+}$ prepared with ZnO revealed that the β-$NaYF_4:Yb^{3+}$, Tm^{3+}/ZnOcomposites have higher photocatalytic activity than the mixture of β-$NaYF_4:Yb^{3+}$,Tm^{3+} and ZnO. The activation of ZnO by energy transfer from upconverting β-$NaYF_4:Yb^{3+}$, Tm^{3+} to ZnO was attributed to the NIR photocatalysis. Moreover, the energy transfer processes determine the overall photoactivity and were found different from that of the physical mixture of β-$NaYF_4:Yb^{3+}$,Tm^{3+} and ZnO nanocomposites [90].

Interestingly, an attempt was made to utilize ultrasonic source conversion to UV light by using $Er^{3+}:YAlO_3$ having the capability to absorb the visible light from the ultrasonic source. The sonocatalytic activity of $Er^{3+}:YAlO_3/TiO_2$-ZnO composite was investigated to degrade acid red B dye by means of transfer of electron and hole between conduction and valence bands of ZnO and TiO_2, respectively, which counts slow electron/hole recombination and increases the sonocatalytic activity of $Er^{3+}:YAlO_3/TiO_2$-ZnO composite [91]. Another efficient material systems that can convert the visible solar light into ultraviolet light are $Er^{3+}:YAlO_3$ coated Co and Fe-doped TiO_2-ZnO films prepared via sol-gel process have been tested for degradation of azo fuchsine. Moreover, the photocatalytic efficiency of $Er^{3+}:YAlO_3$/Co-doped TiO_2-ZnO-doped was superior to that of $Er^{3+}:YAlO_3$/Fe-doped TiO_2-ZnO [92].

Table 4 ZnO/ Luminescent based composites used for the photodegradation of organic pollutants.

Photo-catalyst	Organic pollutants	Photocatalytic efficiency w.r.t time & light source	Content	Concentration of photocatalyst & organic pollutant	Synthesis method	Ref.
$Er^{3+}:Y_3Al_5O_{12}/TiO_2/ZnO$Cs	AF	96.65 %, 60 min solar light irradation	10 wt % $Er^{3+}:Y_3Al_5O_{12}$	1.0 g/L & 10 mg/L (100 mL)	Nitrate-citrate acid & heat-treated method	[89]
β-$NaYF_4$: Yb^{3+},Tm^{3+}/ZnOCs	RhB	65%, 30 min Near infrared irradiation	β-$NaYF_4$:20 % Yb,Tm:0.5%	0.5 mg & 20 mg/L (0.5 mL)	High temperature thermolysis method	[90]
$Er^{3+}:YAlO_3/TiO_2$/ZnOCs	ARB	76.54%, 150 min ultrasonic irradiation	--	1.0 g/L & 10 mg/L (100 mL)	Sol-gel & auto-combustion method	[91]
$Er^{3+}:YAlO_3$Fe, Co/TiO_2/ZnOCs	AF	93.5 & 95.53 %, 60 min visible light irradiation (Xenon lamp 300W)	15 wt.% $Er^{3+}:YAlO_3$/Fe & Co TiO_2/ZnO	10 mg/L & 1000 mg/L (100 mL)	Nitrate-citric acid method	[92]
$Er^{3+}:Y_3Al_5O_{12}$/ZnOCs	AF	84.27%, 120 min solar light irradiation	12 wt % $Er^{3+}:Y_3Al_5O_{12}$	1.0 g/L & 10 mg/L	Nitrate-citric acid method	[93]
Er^{3+}: $YAlO_3$/ Co,Fe ZnOCs	AF	89.8 and 99.2%, 80 min solar light irradiation	15.0 & 25.0 wt.% $Er^{3+}:YAlO_3$	1000 mg/L & 10 mg/L	Sol-gel method	[94]

Azo fuchsine=AF, Acid red-B=ARB, Rhodamine-B=RhB, Cs=Composites or nanocomposites.

6. Photocatalytic ZnO-transition metals, rare earth metals, and non-metals based composites

6.1 Transition metals-doped ZnO composites

Doping of transition metals including noble metals (such as, Mn, Fe, Cu, Cd, Pt, Hf, Ag and Au) in semiconductors is an effective approach to enhance the visible light absorption by generating energy states within the bandgap, which act as intermediate steps for electrons in their transitions between the valence and conduction bands due to photoexcitation [95-102]. The capability of a metal to trap photogenerated electrons within semiconductor during photochemical reactions allows the holes to interact with radical or species bound to the surface and promotes degradation of organic pollutants [103]. Photoactive surfaces such as oxygen vacancies, crystal interfaces and higher light scattering of metal are responsible for the enhancement of the photocatalytic activity of semiconductors [104,105]. In addition, the metal-doped semiconductors experience local electrical fields that enhance the surface plasmon activity, indeed is significant for photocatalytic applications [106].

Photocatalytic activities of the Ag-doped ZnO nanoparticles and undoped ZnO prepared through precipitation method were studied and compared for degradation of MO. The results revealed that Ag-doped ZnO nanoparticles containing 1 and 8% Ag content showed enhanced photocatalytic activity than the undoped ZnO nanoparticles and the time for the complete decomposition decreased to only 60 min [107]. Multivalent Cu-doped ZnO nanoparticles synthesized by green-chemical approaches with a yield of ~100% showed superior photocatalytic activity for degradation of indigo carmine, orange G, rhodamine 6G and MB [108]. On the other hand, 94% of acid red 1 dye degraded (at optimized conditions) using Fe-doped ZnO as a photocatalyst which was prepared by solution combustion synthesis method. Various parameters such as the amount of Fe-doped, calcination temperature, calcination time, pH of the dye solution, catalyst dose, hydrogen peroxide concentration, reaction temperature and initial concentration of the dye solution were optimized for obtaining superior photocatalytic activity [109]. In another study, the Au-ZnO nanopyramids with controlled morphologies exhibited admirable photocatalytic activity for the photodegradation of RhB which was associated with the exposed crystal facets, crystallinity, and the creation of hybrid nanostructure.The most exposed polar {001} facts better crystallinity as well as the formation of a hybrid structure with the well-defined interface was ascribed for better photocatalytic activities [110].

Table 5 ZnO/ transition metal-doped composites used for the photodegradation of organic pollutants.

Photo-catalyst	Organic pollutants	Photocatalytic efficiency w.r.t time & light source	Content	Concentration of photocatalyst & organic pollutant	Synthesis method	Ref.
Au/ ZnOCs	MB	99% , 30 min UV-visible light irradation (UV lamp & tungsten lamp 60 W)	--	50 mg/16 mL & $5x10^{-5}$ M (4.0 mL)	Surfactant mediated method	[102]
Ag/Zn OCs	MO	100%, 60 min UV light irridation (UV lamp 100 W)	1 and 8 % Ag	50 mg & 50 mg /50 mL	Precipitatio n method	[107]
Cu/Zn OCs	IC, O-G, RhB-6G &MB	UV-visible light irradiation (high-pressure mercury vapor lamp 125 W & metal halide lamp 400 W)	1.5 M % Cu:Zn O	0.6 g/L & 7.8×10^{-2}, 5.3×10^{-2}, 5.2×10^{-2} & 5.5×10^{-2} mM	Green-chemical method	[108]
Fe/ZnO Cs	AR-1	94%,180 min UV light irradation (UV bulb 125 W)	2.5 wt.% Fe/ZnO	1.25 g/L & 50 mg/L (250 mL)	Solution combustion synthesis method	[109]
Au/Zn OCs	RhB	~100 %, 30 min UV light irridation (UV lamp 175 W)	---	10 mg & 1 mg/L (100 mL)	--	[110]
Mn/ ZnOCs	O-II	100%, 210 min Solar light irradiation (Sylvania LuxLine FHO T5 neon tubes)	3% ZnO:M n	60 mg & 10 mg/L (30 mL)	Solvo-thermal method	[111]
Fe/ZnO Cs	MO& MB	2 h UV light irradiation (Two UV lamps 20 W)	Fe conc. 21 %	0.6 g/L & 20 mg/100 mL	co-precipitatio n method	[112]
Fe/ZnO Cs	RhB	89%,180 min UV-visible light (high-pressure mercury lamp 500 W&Xenon lamp500 W)	2% Fe	20 mg &10 mg/L (20 mL)	Hydro-thermal process	[113]
Cd/Zn OCs	MB&M O	65& 45 %, 300 min	3 % Cd/ ZnO	(0.5 g/L) &10^{-5} M (100 mL)	Sono-chemical method	[114]

Ag/ZnOCs	MO, MB&4-NP	~100 %, 5, 4 & 6 h visible light irradiation	--	2 mg & 10 mg/L (20 mL), 10 mg/L (20 mL), 5 mg/L (20 mL)	--	[115]
Ag/ZnO/CdOCs	AB-1 &AV-7	~100 %, 45 & 50 min Sunlight irradation	--	3×10^{-4} M, 5×10^{-4} M & 3 g/L	Co-precipitation method	[116]
Pt/ZnOCs	AO-II	86%,5 h UV light irradiation (UV lamp)	1% Pt/ZnO	0.05 g/100 mL & 0.020 g/L	Solvo-thermal method	[117]
Hf/ZnOCs	MB	85%, 120 min Sunlight irradation	2M % Hf/ZnO	25 mg/100 mL & 250 mg/L	Sol-gel method	[100]

Acid black-1=AB-1, Acid violet 7=AV-7, Acid orange II=AO-II, 4-Nitrophenol=4-NP, Acid red 1=AR-1, Indigo carmine=IC, Orange-G=O-G, Orange-II=O-II, Rhodamine-B=RhB, Rhodamine-6G=RhB-6G, Cs=Composites or nanocomposites.

6.2 Rare earth metals-doped ZnO composites

Many researchers have proposed that rare earth metals are efficient dopant for ZnO nanoparticles for photocatalytic activity because of their luminescence properties delay time, and band edge emission with respect to particle size variation. The rare earth metals (Sm, Gd, Dy)-doped ZnO nanoparticles and their photocatalytic degradation efficiencies for acid blue 92, acid orange 7 and acid red 17 have been investigated [118-120]. In another study, La-doped ZnO nanorods synthesized by using different La contents via a microwave-assisted method showed that degradation efficiency of metasystox (pesticide) over La-doped ZnO increases up to 0.5 mol% doping then decreases for higher doping levels [121]. Whereas, Nd-doped ZnO showed highest photocatalytic activity for degradation of phenol with the optimal Nd content was 2.0% under visible light irradiation [122].

Table 6 ZnO/rare earth metal-doped composites used for the photodegradation of organic pollutants.

Photo-catalyst	Organic pollutants	Photocatalytic Efficiency w.r.t time & light source	Content	Concentration of photocatalyst & organic pollutant	Synthesis method	Ref.
Sm/ZnO Cs	AB-92	90.10%, 150 min ultrasonic irradiation	6% Sm	1 g/L & 10 mg/L	Sonochemical method	[118]
Gd/ZnO Cs	AO-7	90%, 90 min ultrasonic irradiation	5% Gd	100 mg & 5 mg/L (100 mL)	Sonochemical method	[119]
Dy/ZnO Cs	AR-17	100%, 180 min ultrasonic irradiation	3% Dy	1 g/L & 5 mg/L	Sonochemical method	[120]
La/ZnO Cs	M	90 %, 150 min UV and sunlight irradation (high pressure mercury lamp)	0.5 mol % La	100 mg & 1.4×10^{-3} M (100 mL)	Microwave assisted method	[121]
Nd/ZnO Cs	P	visible light irradiation	2.0%	100 mg & 20 mg/L (100 mL)	Chemical precipitation method	[122]
Ce/ZnO Cs	MO	100% , 80 min UV-visible light irradiation	0.2 M Ce	10 ppm MO	Hydrothermal process	[123]

Acid blue-92=AB-92,Acid orange-7=AO-7, Acid red-17=AR-17, Methyl orange=MO, Metasystox=M, Phenol=P, Cs=Composites or nanocomposites.

6.3 Non-metals-doped ZnO composites

Doping with nonmetallic elements such as C, S, and N has also been considered to reduce the band gap of semiconductors [124]. The photocatalytic activities of C-ZnO nanocrystals obtained with different morphologies and carbon contents were evaluated for degradation of methylene blue under visible light irradiation [125]. Results showed that incorporation of carbon decreases the energy band gap of ZnO, improves the separation efficiency of its electron-hole pairs, and considerably enhances the visible light photocatalytic activity. On the other hand, nitrogen has been widely used as a dopant to modify the electronic structures of oxide semiconductors due to its comparable size to oxygen and small formation energy required for the substitution of O. Therefore, N doping has been applied to ZnO with the expectation of improving the light absorption property of ZnO in the visible region. In another study, N-doped mesoporous ZnO nanospheres obtained via solvothermal method were found to exhibit higher photocatalytic activity for degradation of RhB compared to pure ZnO nanoparticles [126].

Recently, coupling of nonmetal-doped semiconductor and soft polymeric semiconductor (graphitic carbon nitride g-C_3N_4) materials and the generation of several kinds of g-C_3N_4 composite photocatalysts (semiconductor/g-C_3N_4 coupled with graphene oxide or CNTs etc.) have attracted a great deal of scientific interest in photocatalytic water splitting [127,128] and degradation of environmental pollutants under visible light irradiation due to their suitable band gap to absorb the visible-light and unique properties. This soft polymeric semiconductor can be easily coated on the surface of other materials in comparison to inorganic π-conjugative materials such as CNTs and graphenes [129]. Moreover, it has a well crystallized lamellar structure that assists improved charge transfer than organic π-conjugative materials. LaterN-doped ZnO/g-C_3N_4 hybrid core-shell nanoplates were utilized for RhB degradation. The direct Z-scheme mechanism is proposed for the prolonged lifetime of charge carrier transfer at the interface of N-doped ZnO/g-C_3N_4 by trapping photoinduced electrons and holes under visible-light irradiation [130].

7. Conclusion

This chapter summarizes the potential aspects of ZnO based composites for degradation of various organic pollutants. The potential feature of different kinds of ZnO composites including carbon-based composites (graphene, reduced graphene oxides, CNT and MWCNT), mesoporous materials (silica, alumina, zeolite, and clay), luminescent agents, transition metals and rare earth metals and non-metals have been discussed thoroughly. It has been revealed that the combination of various materials with ZnO semiconductor improved overall photocatalytic efficiency of ZnO based composites by means of various aspects: bandgap modification, enhance charge transfer, reduction in the recombination of photogenerated electrons and holes, efficient use of sunlight and availability of active sites for the adsorption of organic pollutants by providing a high surface area.

References

[1] I.A. Balcioglu, Y. Inel, Photocatalytic degradation of organic contaminants in semiconductor suspensions added H_2O_2, J. Environ. Sci. Health., Part A 31(1996) 123-138.

[2] C.C. Tsao, J.E. Campbell, M. Mena-Carrasco, S.N. Spak, G.R. Carmichael Y. Chen, Increased estimates of air-pollution emissions from Brazilian sugar-cane ethanol, Nat. Clim. Change 2 (2011) 53-57. https://doi.org/10.1038/nclimate1325

[3] S. Ghasemi, S. Rahimnejad, S.R. Setayesh, S. Rohani, M.R. Gholami, Transition metal ions effect on the properties and photocatalytic activity of nanocrystalline TiO_2 prepared in an ionic liquid, J. Hazard. Mater. 172 (2009)1573-1578. https://doi.org/10.1016/j.jhazmat.2009.08.029

[4] S.P. Kamble, S.P. Deosarkar, S.B. Sawant, J.A. Moulijn, V.G. Pangarkar, Photocatalytic degradation of 2,4-dichlorophenoxyacetic acid using concentrated solar radiation: Batch and continuous operation, Ind. Eng. Chem. Res. 43 (2004) 8178-8187. https://doi.org/10.1021/ie0494263

[5] S. Bull, K. Fletcher, A. Boobis, J. Batterrshill, Evidence for genotoxicity of pesticides in pesticide applicators, Mutagenesis 21 (2006) 93-103. https://doi.org/10.1093/mutage/gel011

[6] K.C. Kemp, H. Seema, M. Saleh, N.H. Le, K. Mahesh, V. Chandra, K.S. Kim, Environmental applications using graphene composites: Water remediation and gas adsorption, Nanoscale 5 (2013) 3149-3171. https://doi.org/10.1039/c3nr33708a

[7] H. Nakamura, Recent organic pollution and its biosensing methods, Anal. Methods 2 (2010) 430-444. https://doi.org/10.1039/b9ay00315k

[8] N.R. Khalid, E. Ahmed, Zhanglian Hong, M. Ahmad, Synthesis and photocatalytic properties of visible light responsive La/TiO_2-graphene composites, Appl. Surf. Sci. 263 (2012) 254-259. https://doi.org/10.1016/j.apsusc.2012.09.039

[9] S. Wang, H. Sun, H.A Ng, M. Tade, Adsorptive remediation of environmental pollutants using novel graphene-based nanomaterials, Chem. Eng. J. 226 (2013) 336-347. https://doi.org/10.1016/j.cej.2013.04.070

[10] Z. Liu, W. Xu, J. Fang, X. Xu, S. Wu, X. Zhu, Z. Chen, Decoration of BiOI quantum size nanoparticles with reduced graphene oxide in enhanced visible-light-driven photocatalytic studies, Appl. Surf. Sci. 259 (2012) 441-447. https://doi.org/10.1016/j.apsusc.2012.07.063

[11] T. Thomas, N. Kottam, Combining "chimie douce" and green principles for the developing world: Improving industrial viability of photocatalytic water remediation, Chem. Eng. Sci. 102 (2013) 283-288. https://doi.org/10.1016/j.ces.2013.08.004

[12] G. Sharma, S. Bhogal, M. Naushad, Inamuddin, A. Kumar, F.J. Stadler, Microwave assisted fabrication of La/Cu/Zr/carbon dots trimetallic nanocomposites with their adsorptional vs photocatalytic efficiency for remediation of persistent organic pollutants, J. Photochem. Photobiol. A Chem. 347 (2017) 235–243. https://doi.org/10.1016/j.jphotochem.2017.07.001

[13] A. Kumar, M. Naushad, A. Rana, Inamuddin, Preeti, G. Sharma, A.A. Ghfar, F.J. Stadler, M.R. Khan, ZnSe-WO3 nano-hetero-assembly stacked on Gum ghatti for photo-degradative removal of Bisphenol A: Symbiose of adsorption and photocatalysis, Int. J. Biol. Macromol. 104 (2017) 1172–1184. https://doi.org/10.1016/j.ijbiomac.2017.06.116

[14] X.P. Gao, J.L. Bao, G.L. Pan, H.Y. Zhu, P.X. Huang, F. Wu,D. Y. Song, Preparation and Electrochemical Performance of Polycrystalline and Single Crystalline CuO Nanorods as Anode Materials for Li-Ion Battery, J. Phys. Chem. B 108 (2004) 5547-5551. https://doi.org/10.1021/jp037075k

[15] S.P. Kim, M.Y. Choi, H.C. Choi, Photocatalytic activity of SnO_2 nanoparticles in methylene blue degradation, Mater. Res. Bull. 74 (2016) 85-89. https://doi.org/10.1016/j.materresbull.2015.10.024

[16] C.A. Bignozzi, S. Caramori, V. Cristino, R. Argazzi, L. Meda, A. Tacca, Nanostructured photoelectrodes based on WO_3: applications to photooxidation of aqueous electrolytes, Chem. Soc. Rev. 42 (2013) 2228-2246. https://doi.org/10.1039/C2CS35373C

[17] M. Barroso, S.R. Pendlebury, A.J. Cowan, J.R. Durrant, Charge carrier trapping, recombination and transfer in hematite (α-Fe_2O_3) water splitting photoanodes, Chem. Sci. 4 (2013) 2724-2734. https://doi.org/10.1039/c3sc50496d

[18] Q. Yang, C. Hu, S. Wang, Y. Xi, K. Zhang, Tunable synthesis and thermoelectric property of Bi_2S_3 nanowires, J. Phys. Chem. C 117 (2013) 5515-5520. https://doi.org/10.1021/jp307742s

[19] R.M. Asmussen, M. Tian, A. Chen, A new approach to wastewater remediation based on bifunctional electrodes, Environ. Sci. Technol. 43 (2009) 5100-5105. https://doi.org/10.1021/es900582m

[20] R. Nagaraja, N. Kottam, C.R. Girija, B. M. Nagabhushana, Photocatalytic degradation of Rhodamine B dye under UV/solar light using ZnO nanopowder synthesized by solution combustion route. Powder Technol. 215-216 (2012) 91-97. https://doi.org/10.1016/j.powtec.2011.09.014

[21] L. Jing, W. Zhou, G. Tian, H. Fu, Surface tuning for oxide-based nanomaterials as efficient photocatalysts, Chem. Soc. Rev. 42 (2013) 9509-9549. https://doi.org/10.1039/c3cs60176e

[22] D.S. Bhatkhande, V.G Pangarkar, A.A.C.M. Beenackers, Photocatalytic degradation for environmental applications-A review, J. Chem. Technol. Biotechnol. 77 (2002) 102-116. https://doi.org/10.1002/jctb.532

[23] B.K. Kumar, T. Imae, J. Miras, J. Esquena, Synthesis and azo dye photodegradation activity of ZrS_2-ZnO nanocomposites, Sep. Purif. Technol. 132 (2014) 281-288. https://doi.org/10.1016/j.seppur.2014.05.018

[24] J. Yu, S. Zhuang, X. Xu, W. Zhu, B. Feng, J. Hu, Photogenerated electron reservoir in hetero-*p-n* CuO-ZnO nanocomposite device for visible-light-driven photocatalytic reduction of aqueous Cr(VI), J. Mater. Chem. A 3 (2015) 1199-1207. https://doi.org/10.1039/C4TA04526B

[25] M. Alvaro, E. Carbonell, M. Espla, H. Garcia, Iron phthalocyanine supported on silica or encapsulated inside zeolite Y as solid photocatalysts for the degradation of phenol and sulfur heterocycles, Appl. Catal. B Environ. 57 (2005) 37-42. https://doi.org/10.1016/j.apcatb.2004.10.003

[26] A. Zhang, R. Zhang, N. Zhang, S. Hong, M. Zhang, Synthesis and characterization of TiO_2-montmorillonite nanocomposites and their photocatalytic activity, Kinet. Catal. 51 (2010) 529-533. https://doi.org/10.1134/S0023158410040117

[27] N. Sapawe, A.A. Jalil, S. Triwahyono, S.H. Adam, N.F. Jaafar, M.A.H. Satar, Isomorphous substitution of Zr in the framework of aluminosilicate HY by an electrochemical method: Evaluation by methylene blue decolorization, Appl. Catal. B: Environ. 125 (2012) 311-323. https://doi.org/10.1016/j.apcatb.2012.05.042

[28] Q. Wang, Z. Li, J. Wang, P. Ye, Structure and photoluminescent properties of ZnO encapsulated in mesoporous silica SBA-15 fabricated by two-solvent strategy nanoscale, Res. Lett. 4 (2009) 646-654.

[29] X. Zhou, T. Shi, H. Zhou, Hydrothermal preparation of ZnO-reduced graphene oxide hybrid with high performance in photocatalytic degradation, Appl. Surf. Sci. 258 (2012) 6204-6211. https://doi.org/10.1016/j.apsusc.2012.02.131

[30] S. Rabieh, K. Nassimi, M. Bagheri, Synthesis of hierarchical ZnO-reduced graphene oxide nanocomposites with enhanced adsorption photocatalytic performance, Mater. Lett. 162 (2016) 28-31. https://doi.org/10.1016/j.matlet.2015.09.111

[31] A.M. Turek, I.E. Wachs, E.D. Canio, Acidic properties of alumina-supported metal oxide catalysts: An infrared spectroscopy study, J. Phys. Chem. 96 (1992) 5000-5007. https://doi.org/10.1021/j100191a050

[32] S. Kalathil, M.M. Khan, A.N. Banerjee, J. Lee, M.H. Cho, A simple biogenic route to rapid synthesis of Au@TiO_2 nanocomposites by electrochemically active biofilms, J. Nanopart. Res. 14 (2012) 1051-1060. https://doi.org/10.1007/s11051-012-1051-x

[33] X. Zong, C. Sun, H, Yu, Z.G. Chen, Z. Xing, D. Ye, G. Q. (Max) Lu, X. Li, L. Wang, Activation of photocatalytic water oxidation on N-doped ZnO bundle-like nanoparticles under visible light, J. Phys. Chem. C 117 (2013) 4937-4942. https://doi.org/10.1021/jp311729b

[34] Q. Ding, Y.-E. Miao, T. Liu, Morphology and photocatalytic property of hierarchical polyimide/ZnO fibers prepared via a direct ion-exchange process, ACS Appl. Mater. Interfaces 5 (2013) 5617-5622. https://doi.org/10.1021/am4009488

[35] L. Saikia, D. Bhuyan, M. Saikia, B. Malakar, D.K. Dutta, P. Sengupta, Photocatalytic performance of ZnO nanomaterials for self-sensitized degradation of malachite green dye under solar light, Appl. Catal. A 490 (2015) 42-49. https://doi.org/10.1016/j.apcata.2014.10.053

[36] L. Fang, B. Zhang, W. Li, X. Li, T. Xin, Q. Zhang, Controllable synthesis of ZnO hierarchical architectures and their photocatalytic property, Superlattices Microstruct. 75 (2014) 324-333. https://doi.org/10.1016/j.spmi.2014.03.001

[37] K. Byrappa, A.K. Subramani, S. Ananda, K.M. Rai, M.H. Sunitha, B. Basavalingu, K. Soga, Impregnation of ZnO onto activated carbon under hydrothermal conditions and its photocatalytic properties, J. Mater. Sci. 41 (2006) 1355-1362. https://doi.org/10.1007/s10853-006-7341-x

[38] C.N.R. Rao, A.K. Sood, K.S. Subrahmanyam, A. Govindaraj, Graphene: The new two-dimensional nanomaterial, Angew. Chem. Int. Ed. 48 (2009) 7752-7777. https://doi.org/10.1007/s10853-006-7341-x

[39] J. Liu, H. Bai, Y. Wang, Z. Liu, X. Zhang, D.D. Sun, Self-assembling TiO_2 nanorods on large graphene oxide sheets at a two-phase interface and their anti-recombination in photocatalytic applications, Adv. Funct. Mater. 20 (2010) 4175-4181. https://doi.org/10.1002/adfm.201001391

[40] C. Zhu, S. Guo, P. Wang, L. Xing, Y. Fang, Y. Zhai, S. Dong, One-pot, water-phase approach to high-quality graphene/TiO_2 composite nanosheets, Chem. Commun. 46 (2010) 7148-7150. https://doi.org/10.1039/c0cc01459a

[41] S. Gayathri, P. Jayabal, M. Kottaisamy, V. Ramakrishnan, Synthesis of ZnO decorated graphene nanocomposite for enhanced photocatalytic properties, J. Appl. Phys. 115 (2014)173504-9. https://doi.org/10.1063/1.4874877

[42] S. An, B.N. Joshi, M.W. Lee, N.Y. Kim, S.S. Yoon, Electrospun graphene-ZnO nanofiber mats for photocatalysis applications, Appl. Surf. Sci. 294 (2014) 24- 28. https://doi.org/10.1016/j.apsusc.2013.12.159

[43] Y. Bu, Z. Chen, W. Li, B. Hou, Highly efficient photocatalytic performance of graphene-ZnO quasi-shell-core composite material, ACS Appl. Mater. Interfaces 5 (2013) 12361-12368. https://doi.org/10.1021/am403149g

[44] S. Thangavel, K. Krishnamoorthy, V. Krishnaswamy, N. Raju, S.J. Kim, G. Venugopal, Graphdiyne-ZnO nanohybrids as an advanced photocatalytic material, J. Phys. Chem. C 119 (2015) 22057-22065. https://doi.org/10.1021/acs.jpcc.5b06138

[45] W.-J. Ong, S.-Y. Voon, L.-L. Tan, B.T. Goh, S.-T. Yong, S.-P. Chai, Enhanced daylight-induced photocatalytic activity of solvent exfoliated graphene (SEG)/ZnO hybrid nanocomposites towards degradation of Reactive Black 5, Ind. Eng. Chem. Res. 53 (2014) 17333-17344. https://doi.org/10.1021/ie5027088

[46] D. Kale, P. Thakur, Highly efficient photocatalytic degradation and mineralization of 4-nitrophenol by graphene decorated ZnO, J Porous Mater. 22 (2015) 797-806. https://doi.org/10.1007/s10934-015-9953-5

[47] P. Roy, A.P. Periasamy, C-T. Liang, H-T. Chang, Synthesis of graphene-ZnO-Au nanocomposites for efficient photocatalytic reduction of nitrobenzene, Environ. Sci. Technol. 47 (2013) 6688-6695. https://doi.org/10.1021/es400422k

[48] S. Rabieh, K. Nassimi, M. Bagheri, Synthesis of hierarchical ZnO-reduced graphene oxide nanocomposites with enhanced adsorption photocatalytic Performance, Mater. Lett. 162 (2016) 28-31. https://doi.org/10.1016/j.matlet.2015.09.111

[49] B. Weng, M.-Q. Yang, N. Zhang Y.-J. Xu, Toward the enhanced photoactivity and photostability of ZnO nanospheres via intimate surface coating with reduced graphene oxide, J. Mater. Chem. A 2 (2014) 9380-9389. https://doi.org/10.1039/c4ta01077a

[50] E.H. Umukoro, M.G. Peleyeju, J.C. Ngila, O.A. Arotiba, Photocatalytic degradation of acid blue 74 in water using Ag-Ag_2O-ZnO nanostructures anchored on graphene

oxide, Solid State Sci. 51 (2016) 66-73. https://doi.org/10.1016/j.solidstatesciences.2015.11.015

[51] M. Ahmad, E. Ahmed, Z.L. Hong, J.F. Xu, N.R. Khalid, A. Elhissi, W. Ahmed, A facile one-step approach to synthesizing ZnO/graphene composites for enhanced degradation of methylene blue under visible light, Appl. Surf. Sci. 274 (2013) 273-281. https://doi.org/10.1016/j.apsusc.2013.03.035

[52] J. Wang, T. Tsuzuki, B. Tang, X. Hou, L. Sun, X. Wang, Reduced graphene oxide/ZnO composite: Reusable adsorbent for pollutant management, ACS Appl. Mater. Interfaces 4 (2012) 3084-3090. https://doi.org/10.1021/am300445f

[53] S. Ameen, M.S. Akhtar, H.-K. Seo, H.S. Shin, Advanced ZnO-graphene oxide nanohybrid and its photocatalytic Applications, Mater. Lett. 100 (2013) 261-265. https://doi.org/10.1016/j.matlet.2013.03.012

[54] A. Tayyebi, M. Outokesh, M. Tayebi, A. Shafikhani, S.S. Şengor, ZnO Quantum Dots-Graphene composites: Formation mechanism and enhanced photocatalytic activity for degradation of methyl orange dye, J. Alloys Compounds 663 (2016) 738-749. https://doi.org/10.1016/j.jallcom.2015.12.169

[55] R. Cai, J. Wu, L. Sun, Y. Liu, T. Fang, S. Zhu, S. Li, Y.Wang, L.Guo, C. Zhao, A. Wei, 3D graphene/ZnO composite with enhanced photocatalytic activity, Mater. Des. 90 (2016) 839-844. https://doi.org/10.1016/j.matdes.2015.11.020

[56] D. Amaranatha Reddy, R. Ma, T.K. Kim, Efficient photocatalytic degradation of methylene blue by heterostructured ZnO-RGO/RuO_2 nanocomposite under the simulated sunlight irradiation, Ceram. Int. 41 (2015) 6999-7009. https://doi.org/10.1016/j.ceramint.2015.01.155

[57] N. Raghavan, S. Thangavel, G. Venugopal, Enhanced photocatalytic degradation of methylene blue by reduced graphene-oxide/titanium dioxide/zinc oxide ternary nanocomposites, materials science in semiconductor processing, Mater. Sci. Semicond. Process. 30 (2015) 321-329. https://doi.org/10.1016/j.mssp.2014.09.019

[58] J. Qin, R. Li, C. Lu, Y. Jiang, H. Tang, X. Yang, Ag/ZnO/graphene oxide heterostructure for the removal of rhodamine B by the synergistic adsorption-degradation effects, Ceram. Int. 41 (2015) 4231-4237. https://doi.org/10.1016/j.ceramint.2014.11.046

[59] M. Ahmad, E. Ahmed, Z.L. Hong, N.R. Khalid, W. Ahmed, A. Elhissi, Graphene-Ag/ZnO nanocomposites as high-performance photocatalysts under visible light

irradiation, J. Alloys Compd. 577 (2013) 717-727. https://doi.org/10.1016/j.jallcom.2013.06.137

[60] W.J. Lee, M.L. Ju, S.T. Kochuveedu, T.H. Han, Y.J. Hu, M. Park, Biomineralized N-doped CNT/TiO_2 core/shell nanowires for visible light photocatalysis, ACS Nano 6 (2011) 935-943. https://doi.org/10.1021/nn204504h

[61] G. Yu, T.T. Ma, S.W. Liu, Enhanced photocatalytic activity of mesoporous TiO2 aggregates by embedding carbon nanotubes as anelectron-transfer channel, Phys. Chem. Chem. Phys. 13 (2011) 3491-3501. https://doi.org/10.1039/C0CP01139H

[62] G. Zhu, H. Wang, G. Yang, L. Chen, P. Guo, L. Zhang, A facile synthesis of ZnO/CNTs hierarchical microsphere composites with enhanced photocatalytic degradation of methylene blue, RSC Adv. 5 (2015) 72476-72481. https://doi.org/10.1039/C5RA11873E

[63] M. Ahmad, E. Ahmed, Z.L. Hong, W. Ahmed, A. Elhissi, N. R. Khalid, Degradation of Rhodamine B using ZnO/CNTs composites photocatalysts, Ultrason. Sonochem. 21 (2014) 761-773. https://doi.org/10.1016/j.ultsonch.2013.08.014

[64] N. Roozban, S. Abbasi, M. Ghazizadeh, Statistical analysis of the photocatalytic activity of decorated multi-walled carbon nanotubes with ZnO nanoparticles, J. Mater. Sci.: Mater. Electron. 28 (2017) 6047-6055. https://doi.org/10.1007/s10854-016-6280-9

[65] C.S. Chen, T.G. Liu, L.W. Lin, X.D. Xie, X.H. Chen, Q.C. Liu, B. Liang, W.W. Yu, C.Y. Qiu, Multi-walled carbon nanotube-supported metal-doped ZnO nanoparticles and their photocatalytic property, J. Nanopart Res. 15 (2013)1295. https://doi.org/10.1007/s11051-012-1295-5

[66] P. Liu, Y. Guo, Q. Xu, F. Wang, Y. Li, K. Shao, Enhanced thephotocatalytic performance of ZnO/multi-walled carbon nanotube nanocomposites for dye degradation, Ceram. Int. 40 (2014) 5629-5633. https://doi.org/10.1016/j.ceramint.2013.10.157

[67] Y. Yan, T. Chang, P. Wei, S.-Z. Kang J. Mu, Photocatalytic activity of nanocomposites of ZnO and multi-walled carbon nanotubes for dye degradation, J. Dispersion Sci. Technol. 30 (2009)198-203. https://doi.org/10.1080/01932690802498310

[68] M. Ahmad, E. Ahmed, Z.L. Hong, X.L. Jiao, T. Abbas, N.R. Khalid, Enhancement in visible-light-responsive photocatalytic activity by embedding Cu-doped ZnO

nanoparticles on multi-walled carbon nanotubes, Appl. Surf. Sci. 285 (2013) 702-712. https://doi.org/10.1016/j.apsusc.2013.08.114

[69] P. Liu, L. Zhang, Adsorption of dyes from aqueous solutions or suspensions with clay nano-adsorbents. Sep. Purif. Technol. 58 (2007) 32-39. https://doi.org/10.1016/j.seppur.2007.07.007

[70] X. Meng, Z. Qian, H. Wang, X. Gao, S. Zhang, E.M. Yang, Sol-gel immobilization of SiO_2/TiO_2 on hydrophobic clay and its removal of methyl orange from water, J. Sol. Gel. Sci. Technol. 46 (2008) 195-200. https://doi.org/10.1007/s10971-008-1677-4

[71] G. Zhu, S. Qiu, J. Yu, Y. Sakamoto, F. Xiao, R. Xu, O. Terasaki, Synthesis and characterization of high-quality zeolite LTA and FAU single nanocrystals, Chem. Mater. 10 (1998) 1483-1486. https://doi.org/10.1021/cm980061k

[72] A. Nezamzadeh-Ejhieh, M. Khorsandi, A comparison between the heterogeneous photodecolorization of an azo dye using Ni/P zeolite and NiS/P zeolite catalysts, Iranian J. Catal. 1 (2011) 99-104.

[73] N. Mohaghegh, M. Tasviri, E. Rahimi, M.R. Gholami, Nanosized ZnO composites: Preparation, characterization and application as photocatalysts for degradation of AB92 azo dye, Mater. Sci. Semicond. Process. 21(2014)167-179. https://doi.org/10.1016/j.mssp.2013.12.023

[74] J.C. Joo, C.H. Ahn, D.G. Jang, Y.H. Yoon, J.K. Kim, L. Campos, H. Ahn, Photocatalytic degradation of trichloroethylene in aqueous phase using nano-ZNO/Laponite composites, J. Hazard. Mater. 263 (2013) 569-574. https://doi.org/10.1016/j.jhazmat.2013.10.017

[75] F.C. Doria, A.C. Borges, J.K. Kim, A. Nathan, J.C. Joo, L.C. Campos, Removal of metaldehyde through photocatalytic reactions using nano-sized zinc oxide composites, Water Air Soil Pollut. 224 (2013) 1434. https://doi.org/10.1007/s11270-013-1434-3

[76] J.K. Kim, J, Alajmy, A.C. Borges, J.C. Joo, H. Ahn, L.C. Campos, Degradation of humic acid by aphotocatalytic reaction using nano-sized ZnO/laponite composite (NZLC), Water, Air, Soil Pollut. 224 (2013)1749. https://doi.org/10.1007/s11270-013-1749-0

[77] N. Sapawe, A.A. Jalil, S. Triwahyono, One-pot electro-synthesis of ZrO_2–ZnO/HY nanocomposite for photocatalytic decolorization of various dye-contaminants, Chem. Eng. J. 225 (2013) 254-265. https://doi.org/10.1016/j.cej.2013.03.121

[78] L. Shi-Qian, Z. Pei-Jiang, Z. Wan-shun, C. Sheng, P. Hong, Effective photocatalytic decolorization of methylene blue utilizing ZnO/rectorite nanocomposite under simulated solar irradiation, J. Alloys Compd. 616 (2014) 227-234. https://doi.org/10.1016/j.jallcom.2013.10.079

[79] J. Esmaili-Hafshejani A. Nezamzadeh-Ejhieh, Increased photocatalytic activity of Zn(II)/Cu(II) oxides and sulfides by coupling and supporting them onto clinoptilolite nanoparticles in the degradation of benzophenone aqueous solution J. Hazard. Mater. 316 (2016) 194-203. https://doi.org/10.1016/j.jhazmat.2016.05.006

[80] A. Nezamzadeh-Ejhieh, F. Khodabakhshi-Chermahini, Incorporated ZnO onto nano clinoptilolite particles as the active centers in the photodegradation of phenylhydrazine, J. Ind. Eng. Chem. 20 (2014) 695-704. https://doi.org/10.1016/j.jiec.2013.05.035

[81] A. Nezamzadeh-Ejhieh, S. Khorsandi, Photocatalytic degradation of 4-nitrophenol with ZnO supported nano-clinoptilolite zeolite J. Ind. Eng. Chem. 20 (2014) 937-946. https://doi.org/10.1016/j.jiec.2013.06.026

[82] X. Li, H. Yang, Pd hybridizing ZnO/kaolinite nanocomposites: Synthesis, microstructure, and enhanced photocatalytic property, Appl. Clay Sci. 100 (2014) 43-49. https://doi.org/10.1016/j.clay.2014.05.007

[83] I. Fatimah, S. Wang, D. Wulandari, ZnO/montmorillonite for photocatalytic and photochemical degradation of methylene blue, Appl. Clay Sci. 53 (2011) 553-560. https://doi.org/10.1016/j.clay.2011.05.001

[84] N.N. Zu, H.G. Yang, Z.W. Dai, Different processes responsible for blue pumped, ultraviolet and violet luminescence in high-concentrated Er^{3+}:YAG and low concentrated Er^{3+}:YAP crystals, Physica B: Condens. Matter. 403 (2008) 174-177. https://doi.org/10.1016/j.physb.2007.08.097

[85] J. Wang, Y.P. Xie, Z.H. Zhang, J. Li, X. Chen, L.Q. Zhang, R. Xu, X.D. Zhang, Photocatalytic degradation of organic dyes with Er^{3+}:$YAlO_3$/ZnO composite under solar light, Sol. Energy Mater. Sol. Cells 93 (2009) 355-361. https://doi.org/10.1016/j.solmat.2008.11.017

[86] F. Shi, J.S. Wang, D.S. Zhang, G.S. Qin, W.P. Qin, Greatly enhanced size-tunable ultraviolet upconversion luminescence of monodisperse beta-$NaYF_4$:Yb,Tm nanocrystals, J. Mater. Chem. 21 (2011) 13413-13421. https://doi.org/10.1039/c1jm11480h

[87] C.H. Li, F. Wang, J.A. Zhu, J.C. Yu, $NaYF_4$ Yb, Tm/CdS composite as a novel near-infrared-driven photocatalyst, Appl. Catal. B 100 (2010) 433-439. https://doi.org/10.1016/j.apcatb.2010.08.017

[88] W.P. Qin, D.S. Zhang, D. Zhao, L.L. Wang, K.Z. Zheng, Near-infrared photocatalysis based on YF_3:Yb^{3+}, Tm^{3+}/TiO_2 core/shell nanoparticles, Chem. Commun. 46 (2010) 2304-2306. https://doi.org/10.1039/b924052g

[89] L. Yin, Y. Li, J. Wang, Y. Zhai, J. Wang, Y. Kong, B. Wang, X. Zhang, Preparation of Er^{3+}:$Y_3Al_5O_{12}$/TiO_2-ZnO composite and application of solar energy in photocatalytic degradation of organic dyes, Environ. Prog. Sustain. Energ. 32 (2013) 697-704. https://doi.org/10.1002/ep.11688

[90] X. Guo, W. Song, C. Chen, W. Di, W. Qin, Near-infrared photocatalysis of β-$NaYF_4$:Yb^{3+},Tm^{3+}@ZnO composites, Phys. Chem. Chem. Phys. 15 (2013) 14681-14688. https://doi.org/10.1039/c3cp52248b

[91] J. Gao, R. Jiang, J. Wang, P. Kang, B. Wang, Y. Li, K. Li, X. Zhang, The investigation of sonocatalytic activity of Er^{3+}:$YAlO_3$/TiO_2-ZnO composite in azo dyes degradation, Ultrason. Sonochem. 18 (2011) 541-548. https://doi.org/10.1016/j.ultsonch.2010.09.012

[92] C. Lu,Y. Chen, H. Zhang, L. Tang, S. Wei, Y. Song, J. Wang, Preparation of Er^{3+}:$YAlO_3$/Fe- and Co-doped-ZnO coated composites and their visible-light photocatalytic activity in degradation of some organic Dyes, Res. Chem. Intermed. 42 (2016) 4651-4668. https://doi.org/10.1007/s11164-015-2306-9

[93] L.N. Yin, Y. Li, J. Wang, Y.M. Kong, Y. Zhai, B.X. Wang, K. Lia, X.D. Zhang, The improvement of solar photocatalytic activity of ZnO by doping with Er^{3+}:$Y_3Al_5O_{12}$ during dye degradation, Russ. J. Phys. Chem. A 86 (2012) 2049-2056. https://doi.org/10.1134/S0036024412130262

[94] Y. Chen, C. Lu, L. Tang, Y. Song, S. Wei, Y. Rong, Z. Zhang, J. Wang, Photocatalytic degradation of organic dyes by Er^{3+}:$YAlO_3$/Co- and Fe-Doped ZnO Coated Composites under Solar Irradiation,Russ. J. Phys. Chem. A 90 (2016) 2654-2664.

[95] F. Achouri, S. Corbel, L. Balan, K. Mozet, E. Girot, G. Medjahdi, M.B. Said, A. Ghrabi, R. Schneider, Porous Mn-doped ZnO nanoparticles for enhanced solar and visible light photocatalysis, Mater. Des. 101 (2016) 309-316. https://doi.org/10.1016/j.matdes.2016.04.015

[96] R. Saleh, N.F. Djaja, UV light photocatalytic degradation of organic dyes with Fe-doped ZnO nanoparticles, Superlattices Microstruct. 74 (2014) 217-233. https://doi.org/10.1016/j.spmi.2014.06.013

[97] N.M. Jacob, G. Madras, N. Kottam, T. Thomas, Multivalent Cu-Doped ZnO nanoparticles with full solar spectrum absorbance and enhanced photoactivity, Ind. Eng. Chem. Res. 53 (2014) 5895-5904. https://doi.org/10.1021/ie404378z

[98] A. Phuruangrat, S. Madahin, O. Yayapao, S. Thongtem, T. Thongtem, Photocatalytic degradation of organic dyes by UV light, catalyzed by nanostructured Cd-doped ZnO synthesized by a sonochemical method, Res Chem Intermed. 41 (2015) 9757-9772. https://doi.org/10.1007/s11164-015-1963-z

[99] C. Yu, K. Yang, Y. Xie, Q. Fan, J.C. Yu, Q. Shuaand, C. Wang, Novel hollow Pt-ZnO nanocomposite microspheres with hierarchical structure and enhanced photocatalytic activity and stability, Nanoscale 5 (2013) 2142-2151. https://doi.org/10.1039/c2nr33595f

[100] M. Ahmad, E. Ahmed, Z.L. Hong, Z. Iqbal, N.R. Khalid, T. Abbas, I. Ahmad, A.M. Elhissi, W.Ahmed, Structural, optical and photocatalytic properties of hafnium doped zinc oxide nanophotocatalyst, Ceram. Int. 39 (2013) 8693-8700. https://doi.org/10.1016/j.ceramint.2013.04.051

[101] S.A. Ansari, M.M. Khan, M.O. Ansari, J. Lee, M.H. Cho, Biogenic synthesis, photocatalytic, and photoelectrochemical performance of Ag-ZnO nanocomposite, J. Phys. Chem. C 117 (2013) 27023-27030. https://doi.org/10.1021/jp410063p

[102] P. Fageria, S. Gangopadhyayb, S. Pande, Synthesis of ZnO/Au and ZnO/Ag nanoparticles and their photocatalytic application using UV and visible light, RSC Adv. 4 (2014) 24962- 24972. https://doi.org/10.1039/C4RA03158J

[103] S. Ekambaram, Y. Iikubo, A. Kudo, Combustion synthesis and photocatalytic properties of transition metal-incorporated ZnO, J. Alloy. Compd. 433 (2007) 237-240. https://doi.org/10.1016/j.jallcom.2006.06.045

[104] R. Wang, J.H. Xin, Y. Yang, H.Liu, L. Xu, J. Hu, The Characteristics and photocatalytic activities of silver doped ZnO nanocrystallites, Appl. Surf. Sci. 227 (2004) 312-317. https://doi.org/10.1016/j.apsusc.2003.12.012

[105] M.K. Seery, R. George, P. Floris, S.C. Pillai, Silver doped titanium dioxide nanomaterials for enhanced visible light photocatalysis, J. Photochem. Photobiol. A 189 (2007) 258-263. https://doi.org/10.1016/j.jphotochem.2007.02.010

[106] J.-J. Wu, C.-H. Tseng, Photocatalytic properties of nc-Au/ZnO nanorod composites, Appl. Catal. B 66 (2006) 51-57. https://doi.org/10.1016/j.apcatb.2006.02.013

[107] O.A. Yıldırım, H.E. Unalan, C. Durucan, Highly efficient room temperature synthesis of silver-doped zinc oxide (ZnO:Ag) nanoparticles: Structural, optical, and photocatalytic properties, J. Am. Ceram. Soc. 96 (2013) 766-773. https://doi.org/10.1111/jace.12218

[108] N.M. Jacob, G. Madras, N. Kottam, T. Thomas, Multivalent Cu-doped ZnO nanoparticles with full solar spectrum absorbance and enhanced photoactivity, Ind. Eng. Chem. Res. 53 (2014) 5895-5904. https://doi.org/10.1021/ie404378z

[109] Priyanka, V.C. Srivastava, Photocatalytic oxidation of dye bearing wastewater by iron doped zinc oxide, Ind. Eng. Chem. Res. 52 (2013)17790-17799. https://doi.org/10.1021/ie401973r

[110] Y. Chen, D. Zeng, K. Zhang, A. Lu, L. Wang, D.-L. Peng, Au-ZnO hybrid nanoflowers, nanomultipods, and nanopyramids: One-pot reaction synthesis and photocatalytic properties, Nanoscale 6 (2014) 874-881. https://doi.org/10.1039/C3NR04558G

[111] F. Achouri, S. Corbel, L. Balan, K. Mozet, E. Girot, G. Medjahdi, M.B. Said, A. Ghrabi, R. Schneider, Porous Mn-doped ZnO nanoparticles for enhanced solar and visible light photocatalysis, Mater. Des. 101 (2016) 309-316. https://doi.org/10.1016/j.matdes.2016.04.015

[112] R. Saleh, N.F. Djaja, UV light photocatalytic degradation of organic dyes with Fe-doped ZnO nanoparticles, Superlattices Microstruct. 74 (2014) 217-233. https://doi.org/10.1016/j.spmi.2014.06.013

[113] S. Yi, J, Cui, S. Li, L, Zhang, D. Wang, Y. Lin, Enhanced visible-light photocatalytic activity of Fe/ZnO for rhodamine B degradation and its photogenerated charge transfer properties, Applied Surface Science 319 (2014) 230-236. https://doi.org/10.1016/j.apsusc.2014.06.151

[114] A. Phuruangrat, S. Madahin, O. Yayapao, S. Thongtem, T. Thongtem, Photocatalytic degradation of organic dyes by UV light, catalyzed by nanostructured Cd-doped ZnO synthesized by a sonochemical method, Res Chem Intermed. 41 (2015) 9757-9772. https://doi.org/10.1007/s11164-015-1963-z

[115] S.A. Ansari, M.M. Khan, M.O. Ansari, J. Lee, M.H. Cho, Biogenic synthesis, photocatalytic, and photoelectrochemical performance of Ag-ZnO Nanocomposite, J. Phys. Chem. C 117 (2013) 27023-27030. https://doi.org/10.1021/jp410063p

[116] S. Balachandran, S.G. Praveen, R. Velmuruganc, M. Swaminathan, Facile fabrication of highly efficient, reusable heterostructured Ag-ZnO-CdO and its twin applications of dye degradation under natural sunlight and self-cleaning, RSC Adv. 4 (2014) 4353- 4362. https://doi.org/10.1039/C3RA45381B

[117] C. Yu, K. Yang, Y. Xie, Q. Fan, J.C. Yu, Q. Shuaand, C. Wanga, Novel hollow Pt-ZnO nanocomposite microspheres with hierarchical structure and enhanced photocatalytic activity and stability, Nanoscale 5 (2013) 2142- 2151. https://doi.org/10.1039/c2nr33595f

[118] A. Khataee, S. Saadi, B. Vahid, S.W. Joo, B.-K. Min, Sonocatalytic degradation of Acid Blue 92 using sonochemically prepared samarium doped zinc oxide nanostructures, Ultrason. Sonochem. 29 (2016) 27-38. https://doi.org/10.1016/j.ultsonch.2015.07.026

[119] A. Khataee, R.D.C. Soltani, A. Karimi, S.W. Joo, Sonocatalytic degradation of a textile dye over Gd-doped ZnO nanoparticles synthesized through sonochemical process, Ultrason. Sonochem. 23 (2015) 219-230. https://doi.org/10.1016/j.ultsonch.2014.08.023

[120] A. Khataee, R.D.C. Soltani, Y. Hanifehpour, M. Safarpour, H.G. Ranjbar, S.W. Joo, Synthesis and characterization of dysprosium-doped ZnO nanoparticles for photocatalysis of a textile dye under visible light irradiation, Ind. Eng. Chem. Res. 53 (2014) 1924-1932. https://doi.org/10.1021/ie402743u

[121] P.V. Korake, R.S. Dhabbe, A.N. Kadam, Y.B. Gaikwad, K.M. Garadkar, Highly active lanthanum doped ZnO nanorods for photodegradation of metasystox, J. Photochem. Photobiol. B 130 (2014) 11-19. https://doi.org/10.1016/j.jphotobiol.2013.10.012

[122] J.-C. Sin, S.-M. Lam, K.-T. Lee, A.R. Mohamed, Preparation of rare-earth-doped ZnO hierarchical micro/nanospheres and their enhanced photocatalytic activity under visible light irradiation, Ceram. Int. 40 (2014) 5431-5440. https://doi.org/10.1016/j.ceramint.2013.10.128

[123] C.-J. Chang, C.-Y. Lin, M.-H. Hsu, Enhanced photocatalytic activity of Ce-doped ZnO nanorods under UV and visible light, J. Taiwan Inst. Chem. Eng. 45 (2014) 1954-1963. https://doi.org/10.1016/j.jtice.2014.03.008

[124] W. Yu, J. Zhang, T. Peng, New insight into the enhanced photocatalytic activity of N-, C- and S-doped ZnO photocatalysts, Appl. Catal. B 181 (2016) 220-227. https://doi.org/10.1016/j.apcatb.2015.07.031

[125] X. Zhang, J. Qin, R. Hao, L. Wang, X. Shen, R. Yu, S. Limpanart, M. Ma, R. Liu, Carbon-doped ZnO nanostructures: Facile synthesis and visible light photocatalytic applications, J. Phys. Chem. C, 119 (2015) 20544-20554. https://doi.org/10.1021/acs.jpcc.5b07116

[126] D. Zhang, J.Y. Gong, J.J. Ma, G.Q. Hana, Z.W. Tong, A facile method for synthesis of N-doped ZnO mesoporous nanospheres and enhanced photocatalytic activity, Dalton Trans. 42 (2013) 16556-16561. https://doi.org/10.1039/c3dt52039k

[127] W.-K. Jo, N. Clament Sagaya Selvam, Enhanced the thevisible light-driven photocatalytic performance of ZnO-g-C_3N_4 coupled with graphene oxide as a novel ternary nanocomposite, J. Hazard. Mater. 299 (2015) 462-470. https://doi.org/10.1016/j.jhazmat.2015.07.042

[128] J. Li, M. Zhou, Z. Ye, H. Wang, C. Ma, P. Huo, Y. Yan, Enhanced photocatalytic activity of g-C_3N_4-ZnO/HNTs composite heterostructure photocatalysts for degradation of tetracycline under visible light irradiation, RSC Adv. 5 (2015) 91177-91189. https://doi.org/10.1039/C5RA17360D

[129] F. Dong, Z. Zhao, T. Xiong, Z. Ni, W. Zhang, Y. Sun, W.K. Ho, ACS Appl. Mater. Interface 5 (2013) 11392-11401. https://doi.org/10.1021/am403653a

[130] S. Kumar, A. Baruah, S. Tonda, B. Kumar, V. Shanker, B. Sreedharc, Cost-effective and eco-friendly synthesis of novel and stable N-doped ZnO/g-C_3N_4 core-shell nanoplates with excellent visible-light responsive photocatalysis, Nanoscale 6 (2014) 4830-4842. https://doi.org/10.1039/c3nr05271k

Chapter 10

TiO_2 Based Nanocomposite for Photocatalytic Degradation of Organic Pollutants

Dhiraj Sud[1*], Nidhi Sharotri[2]

[1] Department of Chemistry, Sant Longowal Institute of Engineering and Technology, Punjab, Pincode 148106, India

[2] Department of Chemistry, Sri SaiUniversity, Palampur, Himachal Pradesh, Pincode 176061, India

suddhiraj@yahoo.com [1*]; nidhisliet11@gmail.com [2]

Abstract

Organic contaminants such as synthetic dyes, phenols, pesticides, fertilizers, herbicides, and surfactants, etc. present in aqueous streams are of foremost concern with respect to the health of the general public. Titania based heterogeneous photocatalytic oxidation has received the attention of researchers for many years as an alternative method for purification of both air and water systems. The major issues of immense importance from industrial viewpoint have been the recovery and recycling of the photocatalyst. The present chapter discusses the utility of TiO_2 based nanocomposite materials in photocatalytic degradation of organic pollutants.

Keywords

Nanocomposites, Titania, Photocatalysis, Organic Pollutants, Wastewater

Contents

1. Introduction

Organic contaminants such as synthetic dyes, phenols, pesticides, fertilizers, herbicides, and surfactants, etc. found in wastewater are of major concern with respect to the health of general public. A number of physicochemical methods, based on precipitation, coagulation, adsorption and membrane filtration, have been applied for the removal of the organic impurities from the waste streams, however, each process has its own limitations. Chemical oxidation technologies seem to have the most potential for the future. Therefore, the elimination of these pollutants from the environment has become a prime concern for researchers.

Synthetic dyes: Synthetic dyes are frequently used in various sections of the textile industry, leather tanning industry, paper production industry, food technology and in agricultural research, light-harvesting arrays, photoelectrochemical cells and also in hair colorings. Discharges from various industries usually contain considerable quantities of synthetic organic dyes and due to their large-scale production and extensive application, synthetic dyes can cause considerable nonaesthetic pollution and serious health-risk factors. Dyes contain chromophores such as -C=C-, -C=N-, -C=O, -N=N-, $-NO_2$, quinoid rings, and the auxochromes are $-NH_3$, -COOH, $-SO_3H$ and –OH groups which cause or intensify the color.

Dyes are classified on the basis of the chemical structure or according to the methods of their application onto the substrate. Based on chemical structures, the dyes are classified into 20-30 different classes such as the azo, anthraquinone, sulfur, indigoid, triphenylmethyl (trityl), and phthalocyanine derivatives. Azo dyes include the largest class of organic colorants recorded in the Colour Index (60-70% of the total) and constitute the majority of the dyes used by the textile-processing industries. Anthraquinone dyes are the second largest class (~15%), followed by triarylmethanes (~3%) and phthalocyanines (~2%) of the entries in the Colour Index.

Pesticides: Pesticides are the synthetic compounds or mixtures aimed for preventing, destroying, controlling, pest causing human or animal disease, unwanted species of plants, and animals causing harm during or otherwise interfering with the production, storage, processing, and transportation, or marketing of food or agricultural commodities. Pesticides are classified according to their biological activity as fungicides, herbicides, bactericides, rodenticides, insecticides, nematicides, and molluscicides. Some of the widely used pesticides are organophosphate pesticides, organochlorine pesticides, carbamates, urea, anilides, pyrethroid, and miscellaneous compounds [1].

Phenols and their derivatives: Phenols and their derivatives are the most important organic contaminants found in wastewater, released from industries such as petroleum refining, steel, dyestuff, synthetic resins, byproducts of agricultural chemicals, paper and pulp mills, tanning and fiberboard product to the environment. These compounds are also used as fungicides, bactericides, antiseptics, disinfectants, and wood and glue preservatives.

Pharmaceutical compounds: The emerging environmental problem is the presence of pharmaceutical residues in municipal, surface, ground- and even drinking water and has a potential impact on human health and the environment even at very low concentrations. Pharmaceutical residues usually detected in the environment at low concentration (ng/L) can induce toxic effects. Mostly, pharmaceutical residues are not toxic but may have some unwanted side effects. A well-known example is the identification of disruption of the endocrine system in the feminization of male sex organs in fish and snails [2]. Based on the manufacturing processes, pharmaceutical industries can be subdivided into the following five major subcategories [3]: (1) fermentation plants; (2) synthesized organic chemicals plants; (3) fermentation/synthesized organic chemicals plants (generally moderate to large plants); (4) natural/biological product extractions (antibiotics/vitamins/enzymes, etc.) and (5) drug mixing, formulation, and preparation plants (tablets, capsules, and solutions, etc.).

2. Heterogeneous photocatalysis

Heterogeneous photocatalysis has emerged as a potent destructive technology for the degradation of organic pollutants in a slurry or immobilized mode using semiconductors as a photocatalyst in the presence of ultraviolet (UV) /solar light. The heterogeneous photocatalytic systems are found to be more efficient than the homogeneous systems [4]. The heterogeneous photocatalytic oxidation leads to the complete removal of organic compounds [5,6] and also undergoes partial decomposition of non-biodegradable contaminants to biodegradable intermediates [7]. Titania assisted heterogeneous photocatalytic oxidation has received the attention of researchers for many years as an alternative method for purification of both air and water systems.

2.1 Titanium dioxide as photocatalyst

TiO_2 is preferred as photocatalyst due to its biologically and chemically inertness, chemical and photo-corrosion resistance, non-toxicity, hydrophilicity and cost-effectiveness. Additionally, it is more stable than other photocatalysts in ambient conditions and can be reused again [8,9]. Furthermore, the favorable energy band gap facilitates the extensive applications of TiO_2 in the area photocatalysis, batteries, photovoltaics, UV blocker, the coating material in textile, paints, cosmetics, paper and biomedical sciences.

2.2 Electronic structure of TiO_2

Molecular orbitals are formed by contribution of atomic orbitals involved. This led to the formation of conduction bands by overlapping of the lowest unoccupied molecular orbitals (LUMO) and valence band by overlapping the highest occupied molecular orbital (HOMO). The energy difference in these bands is known as the band gap (Eg). The energy band gap of TiO_2 of anatase form is 3.2 eV and that of rutile is 3.0 eV corresponding to an absorbance maxima at, λ= 388 and 415 nm respectively. Titanium dioxide is regarded as an n-type semiconductor due to the presence of oxygen vacancies in the lattice. These vacancies are formed upon the release of two electrons and molecular oxygen leaving a positive (+2) oxide ion vacancy. When electrons of energy lower than the conduction band are present, the result is an n-type semiconductor. On the other hand, if a material is added with fewer electrons than the host, positive holes are added above the valence band resulting in a p-type semiconductor.

2.3 Physical properties of TiO_2

Titania is known as the world's fourth most abundant metal and ninth most abundant element. It does not exist in its elemental state, but is mainly found in mineral form like rutile, ilmenite, leucoxene, anatase, brookite, perovskite and spene; titanates and as ores. The primary source and the most stable form of titanium dioxide is rutile ore. Rutile is found as one of the known polymorphs of titanium dioxide (TiO_2) and the other polymorphs are anatase and brookite. All three prominent forms (brookite, rutile, and anatase) contain TiO_6 octahedra, where six oxygen (O^{2-}) atoms are coordinated to titanium (Ti^{4+}) atom (Fig. 1). The anatase form has a tetragonal structure formed by the sharing of the corner (vertices) of octahedra. In rutile, the TiO_6 octahedra share two edges to form linear chains with tetragonal structure and are slightly distorted. In case of the brookite form, both edges and corners are shared to form an orthorhombic structure [10-13]. The distortion of anatase TiO_6 is slightly larger than rutile phase. The density of anatase (3.79) is less as compared to the rutile phase (4.23).

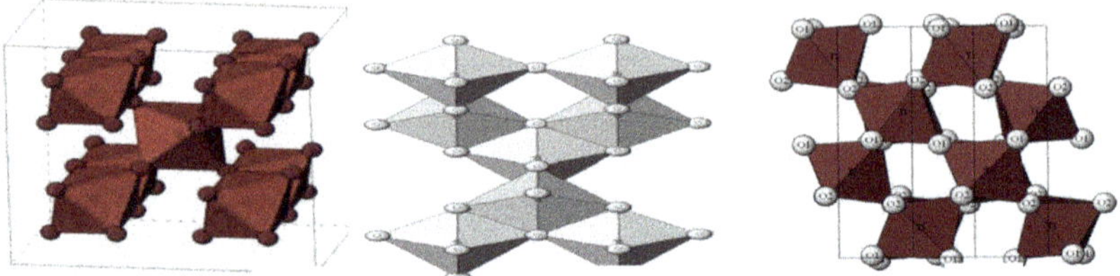

a b c

Figure 1 Crystalline structures of titanium dioxide (a)-rutile, (b)-anatase and (c)-brookite [10].

The rutile phase is the thermodynamically favorable at all temperatures and pressures because of its linear chains than the zig-zag anatase form (ca. 1.2–2.8 kcal mol^{-1}). Phase transformation (anatase to rutile) occurs at higher temperature 700–1000°C depending on the crystallite size and impurity content. At lower temperature, anatase is the more stable form.

Table 1 Physical and chemical properties of anatase and rutile phases.

Property	**Anatase**	**Rutile**	**Brookite**	**Ref.**
Molecular Weight (g/mol)	79.88	79.88	79.88	
Melting point (°C)	1843	1843	1843	
Boiling Point (°C)	2500 ~ 3000	2500 ~ 3000	2,972	
Crystal structure	Tetragonal	Tetragonal	Orthorhombic	[14]
Light absorption (nm)	< 390	< 415	< 415	[15-17]
Atoms per unit cell (Z)	4	2	8	[18-19]
Lattice parameters (nm)	a = 0.3785, c = 0.9514	a = 0.4594, c = 0.29589	a= 9.184 b=5.447 c= 5.154	[18-19]
Solubility in H_2O	Insoluble	Insoluble	Insoluble	[14]
Density (g/cm^3)	3.79	4.23	3.99	[18-19]
Ti–O bond length (Å)	1.94 (4), 1.97 (2)	1.95 (4) 1.98 (2)	1.87-2.04	
Volume/molecule (g cm^3)	34.061	31.2160	32.172	
Density (kg/m^3)	3894	4250	3990	[18-19]

Band gap energies for anatase and rutile have been estimated and found to be 3.2 and 3.0 eV, respectively. The valence band of TiO_2 consists of 2p orbital of hybridized oxygen with a 3d orbital of titanium, while the conduction band consists of 3d orbital of titanium. The smaller bandgap of rutile, however, implies that a higher fraction of the solar radiation can induce band gap excitation (threefold increase in photocurrents). The structural, physical and chemical properties of anatase and rutile forms are given in Table 1.

2.4 Mechanism of TiO_2 assisted heterogeneous photocatalysis

The basic principle of heterogeneous photocatalysis involves the photoinduced reactions which are accelerated by the presence of semiconductor catalysts [20]. The detailed mechanism of the process as discussed by various scientists [21-23] is summarized in the preceding section.

(i) Photoexcitation of the semiconductor

When a photon energy is higher or equal to the threshold energy is absorbed by a semiconductor, an electron gets promoted from the valence band (VB) to the conduction band (CB) with the immediate creation of a hole in the valance band (hVB+). The conduction band energy level is known as the reduction potential of photoelectrons and the valance band energy level is helpful in determining the oxidizing ability of photo holes, and reflects the capability to promote reduction and oxidation reactions [24-26].

$$TiO_2 + hv\ (UV) \rightarrow TiO_2\ (e_{CB}^- + h_{VB}^+) \qquad (1)$$

where hvis photon of energy required to excite the electron from the valence band (VB) region to the conduction band (CB) region.

(ii) Formation of hydroxyl radical

The eCB−and the hVB+ can migrate to the catalyst surface and undergo redox reaction on the surface of the catalyst. The photogenerated holes helps to oxidize the organic molecule or react with OH−or H_2O to generate hydroxyl (OH•) radicals as:

$$TiO_2\ (h_{VB}^+) + H_2O \rightarrow TiO_2 + H^+ + OH^{\bullet} \qquad (2)$$

$$TiO_2\ (h_{VB}^+) + OH^- \rightarrow TiO_2 + OH\bullet \qquad (3)$$

The photogenerated electrons reacts with the oxygen to generate superoxide radical anion ($O_2^{-\bullet}$) which further lead to the additional formation of OH• radical [27-29].

$$TiO_2\,(e_{CB}^-) + O_2 \rightarrow TiO_2 + O_2^{\bullet -} \tag{4}$$

$$O_2^{\bullet -} + H^+ \rightarrow HO_2^{\bullet} \tag{5}$$

$$HO_2^{\bullet} + H^+ + TiO_2\,(e_{CB}^-) \rightarrow H_2O_2 + TiO_2 \tag{6}$$

$$H_2O_2 \rightarrow 2\,{}^{\bullet}OH \tag{7}$$

(iii) Recombination reaction

The recombination of e_{CB}^- and the h_{VB}^+ on the surface or in the bulk of the catalyst in a few nanoseconds results in the dissipation of energy as heat which must be prevented to have favorable photocatalysed reactions [30-31].

$$e_{CB}^- + h_{VB}^+ \rightarrow e_{CB} + \text{heat} \tag{8}$$

(iv) Degradation of organic pollutants

Organic pollutants adsorbed on TiO_2 catalyst surface undergo oxidation by an ${}^{\bullet}OH$ radical, which is a very strong oxidizing agent (standard redox potential +2.8V) can oxidize organic pollutant to the mineralized products (Fig. 2).

$$\text{Organic pollutant} + h_{VB}^+ \rightarrow \text{oxidation products} \tag{9}$$

$$\text{Organic pollutant} + e_{CB}^- \rightarrow \text{reduction products} \tag{10}$$

$$\text{Organic pollutant} + OH^{\bullet} \rightarrow \text{degradation products} \tag{11}$$

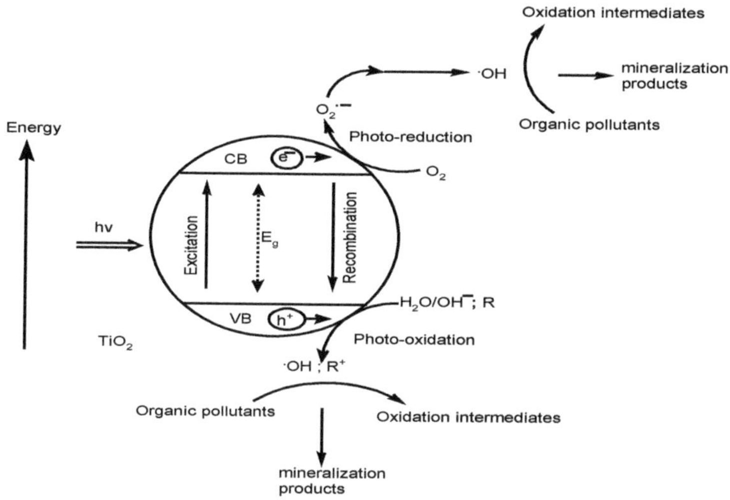

Figure 2 Schematic diagram for the photocatalytic process over TiO_2 [29].

Titanium dioxide (TiO_2) has been extensively used as a most efficient photocatalyst in widespread environmental applications [32-33].

2.5 Modification of TiO_2 photocatalysts

The titanium dioxide has the desirable properties of being chemically stable, readily available and widely used as photocatalyst as it absorbs the UV light because of its larger band gap (3.2 eV) [34].

The major disadvantage of the wide band gap is that it hinders the practical application of titanium dioxide in the visible light. Different strategies such as surface sensitization, bandgap modification by creating oxygen vacancies and doping of photocatalyst metals and nonmetals have been discovered for the synthesis of visible light responsive TiO_2 [35-39].

Other major issues to address are to control recombination reaction and the recovery as well as recycling of the catalyst.

2.6 Nanocomposites and their classification

The word "composite" means "made of two or more different parts". A composite is a combination of two or more different materials that are mixed in an effort to obtain better characteristic properties than that of an individual component. Mostly composite materials consist of various discontinuous phases distributed in single continuous phase. The continuous phase is known as"matrix" whereas the discontinuous phase is known to be a "reinforcement, or reinforcing the material.

A nanocomposite is a composite material, in which one of the components is in nano the range i.e. around 10^{-9} m. The matrix comprised of resin and filler, which is used to improve the characteristic properties of the resin as well as the production cost. The filler-resin system acts as a homogeneous material, whereas the composites are made of a matrix and reinforcing the material. The reinforcement induces greater mechanical performance to composite material thereby external mechanical load is increased by the matrix and this protects the fibers against external attack. The type of the reinforcement matrix association depends upon desired characteristic properties of the product such as high mechanical characteristics, good thermal stability, and resistance to corrosion, etc. Most developed nanocomposites with demonstrable technological importance are composed of two phases, and can be microstructurally classified into three principal types:

(a) Nanolayered composite composed of alternating layers of nanoscale dimension

(b) Nanofilamentary composites composed of a matrix with embedded (and generally aligned) nanoscale diameter filaments

(c) Nanoparticulates composites composed of a matrix with embedded nanoscale particles.

3. TiO_2 based nanocomposites for photocatalytic degradation of organic pollutants

3.1 Organic Polymer based TiO_2 nanocomposites for degradation of organic pollutants

Poly(vinyl alcohol)(PVA)/poly(acrylic acid)(PAAC)/TiO_2/graphene oxide nanocomposite hydrogels were prepared by using radical polymerization and condensation reaction for the photocatalytic treatment of wastewater [40]. Graphene oxide (GO) was used as an additive to increase photocatalytic activity as PVA/PAAC/TiO_2 nanocomposite hydrogels. Both TiO_2 and graphene oxide were immobilized in PVA/PAAC hydrogel matrix for an easier recovery after the wastewater treatment. The photocatalytic activities of PVA/PAAC/TiO_2/GO nanocomposite hydrogels were assessed by degradation of pollutants. The improved removal of pollutants was due to the two-step mechanism based on the adsorption of pollutants by nanocomposite hydrogel and the effective decomposition of pollutants by TiO_2 and GO. The highest swelling of nanocomposite hydrogel was observed at pH 10 indicating that poly(vinyl alcohol)/poly(acrylic acid)/TiO_2/GO nanocomposite hydrogels were suitable as a promising system for the treatment of alkaline wastewater. The highest efficiency of dye decomposition (i.e 88% of Coomassie brilliant blue R-250 (CBB) and 95% of methylene blue (MB) was observed at pH 10 due to the pH-sensitive swelling behavior of the PVA/PAAc/TiO_2/GO nanocomposite hydrogels. The combination of TiO_2 and GO strongly contributed to the efficient electron transfer between TiO_2 and GO. GO strongly prevented TiO_2 photocatalyst from the recombination of electron-hole formed under UV irradiation resulting in the improvement of photocatalytic activity. The synergetic contributions of pH-sensitive swelling characteristics of PVA/PAAc hydrogel matrix and GO having functional groups were beneficial for the efficient photocatalytic decomposition of basic water pollutants with PVA/PAAc/TiO_2 /GO nanocomposite hydrogels.

The pH-sensitive PVA/PAAc hydrogels containing TiO_2 photocatalyst were prepared by radical polymerization and electrospinning methods [41]. The anatase phase of TiO_2 was maintained in the TiO_2/PVA/PAAc composite hydrogels without any change in structure during the preparation process. The TiO_2/PVA/PAAc nanofibers showed faster and higher swelling property due to the larger surface area when the pH of the medium was changed from acidic to alkaline condition.The swelling property of pH-sensitive hydrogels was attributed to the ionization of PAAc component in the PVA/PAAc hydrogel without hindrance even in the presence of PVA. The photodegradation of dye was effectively improved by adding TiO_2 photocatalyst in PVA/PAAc hydrogel nanofiber supports. Thus both hydrogel matrix and fibers help in the absorption of dye without their decomposition activity. The enhancement in decomposition efficiency of dye can be explained by the equal distribution of TiO_2 and with the much higher surface to volume ratio of nano-sized fibrous supports. The faster and larger swelling properties of nanocomposite hydrogel nanofibers played an important role in the improved photo degradation ability of TiO_2.

TiO_2/hydrogel nanocomposites were synthesized in order to estimate their capacity to photodegrade different textile azo dyes: acid dyes C.I. acid red 18, C.I. acid blue 113, reactive dyes C.I. reactive yellow 17, C.I. reactive black 5 and direct dye C.I. direct blue 78 from aqueous solution [42]. Two different types of TiO_2 nanoparticles were immobilized: colloidal TiO_2 nanoparticles synthesized by acidic hydrolysis of $TiCl_4$ and the commercial Degussa P25 TiO_2 nanoparticles. SEM analysis, swelling and rheology studies revealed the dependency on pore size, swelling behavior and mechanical properties of the immobilized TiO_2 nanoparticles. Under illumination, nanocomposite with immobilized colloidal TiO_2 nanoparticles successfully removed and degraded the dyes (AR18, AB113, RB5 and DB78) with the highest removal (55%) of RY17 dye. Reproducibility study was performed for the dye AR18 and after four cycles of illumination, the removal rate was 75%, indicating that prepared TiO_2/hydrogel nanocomposites could be reused without reduction in photocatalytic efficiency

3.2 Inorganic Polymer based TiO2 nanocomposites for degradation of organic pollutants

TiO_2– Montmorillonite(Mt)/polythiophene (PTP) 20%-sodium dodecyl sulfate (SDS) was prepared by in- situ chemical oxidative polymerization to show improved visible light absorption. The synthesized product was analyzed by FTIR, XRF, XRD, and FESEM, and the results indicated that the montmorillonite was successfully modified by TiO_2, followed by PTP, in the presence of surfactant (SDS) [43]. The sonocatalytic, photocatalytic, and sonophotocatalytic degradation processes of R6G were examined in the presence of heterogeneous photocatalysts (TiO_2–Mt/PTP20%-SDS, TiO_2–Mt and TiO_2-P25). The removal of the rhodamine 6G (R6G)from solution was observed and the results confirmed pseudo-first-order kinetics in the presence of all prepared catalysts which were examined by using three processes. The TiO_2–Mt/PTP20%-SDS was a most useful catalyst for R6G degradation. The incorporation of PTP and SDS surfactant into the TiO_2–Mt surface promoted the ease of electron transfer from excited polythiophene polymer to the conduction band of TiO_2. This results from the synergistic effect of the anion surfactant present in the nanocomposite.

TiO_2/clay nanocomposites were synthesized by dispersion of TiO_2 on the surfaces of a natural montmorillonite and a synthetic hectorite with increased sorption ability of TiO_2.This synthetic procedure allowed the formation of delaminated layers for hectorite–TiO_2 samples, whereas in case of montmorillonite–TiO_2 composites, the formation of a more lamellar-like aggregation occurred [44]. The photocatalytic efficiency of the nanocomposites was evaluated by using chloroacetanilide herbicide (dimethachlor) in water as a model compound. Nanocomposites exhibited good photodegradation efficiency and the overall removal efficiency of TiO_2 was better than that of bare TiO_2 produced by the sol-gel method. Because of good sedimentation ability, the nanocomposites could be studied as a hopeful alternative for the removal of organic water contaminants.

TiO_2/montmorillonite composites with intercalation structures containing anatase phase of TiO_2 was prepared [45]. The catalytic properties of TiO_2/montmorillonite composites were tested in the photooxidation of phenol. The total organic carbon (TOC) data showed that the photocatalytic efficiency of TiO_2 can be significantly enhanced by intercalation in montmorillonite support.

3.3 Metal oxide-based TiO_2 nanocomposites for degradation of organic pollutants

The TiO_2/SiO_2@Fe_3O_4 microspheres were synthesized by two sol-gel steps [46]. The as-prepared 9%TiO_2/6%SiO_2@Fe_3O_4 microspheres showed a core-shell structure which was composed of a SiO_2@Fe_3O_4 core of 205 nm in diameter and a porous TiO_2 layer of 5-6 nm in thickness. The anatase form of a TiO_2 layer is composed of primary anatase TiO_2 nanocrystals with a mean size of 3-5 nm and irregular pores with pore sizes of 2-3 nm with highest BET surface area of 373.5 m^2/g. The samples showed a high degree of superparamagnetism and the maximum saturation magnetization (Ms) value of 9%TiO_2/6%SiO_2@Fe_3O_4 microspheres was about 66.4 $A \cdot m^2/kg$. The UV-vis diffuse reflection spectra of synthesized microspheres exhibited redshift towards visible region. The photocatalytic activity for the degradation of methyl orange and methylene blue in the presence of 9%TiO_2/6%SiO_2@Fe_3O_4 was studied. The results confirmed that these two dyes were completely decolorized in 60 min under the irradiation of high-pressure mercury lamp and can be reused for 6 times with maintained high photocatalytic performance.

The novel superparamagnetic Bi_2O_4/Fe_3O_4 nanocomposite was prepared by an in-situ growth method and utilized for photocatalytic removal of Ibuprofen [47]. Ibuprofen (IBU) is one of the persistent organic pollutants (POPs) which can cause adverse effects in humans and wildlife. The effective removal of IBU from water is a worldwide necessity.The structural characterization of the Bi_2O_4/Fe_3O_4 nanocomposite confirms the formation of the diameter of 10 nm with highly assembled nanorods of diameter 120 nm. Under visible light irradiation, the synthesized nanocomposites resulted in a complete photocatalytic degradation of IBU within 2 h, which is 1.7 times higher efficiency than pure Bi_2O_4, and complete mineralization of IBU with an irradiation time of 4 h. The photocatalytic mechanisms of Bi_2O_4/Fe_3O_4 (1:2.5) revealed that the enhanced photocatalytic performance was mainly due to the separation of electron-hole pairs after surface modification of Fe_3O_4, and photogenerated photocatalytic removal of IBU. Additionally, these can be magnetically recycled and shows good reusability without loss of their photocatalytic activity or structural change even after reusing over five cycles.

High-thermostability ordered mesoporous TiO_2/SiO_2 nanocomposites were successfully prepared through evaporation-induced self-assembly method combined with ethylenediamine bounding strategy and subsequent high-temperature calcinations (700°C) [48]. The prepared samples were characterized by thermo gravimetric-differential scanning calorimetry, X-ray diffraction, Raman, transmission electron microscopy and N_2 adsorption. The results indicated that the samples were in the form of ordered mesostructure and SiO_2 inhibited the anatase-to-rutile phase transformation at high temperature. Significantly, the experimental results showed that the high-thermostable ordered mesoporous TiO_2/SiO_2 nanocomposites played bifunctional roles on effectively adsorbing As (III) and completely oxidizing higher toxic As (III) to lower toxic As (V) under various pH values. Moreover, after recycling for ten times the composites kept wonderful photocatalytic and adsorption performances.

Novel heterostructured ZnO/TiO_2 nanocomposites photocatalyst (Z9T) was prepared by sol-gel method and characterized by X-ray diffraction (XRD), Scanning Electron Microscopy (SEM) and UV–vis diffuse reflectance spectroscopy. The degradation efficiency with metal oxide nanocomposite for degradation of pesticides (quinalphos and monocrotophos) was found comparable to TiO_2 [49]. Mesoporous WO_3–TiO_2 nanocomposites at different WO_3 contents (0–5 wt%) have been used for photocatalytic degradation of the herbicide under visible light and UV illumination [50]. Under UV illumination, the overall photocatalytic efficiencies of the 3% WO_3–TiO_2 nanocomposite were 3.5 and 6.6 times higher than that of mesoporous TiO_2 and commercial UV-100 photocatalyst, respectively

Magnetic recoverable $CoFe_2O_4$@TiO_2 decorated reduced graphene oxide nanocomposite was prepared and used for investigating the photodegradation of chlorpyrifos [51]. The photocatalytic degradation of chlorpyrifos followed the pseudo-first-order kinetic model. The results indicated that the nanocomposite exhibited a high efficient photocatalytic activity for the photodegradation of chlorpyrifos. The nanocomposite was separated from the solution by a magnet and reused after the photodegradation of chlorpyrifos. The recyclable of the nanocomposite was economically significant in the industry.

Heterostructured nanoporous Bi_2O_3-TiO_2 (BTO) was synthesized by a greener approach using ultrasonication technique and was calcinated at different temperatures. Doping with Bi_2O_3 led to the lowering of the band gap of TiO_2 from 3.25 eV to 2.5 eV. The kinetic study data of the degradation of quinalphos showed pseudo first order path with a rate constant of 0.01267 1/min. The synthesized BTO mixed oxide nanocomposites showed profound improvement in photocatalytic activity in the visible region as compared to TiO_2 [52].

3.4 Noble metal/ carbon-based TiO_2 nanocomposites for degradation of organic pollutants

The Ag-TiO_2/reduced graphene oxide (r-GO) (ATG) nanocomposite was synthesized by a one-pot solvothermal method[53]. All the ATG samples acquired a large surface area, which can provide more active sites to adsorb more reactant molecules. The excellent properties of r-GO and strong ability to absorb photogenerated electron of Ag nanoparticle enhanced the longer lifetime of an electron-hole pair, which is essential for photocatalytic activity. The photocatalytic activities of all ATG samples were superior to pure TiO_2 both for degrading Rh B dye and the reduction of CO_2 under visible light. The possible mechanism of the photocatalytic process has been discussed. The dyes can be oxidized by strong oxidizing power radical. This study paved a way for one-pot synthesis of graphene-based noble metal-doped semiconductor nanocomposite, which may be used in environmental fields.

An efficient, simple, cost-effective and environmentally benign method was developed for the preparation of the Ag/RGO/TiO_2 nanocomposite using leaf extract of E. helioscopia L. as a stabilizing and reducing agent [54]. The main advantages of this method include the elimination of toxic, expensive and harmful chemicals, the use of water as a solvent, easy-synthesis of the Ag/RGO/TiO_2 nanocomposite under environmental-friendly conditions, easy separation of the catalyst and experimental ease.The photocatalytic activity of the Ag/RGO/TiO_2 nanocomposite was checked by examining the reduction of aqueous solutions of 4-nitrophenol (NP), congo red (CR) and methylene blue (MB) dyes in the presence of $NaBH_4$ at room temperature. The progress of the reaction was monitored by using UV– visible measurements at regular intervals of time. The catalyst was recovered and reused for several times with no significant loss of photocatalytic activity. Table 2 summarizes the research work carried out using TiO_2 based nanocomposites for photocatalytic degradation of organic pollutants

Conclusions

Chemical oxidation technology is the most potential technology for future applications in the field of environmental remediation i.e. the elimination of pollutants from the environment matrix. Titania based heterogeneous photocatalytic oxidation has received the attention of researchers for many years as an alternative method for purification of both air and water system. The major disadvantages that limit the practical application of titanium dioxide include low absorption in the visible region, the efficiency of the process and the recovery as well as recycling of the catalyst. TiO_2 based nanocomposites offer the potential to overcome these bottlenecks. The synthesis and applications of TiO_2 based nanocomposite materials for photocatalytic degradation of organic pollutants has been discussed in the chapter. The environmental applications of TiO_2 based nanocomposites are interesting and endless. However, further insights into the interactions among the hybrid materials need to be revealed using modern analytical techniques. Further, the interplay among the nanocomposites and the targeted pollutants or the substrates needs to be investigated for understanding mechanism and potential applications in wastewater treatment.

Table 2 Summary of the research work using TiO_2 based nanocomposites for photocatalytic degradation of organic pollutants.

S.No.	Name of catalyst	Method of preparation	Photocatalytic activity	Ref.
Organic Polymer based TiO_2 nanocomposites for degradation of organic pollutants				
1	Poly(Vinyl alcohol)/ poly(acrylic acid)/ TiO_2/ graphene oxide nanocomposites	Radical polymerization and condensation reaction	CBB and MB	[40]
2	PVA/PAAc hydrogels TiO_2 photocatalyst	Radical polymerization and electro-spinning method	Dye	[41]
3	TiO_2/ hydrogel nanocomposites	-	Textile azo dyes	[42]
Inorganic Polymer based TiO_2 nanocomposites for degradation of organic pollutants				
4	TiO_2-MT/PTP/ SDS nanocomposites	Chemical oxidative polymerization	Rhodamine 6G	[43]
5	TiO_2/ clay nanocomposites	Precipitation method	Herbicides (dimethachlor)	[44]
6	TiO_2/Montmorillonite nanocomposites	sol-gel method	phenol	[45]
Metal oxidebased TiO_2 nanocomposites for degradation of organic pollutants				
7	TiO_2/ SiO_2/ Fe_2O_4 microsphere	Solvothermal method	Methyl orange and methylene blue	[46]
8	Bi_2O_4/ Fe_3O_4 hybrid nanocomposites	Insitu growth method	Ibuprofen removal	[47]
9	TiO_2/ SiO_2 nanocomposites	Evaporation-induced self-assembly method	Removal of arsenic contaminants	[48]
10	Heterostructured TiO_2	-	Removal of organophosphate pesticides	[49]
11	Mesoporous WO_3-TiO_2 nanocomposites	Sol-gel method	Herbicides (Imazapyr)	[50]
12	$CoFe_2O_4$/ TiO_2 decorated reduced graphene oxide nanocomposites	Coprecipitation method	Chlorpyrifos	[51]
13	Bi_2O_3-TiO_2	Ultrasonication method	Removal of Quinalphos pesticide	[52]
Noble Metal /carbon-based TiO_2 nanocomposites for degradation of organic pollutants				
14	Ag/RGO/TiO_2 nanocomposites	Modified hummer method and ultrasonication	4-NP, Congo red and methyl blue	[53]
15	Ag-TiO_2/G-GO nanocomposites	Solvothermal method	RhB dye and reduction of CO_2	[54]

References

[1] C.D.S. Tomlin, (Ed.), The pesticide manual (15th Ed.). Hampshire, Eng- land: BCPC (2009).

[2] M.W. Luckenbach, P.D. Fur, M.L. Kellogg, P.V. Veld, Potential effects of endocrine disrupting compounds on bivalve populations in Chesapeake Bay: A review of current knowledge and assessment of research needs, Chesapeake Research Consortium CRC Publication No. 10-170 Edgewater, MD. (2009).

[3] C.M. Browner, J.C. Fox, S. Frace, M.B. Rubin, F. Hund, Development document for final effluent limitations guidelines and standards for the pharmaceutical manufacturing point source category, EPA 821-B-98-009; Engineering and Analysis Division, U.S. Environ-mental Protection Agency (EPA): Washington, DC, USA. (1998).

[4] H.D. Mansilla, M.C. Yeber, J. Freer, J. Rodriguez, J. Baeza, Homogeneous and heterogeneous advanced oxidation of a bleaching effluent from the pulp and paper industry, J. Water Sci. Technol. 35 (1997) 273-278.

[5] G. Alhakimi, S. Gebril, L.H. Studnicki, Comparative photocatalytic degradation using natural and artificial UV-light of 4-chlorophenol as a representative compound in refinery wastewater, J. Photochem. Photobiol. A: Chem. 157 (2003) 103-109. https://doi.org/10.1016/S1010-6030(03)00038-8

[6] M. Noorjahan, M.P. Reddy, V.D. Kumari, B. Lavedrine, P. Boule, M. Subrhamanyam, Photocatalytic degradation of H-acid over a novel TiO_2 thin film fixed bed reactor and in aqueous suspensions, J. Photochem. Photobiol. A: Chem. 156 (2003) 179-187. https://doi.org/10.1016/S1010-6030(02)00408-2

[7] I.M. Arabatizis, S. Antonaraki, T. Stergiopoulos, A. Hiskia, E. Papaconstantiou, M.C. Bernard P. Falaras, Preparation, characterization and photocatalytic activity of nanocrystalline thin film TiO_2 catalysts towards 3,5-dichlorophenol degradation, J. Photochem. Photobiol. A: Chem. 149 (2002) 237-245. https://doi.org/10.1016/S1010-6030(01)00645-1

[8] F. Kiriakidou, D.I. Kondarides, X.E. Verykios, The effect of operational parameters and TiO_2-doping on the photocatalytic degradation of azo-dyes, Catal. Today 54 (1999) 119-130. https://doi.org/10.1016/S0920-5861(99)00174-1

[9] B. Sun, and P.G. Smirniotis, Interaction of anatase and rutile TiO2 particles in aqueous photooxidation, Catal. Today 88 (2003) 49-59. https://doi.org/10.1016/j.cattod.2003.08.006

[10] K. Namura, Minerals. Crystal structure gallery. Available Source: http://staff.aist.go.jp/nomura-k/english/itscgallary-e.htm. (2002).

[11] Y. Hu, H. L. Tsai, C.L. Huang, Effect of brookite phase on the anatase-rutile transition in titania nanoparticles, Eur. Ceram. Soc. 23 (2003) 691-696. https://doi.org/10.1016/S0955-2219(02)00194-2

[12] D. Nicholls, Complexes and First-Row Transition Elements; MacMillan Education: Hong Kong (1974). https://doi.org/10.1007/978-1-349-02335-6

[13] Y. Shao, D. Tang, J. Sun, Y. Lee, W. Xiong, Lattice deformation and phase transformation from nano-scale anatase to nano-scale rutile TiO_2 prepared by a sol-gel technique. China Part. 2 (2004) 119.

[14] J. Fisher, T.A. Egerton, Titanium compounds, inorganic, Kirk-Othmerencyclopaedia of chemical technology. Wiley, New York (2001).

[15] K. Madhusudan Reddy, S.V. Manorama, A. Ramachandra Reddy, Bandgap studies on anatase titanium dioxide nanoparticles, Mater. Chem. Phys. 78 (2003) 239. https://doi.org/10.1016/S0254-0584(02)00343-7

[16] N. Serpone, Is the band gap of pristine TiO_2 narrowed by anion- and cation-doping of titanium dioxide in second-generation photocatalysts, J. Phys. Chem. B. 110 (2006) 24287. https://doi.org/10.1021/jp065659r

[17] N. Daude, C. Gout, Electronic band structure of titanium dioxide, J. Phys. Rev. B. 15 (1977) 3229-3235. https://doi.org/10.1103/PhysRevB.15.3229

[18] G. Peters, V. Vill, Index of modern inorganic compounds. Subvolume A. Landolt-Bo¨rnstein numerical data and functional relationships in science and technology, Verlag, Berlin (1989).

[19] J.K. Burdett, T. Hughbanks, G.J. Miller, J.W. Richardson, J.V. Smith, Structural-electronic relationships in inorganic solids: powder neutron diffraction studies of the rutile and anatase polymorphs of titanium dioxide at 15 and 295 K, J. Am. Chem. Soc. 109 (1987) 3639-3646. https://doi.org/10.1021/ja00246a021

[20] O. Carp, C.L. Huisman, A. Reller, Photoinduced reactivity of titanium dioxide, Prog. Solid State Chem. 32 (2004) 33–117. https://doi.org/10.1016/j.progsolidstchem.2004.08.001

[21] J.M. Herrmann, Heterogeneous photocatalysis: fundamentals and applications to the removal of various types of aqueous pollutants, Catal. Today 53 (1999) 115–129. https://doi.org/10.1016/S0920-5861(99)00107-8

[22] N. Serpone, Brief introductory remarks on heterogeneous photocatalysis, Sol. Energy Mater. Sol. Cells. 38 (1995) 369-379. https://doi.org/10.1016/0927-0248(94)00230-4

[23] D.W. Bahnemann, Mechanisms of organic transformations on semiconductor particles, in: Pelizzetti, E.; Schiavello, M. (Eds.), Photochemical Conversion and Storage of Solar Energy; Kluwer Academic Publishers; The Netherlands. 251–276. (1991). https://doi.org/10.1007/978-94-011-3396-8_15

[24] P. Suppan, Chemistry and Light. Royal Society of Chemistry: Cambridge, 107. (1994).

[25] A. Hoffman, E.R. Carraway, M. Hoffman, Photocatalytic production of H_2O_2 and organic peroxides on quantum-sized semiconductor colloids, Environ. Sci. Technol. 28 (1994) 776-785. https://doi.org/10.1021/es00054a006

[26] F. Mahdavi, T.C. Burton, Y. Li, Photoinduced reduction of nitro compounds on semiconductor particles, J. Org. Chem. 58 (1993) 744-746. https://doi.org/10.1021/jo00055a033

[27] W. Choi, A. Termin, M.R. Hoffmann, The role of metal ion dopants in quantum-sized TiO_2: Correlation between photoreactivity and charge carrier recombination dynamics, J. Phys. Chem. B 98 (1994) 13669–13679. https://doi.org/10.1021/j100102a038

[28] K. Vinodgopal, P.V. Kamat, Enhanced rates of photocatalytic degradation of an azo dye using SnO_2/TiO_2 coupled semiconductor thin films, Environ. Sci. Technol. 29 (1995) 841-845. https://doi.org/10.1021/es00003a037

[29] P. Kaur, D. Sud, Photocatalytic degradation of quinalphos in aqueous TiO_2 suspension: Reaction pathway and identification of intermediates by GC/MS, J. Mole. Catal. A: Chem. 362 (2012) 32-38. https://doi.org/10.1016/j.molcata.2012.08.005

[30] Z. Xu, J. Shang, C. Liu, C. Kang, H. Guo, Y. Du, The preparation and characterization of TiO_2 ultrafine particles, Mater. Sci. Eng. B. 63 (1999) 211–214. https://doi.org/10.1016/S0921-5107(99)00084-7

[31] Y. Li, D.S. Hwang, N.H. Lee, S.J. Kim, Synthesis and characterization of carbon-doped titania as an artificial solar light sensitive photocatalyst, Chem. Phys. Lett. 404 (2005) 25-29. https://doi.org/10.1016/j.cplett.2005.01.062

[32] L.B. Reutergarth, M. Iangpashuk, Photocatalytic decolourization of reactive azo dye: A comparison between TiO_2 and us photocatalysis. Chemosphere 35 (1997) 585–596. https://doi.org/10.1016/S0045-6535(97)00122-7

[33] S. Sakthivel, B. Neppolian, M.V. Shankar, B. Arabindoo, M. Palanichamy, V. Murugesan, Solar photocatalytic degradation of azo dye: comparison of

photocatalytic efficiency of ZnO and TiO_2, Sol. Energy Mater. Sol. Cells, 77 (2003) 65-82. https://doi.org/10.1016/S0927-0248(02)00255-6

[34] S. Rehman, R. Ullah, A.M. Butt, N.D. Gohar, Strategies of making TiO_2 and ZnO visible light active, J. Hazard. Mater. 170 (2009) 560–569. https://doi.org/10.1016/j.jhazmat.2009.05.064

[35] D. Li, and H. Haneda, Synthesis of nitrogen-containing ZnO powders by spray pyrolysis and their visible-light photocatalysis in gas-phase acetaldehyde decomposition, J. Photochem. Photobiol. A: Chem. 155 (2003) 171–178. https://doi.org/10.1016/S1010-6030(02)00371-4

[36] P. Wu, R. Xie, K.J. Shang, Enhanced visible-light photocatalytic disinfection of bacterial spores by palladium-modified nitrogen-doped titanium oxide, J. Am. Ceram. Soc. 91 (2008) 2957–2962. https://doi.org/10.1111/j.1551-2916.2008.02573.x

[37] J.A. Cha, S.H. An, H.D. Jang, C.S. Kim, D.K. Song, T.O. Kim, Synthesis and photocatalytic activity of N-doped TiO_2/ZrO_2 visible-light photocatalysts. Adv. Powder Technol. 23 (2011) 717–723. https://doi.org/10.1016/j.apt.2011.09.003

[38] Y. Cong, J. Zhang, F. Chen, M. Anpo, Synthesis and characterization of nitrogen-doped TiO_2 nanophotocatalyst with high visible light activity, J. Phys. Chem. C 111 (2007) 6976–6982. https://doi.org/10.1021/jp0685030

[39] S.M. Gupta, M. Tripathi, A review of TiO_2 nanoparticles, Chinese Sci. Bull. 56 (2011) 1639-1657. https://doi.org/10.1007/s11434-011-4476-1

[40] M. Young-E, J. Gowun, Y. Jumi, K. Hyung, Poly(vinyl alcohol)/poly(acrylic acid)/TiO_2/graphene oxide nanocomposite hydrogels for pH-sensitive photocatalytic degradation of organic pollutants, Mater. Sci. Eng. B 178 (2013) 1097–1103. https://doi.org/10.1016/j.mseb.2013.07.002

[41] Y. Jumi, I.J. Sun, O. Aeri, J. Dong-Hwee, B. Tae-Sung, L. Young-Seak, K. Hyung, pH-sensitive photocatalytic activities of TiO_2/poly(vinyl alcohol)/poly(acrylic acid) composite hydrogels, Mater. Sci. Eng. B 176 (2011) 276–281. https://doi.org/10.1016/j.mseb.2010.11.011

[42] L. Marija, M. Nedeljko, R. Maja, Š. Zoran, R. Marija, K.M. Kalagasidis, The potential application of TiO_2/hydrogel nanocomposite for removal of various textile azo dyes, Sep. Purif. Technol. 122 (2014) 206–216. https://doi.org/10.1016/j.seppur.2013.11.002

[43] N.K. Boutoumi, H. Boutoumi, H. Khalaf, D. Bernard, Synthesis and characterization of TiO_2–montmorillonite/polythiophene-SDS nanocomposites: Application in the sonophotocatalytic degradation of rhodamine 6G, Appl. Clay Sci. 80–81 (2013) 56–62. https://doi.org/10.1016/j.clay.2013.06.005

[44] V. Belessi, D. Lambropoulou, I. Konstantinou, A. Katsoulidis, P. Pomonis, D. Petridis, T. Albanis, Structure and photocatalytic performance of TiO_2/clay nanocomposites for the degradation of dimethachlor, Appl. Catal. B: Environ. 73 (2007) 292–299. https://doi.org/10.1016/j.apcatb.2006.12.011

[45] R. Kun, K. Mogyorósi, I. Dékány, Synthesis and structural and photocatalytic properties of TiO_2/montmorillonite nanocomposites, Appl. Clay Sci. 32 (2006) 99–110. https://doi.org/10.1016/j.clay.2005.09.007

[46] L. Hongfei, J. Shengfu, Z. Yuanyuan, L. Mingand, Y. Hao, Porous TiO_2-coated magnetic core-shell nanocomposites: Preparation and enhanced photocatalytic activity, Chinese J. Chem. Eng. 21 (2013) 569-576. https://doi.org/10.1016/S1004-9541(13)60521-2

[47] D. Xia, M.C. Irene, Synthesis of magnetically separable Bi_2O_4/Fe_3O_4 hybrid nanocomposites with enhanced photocatalytic removal of ibuprofen under visible light irradiation, Water Res. 100 (2016) 393-404. https://doi.org/10.1016/j.watres.2016.05.026

[48] Y. Wang, Z. Xing, Z. Li, X. Wu, G. Wang, W. Zhou, Facile synthesis of high-thermostably ordered mesoporous TiO_2/SiO_2 nanocomposites: An effective bifunctional candidate for removing arsenic contaminations, J. Coll. Interface Sci. 485 (2017) 32–38. https://doi.org/10.1016/j.jcis.2016.09.022

[49] P. Kaur, P. Bansal, D. Sud, Heterostructured nanophotocatalysts for degradation of organophosphate pesticides from aqueous streams, J. Korean Chem. Soc. 37 (2013) 382-388. https://doi.org/10.5012/jkcs.2013.57.3.382

[50] A.A. Ismail, I. Abdelfattah, A. Helal, S.A. Al-Sayari, L. Robben, D.W. Bahnemann, Ease synthesis of mesoporous WO_3–TiO_2 nanocomposites with enhanced photocatalytic performance for photodegradation of herbicide imazapyr under visible light and UV illumination, J. Hazard. Mater. 307 (2016) 43–54. https://doi.org/10.1016/j.jhazmat.2015.12.041

[51] V.K. Gupta, T. Eren, N. Atar, M.L. Yola, C. Parlak, H.K. Maleh, $CoFe_2O_4$@TiO_2 decorated reduced graphene oxide nanocomposite for photocatalytic degradation of chlorpyrifos, J. Mol. Liq. 208 (2015) 122–129. https://doi.org/10.1016/j.molliq.2015.04.032

[52] D. Sud, A. Syal, Investigations on phase transformation, optical characteristics and photocatalytic activity of synthesized heterostructurednanoporous Bi_2O_3-TiO_2, 63 (2016) 776–783.

[53] L. Zhang, C. Ni, H. Jiu, C. Xie, J. Yan, G. Qi, One-pot synthesis of Ag-TiO_2/reduced graphene oxide nanocomposite for high performance of adsorption and photocatalysis, Ceram. Int. 43 (2017) 5450–5456. https://doi.org/10.1016/j.ceramint.2017.01.041

[54] M. Nasrollahzadeh, M. Atarod, B. Jaleh, M. Gandomirouzbahani, In situ green synthesis of Ag nanoparticles on graphene oxide/TiO_2 nanocomposite and their catalytic activity for the reduction of 4-nitrophenol, congo red and methylene blue, Ceram. Int. 42 (2016) 8587–8596. https://doi.org/10.1016/j.ceramint.2016.02.088

Chapter 11

Industrial Water Pollution and Treatment - Can Membranes be a Solution?

Catia Giovanna Lopresto, Debolina Mukherjee, Karen Alejandra Bueno Zabala, Vincenza Calabro, Stefano Curcio, Sudip Chakraborty*

Department of DIMES, University of Calabria, Via P. Bucci, Cubo 42/a, 87036 Rende (CS) – Italy

sudip.chakraborty@unical.it*

Abstract

Industry is one of the main sources of water pollution since a long time and currently increasing day by day. Industries produces pollutants that are not only extremely harmful to people but also to the aquatic environment. Many industrial facilities use freshwater to carry away waste from the industrial plants into rivers, lakes and oceans. During the last century a huge amount of industrial wastewater was discharged into rivers, lakes and coastal areas. This resulted in serious pollution problems in the water bodies and caused negative effects to the eco-system and human's life. There are many types of industrial wastewater pollutants based on different industries and each sector produces its own particular combination of pollutants from the processes. The technology for the treatment of industrial wastewater must be designed more specifically for the particular type of pollutants produced. Membrane has made important contributions to the welfare of people with positive quality of life more than the majority of all other disciplines. In this chapter most of the industrial pollutants and their membrane based treatment mechanisms are described.

Keywords

Membrane, Bioreactor, Industrial Wastewater, Water Pollution, Environmental Impact, Membrane Materials

Contents

1. Introduction

Wastewaters from agro-industries are characterized by high levels of chemical oxygen demand (COD), together with the presence of nitrogen and phosphorus.

The classic technologies used for the treatment of industrial wastewater are insufficient to meet current demand. Advanced technologies in wastewater treatment include membrane processes, ion-exchange processes, adsorption processes, ozonation, UV/ozone/H_2O_2 combinations, Fenton's process, and nitrogen and phosphorus removal technologies. Membrane separation processes associated with a number of advantages, including appreciable energy savings, clean and easy operation, higher effectiveness, and greater flexibility in system design have been preferred. Membrane processes have recently become a great topic of research due to their applicability in wastewater treatment. Decreasing costs of installation and operation of membranes favored the use of membrane processes. Of the membrane processes, microfiltration and ultrafiltration are used mainly for primary treatment purposes while nanofiltration and reverse osmosis are used for final treatment. Specifically, reverse osmosis membranes offer such a high treatment efficiencies that they are used in a wide range of applications including recovery of materials from industrial wastewaters and treatment of sea water for drinking purposes.

The present chapter focuses on the various kinds of pollution present in food and agro alimentary industries. The remediation of the pollutants by application of membrane is also discussed in this chapter. An overview about the different pollutants and treatment technologies including membrane is presented. The use of different membranes in the remediation of agro food pollutants is described with more attention to downstream purification. A large volume of wastewater in the form of either oil-in-water (o/w) or water-in-oil (w/o) emulsions is generated from various process industries such as metallurgical, transportation, food processing and petrochemical as well as petroleum refineries. In the following sections, we will focus on oil-in-water emulsions derived from vegetable oil processing.

2. Industrial pollutants and their impact on water

Agro-industries are major contributors to worldwide industrial pollution. Effluents from many agro-food industries are hazardous to the environment and require appropriate and a comprehensive management approach [1]. The quantity, composition and concentration of different agro-food wastewaters depend on the final products, production processes, equipment used, and composition variations [2].

Table 1 Characteristics of typical agro-food industrial wastewater.

Industry	COD mg/L	BOD mg/L	pH	TS mg/L	TN mg/L	TP mg/L	Temperature °C	Reference
Distillery	85,000 - 1,000,000	42,000 - 51,000	3.1 - 4.7	70,000- 95,000	800- 1,200	180-- 350	85-95	[3]
Sugar factories	8,339- 9,033	1,010- 5,103	4.4 – 9.5	1,344	44-53	1 – 5	29.1	[2,4]
Dairies	346	50	7.3	250 – 2,750	42	3.3	--	[1,5]
Winery	1,800 – 21,000	--	--	150- 200	310- 410	40-60	--	[1]
Food procesing[a]	1,000- 8,000	600- 4,000	--	--	50	3	--	[1]
Olive mil	130,100	--	4.9	9,090	270	110	--	[1,6]

Notes: [a] contains flour, soybean, tomato, pepper, and salt.
TS: Total solids; TN: Total nitrogen; TP: Total phosphorus; BOD: Biochemical oxygen demand; COD: Chemical oxygen demand.

Table 1 summarizes the physicochemical characteristics of agro-food industrial wastewater reported by various authors.

As evident from the data presented in Table 1, there is significant variation in COD (346 – 1,000,000 mg/L) and BOD (50–51,000 mg/L). The pH and total solids (TS) concentration varies in the range of 3,1 – 9,5 and 6,150 – 95,000 mg/L, respectively; and significant amount of nutrients, 42 – 1,200 mg/L of total nitrogen (TN) and 1 – 350 mg/L of total phosphorous are also found in agro-food industry wastewater. The wastewaters impact colour and unpleasant odor in water bodies but some of wastewaters exhibit good biodegradability [7].

One of the main characteristics of agro-food wastewater is the amount of water consumption. The high amount of effluent produced, raises specific problems for the treatment process. These relate to the volume and composition of the wastewater produced and consequently treatment plants must be versatile in relation to the loading regime and at the same time be able to cope with a succession of start-ups and closedowns, including periods of inactivity [8].

The untreated wastewater issued by agro-food industries can cause water and soil pollution and therefore permanent risk of environmental pollution (Fig. 1) [5,9]. Due to high organic loading, the oxygen depleting potential of agro-food wastewater is 100 times that of domestic sewage [10]. The conductivity of the agro-food wastewater exceeds 1,500 μS/cm, which proves excessive mineralization of the wastewater. For turbidity, the value is greater than 50 NTU, proving that the analyzed water is cloudy [5] and its dark color obstructs photosynthesis and causes adverse effect on aquatic life [3].

Wastewaters also alter the physico-chemical characteristics of the receiving aquatic bodies and affect aquatic flora and fauna. When discharged into the environment, wastewaters create serious health hazards to the rural and semi-urban populations. Agro-food effluent has an obnoxious odour and unpleasant colour when released into the environment without proper treatment. Contaminants, such as chloride, sulphate, phosphate, magnesium and nitrate, are discharged with the effluent of various industries, which create a nuisance due to physical appearance, odour and taste. Such harmful water is injurious to plants, animals and human beings [2,4].

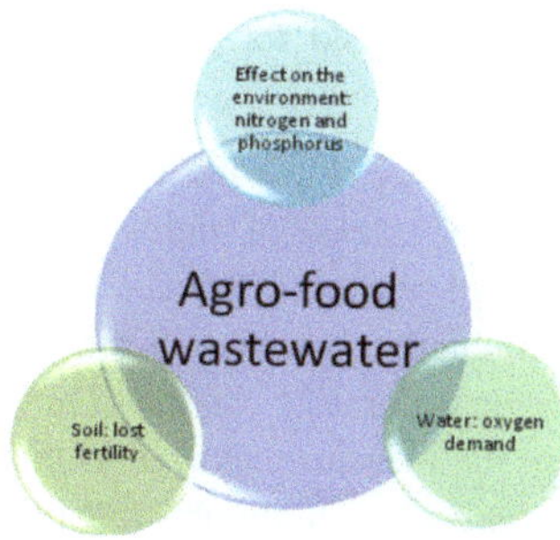

Figure 1 Effects of agro-food wastewater on the environment.

The major chemical pollutants in agro-food wastewater are nitrogen and phosphorus. The presence of nitrogen in discharged wastewater is undesirable because its ecological impacts and adverse effect on public health. A major problem in some plants is a low pH (pH = 6) which arises due to extensive nitrification and low wastewater alkalinity. Nitrogen in the form of ammonia is toxic to fish and exerts an oxygen demand on receiving water by nitrifiers [11].

Surface waters may also contain phosphorus in various compounds, which is an essential constituent of living organisms. In natural conditions, phosphorus concentration in waters is balanced; however, when phosphorus input to waters is higher than that can be assimilated by a population of living organisms, the problem of excess phosphorus content occurs. Since phosphate is the limiting component for growth in most ecosystems and the emission of phosphate in surface waters lead to eutrophication and algae bloom, thus having negative impacts on nature conservation, recreation, and drinking water production [11]. Algal blooms are esthetically undesirable and they alter the native composition and species diversity of aquatic communities, in addition to impair recreational values of surface waters, impede commercial fishing, and pose problems for water treatment. When deprived of oxygen, fishes and other aquatic organisms die, emitting foul odors [12].

Agro-food effluents contain carbohydrate organics, dissolved and suspended solids. The immediate oxygen demands by these effluents cause rapid depletion of dissolved oxygen in receiving streams and produce anaerobic conditions. High solids level would have an adverse impact on aquatic life, render the receiving water unfit for drinking and domestic purposes, reduce crop yields if used for irrigation, and exacerbate corrosion in water systems and pipes [13].

Agro-food wastewaters also affect the soil. The bacteria and fungi which maintain the soil fertility are affected by the release of highly toxic chemicals, causing concerns about the sustainability of continued reuse of treated wastewater in agriculture. The risks posed for human health represent another question at the heart of any discussion on wastewater reuse. These risks cannot be accurately estimated at the moment, but cannot be ignored. The evidences reported in the literature, as well as the critical analyses on the limitations of some experimental approaches highlight the importance of the accumulation and propagation of biological contaminants in soils due to wastewater irrigation. Human and animal pathogens, phytopathogens and antibiotic resistant bacteria and their genes are important biological contaminants that can be transported by wastewater and/or be enriched in soil. Also, numerous other chemical contaminants, such as xenobiotics, pharmaceuticals and metals, can threaten the environmental and human health. The mixture of these contaminants may have unpredictable consequences on both environment and human health [14].

3. Treatment technologies

The summary of opportunities and limitations of anaerobic, aerobic, physico-chemical and membrane treatment methods for agro-food wastewater is presented in Table 2.

Table 2 Opportunities and limitations of anaerobic, aerobic, physicochemical and membrane treatment methods for agro-food wastewater.

Treatment method	**Opportunities**	**Limitations**
Anaerobic	*Energy production is possible due to the generation of methane during degradation of organic matters. *Less sludge production. *Effluent quality in terms of COD is good but nitrogen removal is low.	*Oil and grease are not easily degraded. *Post-treatment of effluent is often required.
Aerobic	*Excellent effluent quality in terms of COD, BOD and nutrient removal. *Aerobic-SBOR has been reported to give high percentage of organics removal.	*Excess sludge produced is high. *Require larger area. *Complete removal of organics is not possible.

	*In the case of aerobic SBR, smaller area is needed as compared to other aerobic activated sludge processes.	*Aerobic SBR treatment systems offer better controlling.
Physico-chemical	*Coagulation/flocculation, adsorption and electrochemical methods are the various physicochemical methods.*Electro-oxidation was better treatment option in comparison to electro-coagulation. *No generation of secondary pollutants take place in electro-oxidation method.	*Chemical coagulation/flocculation process generates secondary pollutants. *In the case of electrocoagulation, treated effluent may be contaminated with electrode material.
Membranes	*Membrane assisted treatments methods such as RO, MF, NF, UF are most suitable to produce high-quality effluents to reuse directly. * Membrane separation can be used in dilute streams. *Effective in removal of recalcitrant contaminants. *Product recovery is possible. *Smaller footprint. *Improve nitrification and denitrification processes.	*Volumes pumped are very high comparing to other food industry branches, the solutions exhibit high viscosity and high osmotic pressure. * Membrane fouling might be severe at the later stage of treatment.

Notes: Reverse osmosis (RO), microfiltration (MF), nanofiltration (NF), ultrafiltration (UF). Adapted from [2,15,16]

Depending on the characteristics of the wastewater and the preferred final disposal routes, wastewater treatment methods can involve a simple single operation or a series of separate operations, involving multiple units. Anaerobic treatment method for concentrated wastewater, in terms of pollutants is a widely used method in the industries. It has several advantages over aerobic processes, which include lesser energy requirement; methane (a source of energy) production due to the degradation of organic; and lesser sludge production, which indirectly reduces the sludge disposal cost [2,17].

In case of aerobic treatment process (Table 2), aerated lagoons, aerated submerged fixed-film culture, and mixed culture activated sludge process have been used for the treatment of agro-food wastewater. However, future studies are hoped to focus on aerobic sequential batch reactor (SBR) treatment, a mixed culture activated sludge process. In the case of aerobic SBR, smaller area is needed as compared to other aerobic activated sludge processes. Both technologies, anaerobic and aerobic are complementary to one another in respect of removing organics and nitrogen completely from agro-food industry wastewater [17].

As agro-food wastewaters have high loading of dissolved solids and suspended solids, physicochemical methods like adsorption and coagulation are well suited for their treatment. Khan et *al* [18] reported coagulation with lime and subsequent adsorption with activated charcoal. BOD and COD removal efficiencies were reported to be 96 and 95%, respectively.

Membrane assisted treatment methods are very capable to produce high-quality effluent to reuse directly. Future research in MBR is likely to focus on reduction in energy demand and membrane fouling during the operation. Therefore, in view of producing good quality treated wastewater for reuse, hybrid system comprising of membranes with aerobic/anaerobic treatment methods and/or physicochemical methods may be promising [2].

4. Agro industrial wastewater

4.1 Olive mill wastewater

Olive oil production is an important agricultural and economic activity in Mediterranean countries. Olive oil processing (traditional press method or the most used three phase continuous process) produces large amounts of olive oil mill wastewaters (OOMW). OOMW are composed of olive pulpwater containing vegetable components and minerals, plus water components used for the olive washing, plant cleaning , olive paste dilution in the continuous process, and the residual olive pulp oil. The organic fraction contains sugars, tannins, polyphenols, polyalcohols, pectins, lipids, proteins and organic acids such as acetic, malic, fumaric, lactic, malonic, citric, tartaric, ossalic and succinic [19]. The composition of OOMW is not constant and exhibits great variability because it depends on a lot of parameters such as kind of olive and ripeness, oil extraction technology, and duration of aging.

OOMW from olive mills represent one of the greatest environmental problems in the Mediterranean area, where olives cultivation is widespread and large volumes of effluents are produced in a concentrated period of a few months. The polluting load depends on the high COD values and the presence of phyto-toxic and antibacterial polyphenolic components. Indeed, in these effluents, the presence of phytotoxic recalcitrant pollutants makes them resistant to biological degradation and thus inhibits the efficiency of biological and conventional processes [20].

OOMW are often illegally discharged or widespread in agricultural fields, with adverse consequences on the field fertility. Straight discharge of OOMW has been reported by several authors to cause strong odor nuisance, soil contamination, plants growth inhibition, underground leaks, water body pollution and hindrance of self-purification processes, as well as severe impacts on the aquatic fauna and ecological status, due to the presence of bio-refractory contaminants, including a wide variety of phenolic compounds, tannins, fatty acids and organohalogenated pollutants. Inorganic compounds including chloride, sulfate and phosphoric salts of potassium, calcium, iron,magnesium, sodium, copper and traces of other elements are also common traits of OOMW [20].

Flexible and efficient treatment plants should assure not only a significant reduction of BOD and COD values but also the possibility of selectively recovering some valuable compounds that could be used in the same production cycle or as raw materials for other processes.

Scientific studies on OOMW treatment principally concentrated on the removal of the polluting load [21]. Physical and physico-chemical processes include thermal processes (evaporation, gasification, combustion) [22], thermochemical processes [23], electro-coagulation [7,24], flocculation/clarification, adsorption, extraction [25], and ozonation [26]. Nevertheless, these conventional treatments are not capable to abate the high concentration of dissolved monovalent and divalent ions present in OOMW, which present hazardous salinity levels according to the guidelines established by the Food and Agricultural Organization (F.A.O.) for irrigation uses [20]. Generally, physico-chemical processes need an integrative biological treatment for the complete degradation of the organic substances [27]. Biological treatments can be divided into aerobic and anaerobic. Anaerobic processes have greater efficiency in the pull down of the pollutant load and produce biogas reusable as energetic resource [28]. Anyway, both in the aerobic and in the anaerobic processes, microorganism growth is particularly difficult for the antibacterial action of polyphenols. Although the efficiency of the coupled physico-chemical and biological treatment of OOMW is better, from an environmental point of view, the discharge of the great amount of sludge produced, with volumes comparable to that of the OOMW treated, is a great problem [29].

Recently, researches have began to consider the recovering of polyphenols, as high value compounds with biological activities [30], the antioxidant, anti-inflammatory activities and radicalic elimination [31], and the antimicrobic activity [32], transforming OOMW from effluents to raw material with high potential economic value [33,34].

In this case, technologies offer high efficiency and moderate investment and maintenance expenses [21,35], even if final treatment of OOMW by membrane processes has not been widely accepted, yet, and limited research papers have been published up to date. Recently, Pulido [20] discussed the state of the art of the different pretreatment methods and integral membrane processes proposed up to 2016 for the reclamation and disposal of the effluents generated in olive mills operating with batch-press and two-phase as well as three-phase processes, comprising microfiltration (MF), ultrafiltration (UF), nanofiltration (NF) and reverse osmosis (RO), along with membrane bioreactors (MBRs) [36], with an insight into the problem of fouling. MF and UF have been used mainly for primary treatment purposes, while NF and RO have been used for final treatment [37,38].

Several novel treatment approaches (Table 3) based on tangential flow membrane filtrations for the selective separation and total recovery of hydroxytyrosol, water and organic substances have been patented by ENEA and Verdiana Company [39]. MF and UF of OOMW without preliminary centrifugation step, operating with pilot plants with fixed process parameters were investigated by [38]. MF/UF permeates can be concentrated in RO. MF and UF permeates or RO concentrate can be used as functional integrators or in pharmacologic compositions. RO permeate can be used as base for the beverage industry. MF and UF concentrates can be used as fertilizers or in the production of biogas.

Membrane distillation (MD) is a non-isothermal process that has attracted the interest of academic and scientific communities during the last few years. It can be used successfully for OOMW with the objective to obtain pure water – which can be useful for irrigation or industrial uses – and a concentrate containing high amounts of polyphenols, which can be extracted later. Compared to other separation processes, MD has many advantages. It exhibits a complete rejection of dissolved, non-volatile species, lower operating pressure than the pressure-driven membrane processes, reduced vapor space, etc. The obtained concentrate of OOMW may represent a source of high added value compounds like sugar and polyphenols and the residue can be utilized in agriculture as a fertilizing agent through its high load of organic compounds [40].

One common problem of membrane filtration of OOMW is the membrane fouling that drastically reduces the efficiency of permeate and also changes its selectivity. Therefore, a pretreatment step is necessary to decrease membrane fouling and to increase filtration efficiency. The centrifuging process was found to be a promising pretreatment method [37]. Several MD processes have been identified (Figs. 2-4).

Hybrid processes, combining biological treatments, chemical processes and membrane technology, were developed too [41,42] in order to reach high removal efficiencies of

COD and mono-phenolic compounds. The important treatments processes based on membrane filtration are listed in Table 3.

Table 3 Several novel treatment approaches based on membrane filtration.

	Filtration System	**Membranes (manufacturer)**	**Material**	**Cut-off**	**Feed**	**Ref.**
MF	Tangential flow	Ceramic Tami	Zirconium oxide	0.8 µm 0.45 µm	Acidified OOMW	[38]
	Tangential flow	Ceramic Tami isoflux	Zirconium oxide	0.45 µm	Acidified OOMW	[38]
	Tangential flow	Polymeric Nadir	Spiral-wound PES	500 Kd	Acidified OOMW	[38]
	/	(Inopor GmbH)	Al_2O_3	/	Pre-filtered OOMW	[43]
	/	Osmonics JX	/	/	Pre-treated OOMW	[44]
	MBR	Carbosep (Novasep Orelis)	Carbon with ZrO_2–TiO_2	0.14 µm	Diluted OOMW	[36]
UF	Tangential flow	Polymeric Osmonics	Spiral-wound PS	80 Kd 20 Kd 6 Kd	MF permeates	[38]
	Tangential flow	Ceramic Tami	Zirconium oxide	1 Kd	MF permeates	[38]
	Cross-flow	JW Osmonics	PVDF	30 kDa	Pre-treated OOMW	[45]
	Cross-flow	MW Osmonics	Ultrafilic	100 kDa	Pre-treated OOMW	[45]
	Cross-flow	Membralox-Ceraver	Ultrafine porous ZrO_2 / coarse porous alumina	0.05 µm	Olive mill solid wastewater	[42]
	Cross-flow	UF-PES-004H (Macrodyn-Nadir)	Permanently hydrophilic PES	4 kDa	Pre-filtered OOMW	[19]
	Cross-flow	C005F (Macrodyn-Nadir)	Regenerated cellulose	5 kDa	Pre-filtered	[19]

					OOMW	
	Cross-flow	C010F (Macrodyn-Nadir)	Regenerated cellulose	10 kDa	Pre-filtered OOMW	[19]
	Cross-flow	P010F (Macrodyn-Nadir)	Permanently hydrophilic PES	10 kDa	Pre-filtered OOMW	[19]
	Cross-flow	Etna 01PP (Alfa Laval)	Composite fluoropolymer	1 kDa	UF permeates	[46]
	Cross-flow	/	Ceramic (zirconia)	/	Pre-filtered OOMW	[47]
	/	HFS (Toray)	PVDF	/	Pre-filtered OOMW	[46]
	/	UC010 (Macrodyn® Nadir)	Cellulose	10 kDa	Centrifuged OOMW	[37]
	/	Osmonics GM	/	/	MF permeates	[44]
	/	Desal-5 – GM4040F (Osmonics	Composite	/	Pre-treated OOMW	[48]
	/	PCI France	/	2-25-100 kDa	Pre-treated OOMW	[41]
	Micellar enhanced	/	PVDF	/	OOMW	[49]
NF	/	NF90 (DOW-Filmtec)	Thin-film	/	UFpermeates	[46]
	/	NP010 (Macrodyn® Nadir)	PES	/	UF permeates	[37]
	/	NP030 (Macrodyn® Nadir)	Permanently PES	/	UF permeates	[37]
	/	NF270 (DOW Filmtec™)	PA	200-300 kDa	UF permeates	[37]

	/	DK2540F Osmonics	/	0.5 nm	Pre-treated OOMW	[50]
	/	/	Polymeric	200	UF permeates	[47]
	/	Osmonics DK	/	/	MF permeates	[44]
	/	Desal-5 – DK4040F (Osmonics	Composite	/	UF permeates	[48]
RO	Tangential flow	Polymeric hydronautics	Composite PA	/	UF permeates	[38]
	/	XLE (DOW Filmtec™) BW30 (DOW Filmtec™)	PA	/	UF permeates	[37]
	/	/	Polymeric	100	UF permeates	[47]
	/	Osmonics SC	/	/	NF permeates	[44]
MD	Direct contact MD	GVHP (Millipore)	PVDF	/	OOMW	[40]
	Direct contact MD	TF200 (Gelman)	PTFE	/	OOMW	[40]
	Direct contact MD	TF200 (Gelman)	PTFE	/	Pre-treated OOMW	[51]
	Direct contact MD	TF200 (Gelman) TF450 (Gelman) TF1000 (Gelman)	PTFE/ PP	/	Micro-filtered OOMW	[52]
	Osmotic distillation	TF200 (Gelman) TF450 (Gelman) TF1000 (Gelman)	PTFE/ PP	/	OOMW	[53]
	Osmotic distillation	/	PP	/	NF permeates	[43]

	Vacuum MD	TF200 (Gelman) TF450 (Gelman) TF1000 (Gelman)	PTFE/PP	/	OOMW	[53]
	Vacuum MD	/	PP or PVDF	/	NF permeates	[43]

Note: PA: polyamide; PES: polyethersulfone; PP: polypropylene; PS: polysulfone; PTFE: polytetrafluoroethylene; PVDF: polyvinylidene difluoride.

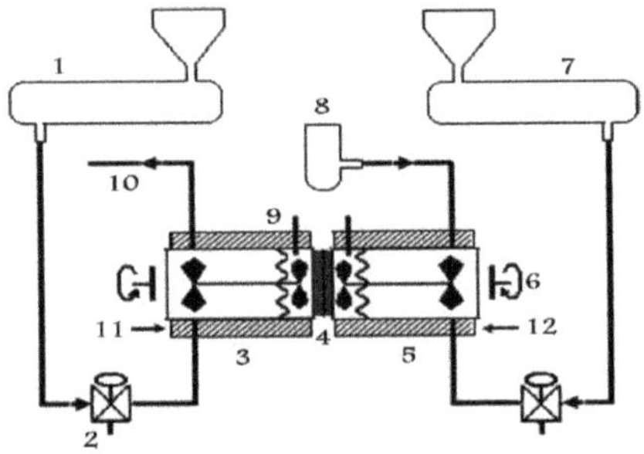

Figure 2 Direct contact membrane distillation system: (1) distilled water reservoir , (2) three-way valve, (3) cold unit, (4) flat sheet membrane module, (5) warm unit, (6) magnetic stirrer (7) feed reservoir, (8) feed tank to keep the feed volume constant (9) temperature sensor, (10) permeate outlet, (11) jacket connected to cold thermostat and (12) jacket connected to hot thermostat [40].

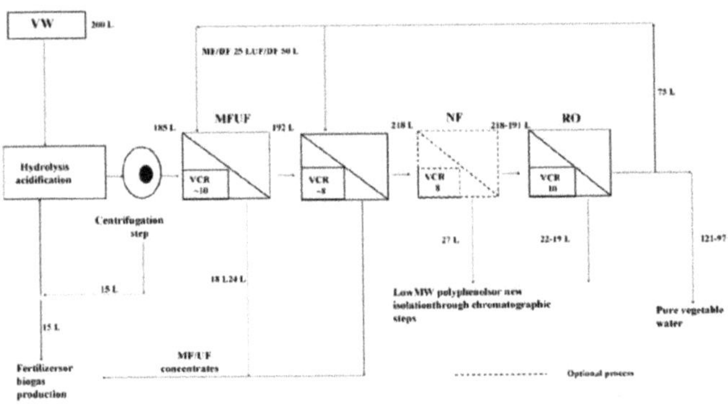

Figure 3 Hypothetical vegetation waters (VW) industrial treatment process [38].

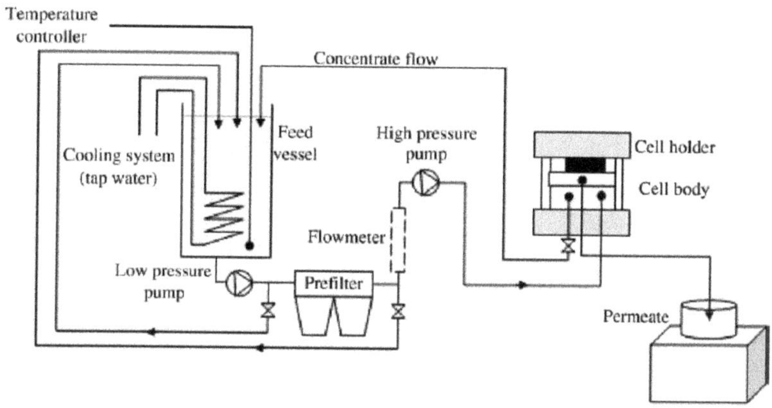

Figure 4 Schematic flow diagram of the experimental set-up [45].

Several integrated membrane processes (a sequence of two UF processes followed by a final NF step or one step of UF followed by NF and RO, or MF followed by NF and osmotic distillation or vacuum membrane distillation) have been also proposed to achieve high levels of purification of OOMWs and a water fraction which can be discharged in aquatic systems or to be reused in the olive oil extraction process [43,46,47,54]. Experimental results confirm that the future direction of the processes for the recovery of antioxidants from OOMW will be toward the utilization of membranes in a sequential design.

Micellar enhanced ultrafiltration (MEUF) treatment method for removal and concentration of polyphenols was developed, using an anionic surfactant (sodium dodecyl sulfate salt) and a hydrophobic poly(vinyldene fluoride) (PVDF) membrane [49]. Some of the schemes of the single or hybrid membrane processes are illustrated below in the form of Figs. 5-8.

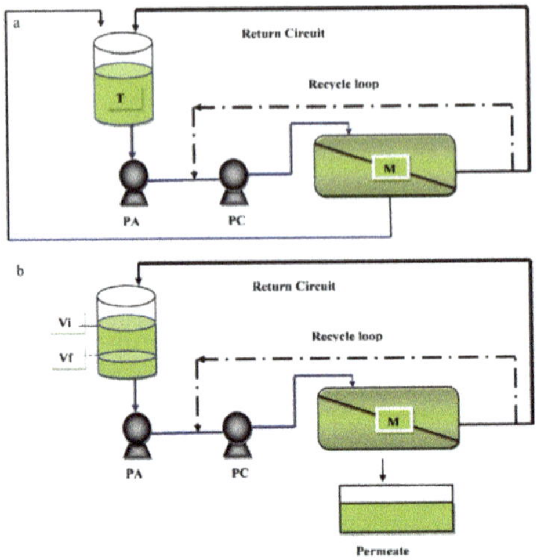

Figure5 Ultrafiltration process T: feed tank, PC: Recycling pump, PA: Feed Pump, M: Ultrafiltration module adapted from reference [42].

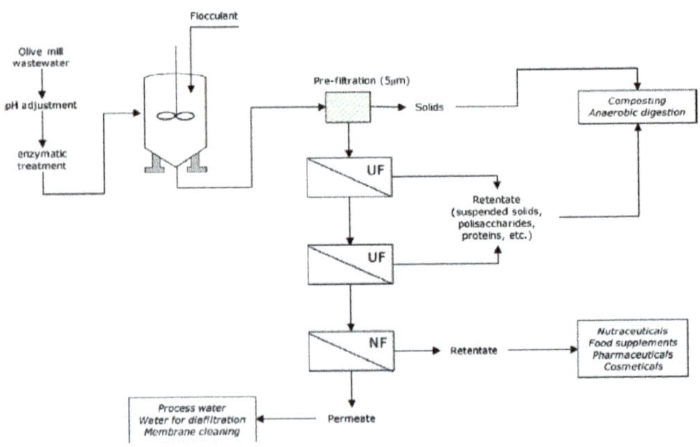

Figure 6 Proposed process scheme for the recovery of water and phenolic compounds from OOMWs [46].

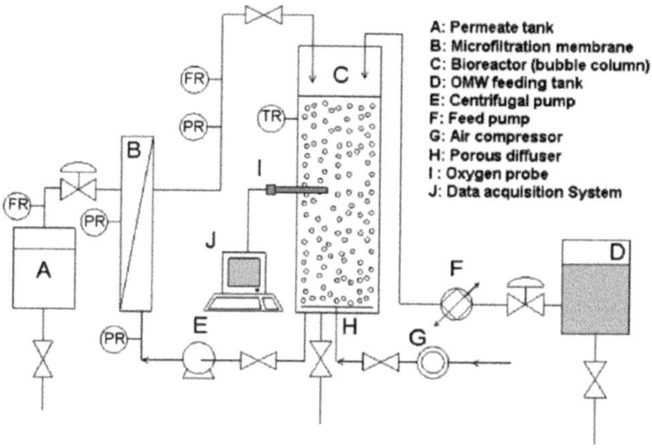

Figure 7 Schematic representation of membrane bioreactor used for OOMW treatment [36].

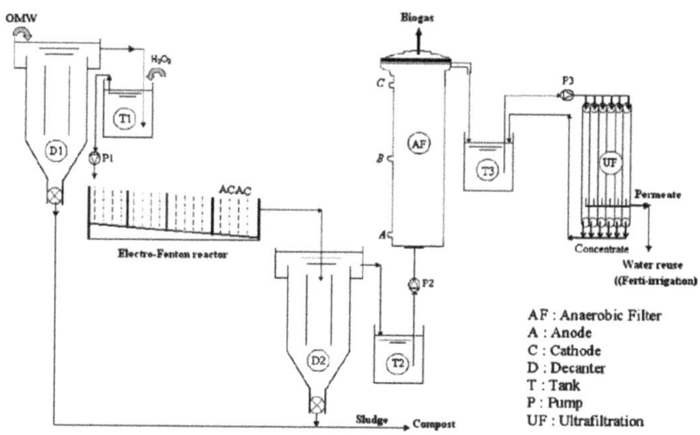

Figure 8 Schematic flow diagram of pilot plant for OOMW treatment [41].

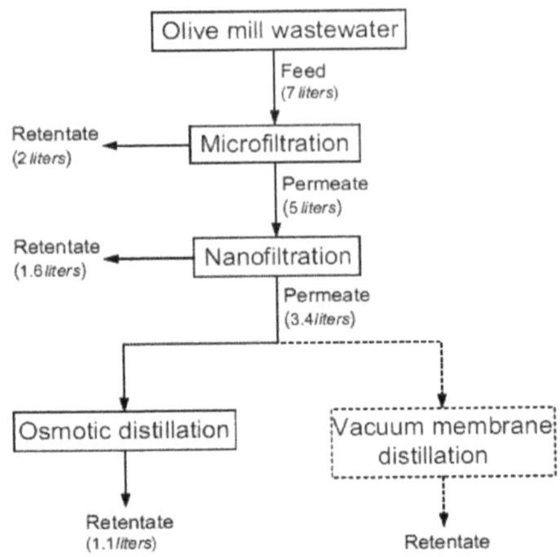

Figure 9 Flowchart of OOMW treatment processes.

In the above flowchart (Fig. 9), the activities carried out for the recovery, purification and concentration of polyphenols from olive mill wastewater are illustrated [43].

3.2 Palm oil mill wastewater

Palm oil is one of the world's most rapidly expanding equatorial crops. Indonesia and Malaysia are the two largest oil palm producing countries, but its by-product–palm oil mill effluent (POME), posed a great threat to the water environment. Palm oil processing, similar to other agricultural and industrial activities, raised environmental issues particularly water pollution, which adversely affects aquatic life and domestic water supply [55]. About half of the water used in the extraction process from the fresh fruit bunch results in highly polluting palm oil mille wastewater (POMW) as evident from Fig. 10. POMW is a colloidal suspension of water (95-96%), oil (0.6-0.7%) and total solids (4-5%). It is non-toxic as no chemicals are added during oil extraction but has an unpleasant odor. It is highly polluting and characterized by low pH (average pH: 3.5–4.2), enhanced biological and chemical oxygen demands ($BOD_{3d,\ 30°C}$: 10–44 g/L, COD_{cr}: 16–100 g/L), increased salt content, and high suspended solids (SS: 5–54 g/L) [56]. The typical POMW characteristics are reported in the following Table 4 [57].

Table 4 Characteristics of palm oil mill effluent [57].

Parameter	Concentration (mg/L)
Ammonical Nitrogen	35
BOD	25000
Boron	7.6
Calcium	439
COD	50000
Copper	0.89
Elements	
Iron	46.5
Magnesium	615
Magnesium	2
Oil & Grease	4000
pH	4.7
Phosphate	180
Potassium	2270
SS	18000
TDS	40500
Total Nitrogen	750
Total volatile solids	34000
Zinc	2.3

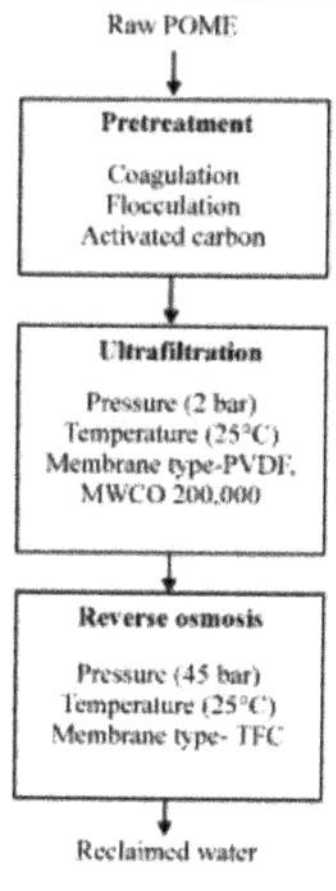

Figure 10 Flowchart of POME treatment using membrane technology [57].

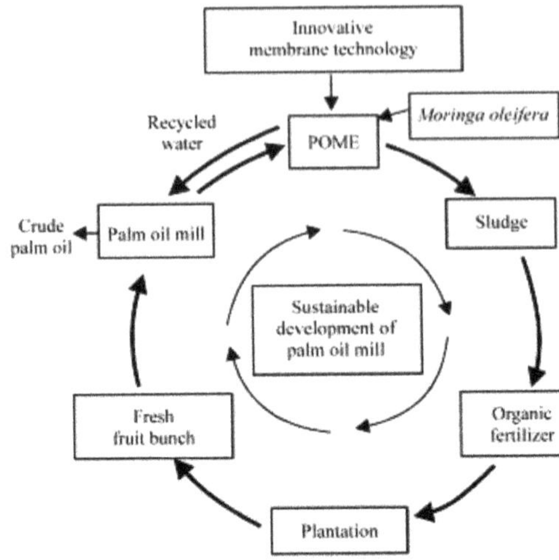

Figure 11 Sustainable development of palm oil mill with zero discharge using membrane technology [57].

The high content of suspended solids and organic matter in the effluent discharge can cause severe pollution of waterways due to oxygen-depletion and other related effects. POMW contains high concentrations of protein, carbohydrate, nitrogenous compounds, lipids and minerals that may be converted into useful materials using microbial processes [58].

There is an urgent need to find a compromising way that will enable the balance between environmental protection and sustainable reuse of the nutrient sources found in the POMW (Fig. 11).

Currently, POMWs are treated by conventional biological processes of anaerobic [59] or aerobic digestion [60]. Several researchers have proposed other biological treatment systems which include aerated lagoon system, conventional anaerobic digester, anaerobic contact process, up-flow anaerobic sludge blanket (UASB) reactor, close tank digester, trickling filter, aerobic lagoon system, aerobic rotating biological contactor and evaporation processes [55,61–63]. However, the proposed biological treatment systems are only confined to lab scale analysis. Moreover, the nutrient sources available in the

POMW cannot be effectively reused as a substrate in fermentation after the conventional treatment process has been adopted.

Membrane separation technology is recognized as an efficient, economical, sustainable and reliable technology that exhibits high potential to be applied in POMW treatment. In the last decade, very few membrane processes were investigated. Ultrafiltration (UF) has been successfully developed from a useful laboratory tool to an industrial process. The main bottleneck of membrane separation is membrane fouling. Therefore, effective techniques of membrane cleaning need to be studied to mitigate membrane fouling.

A two-stage pilot-scale plant was investigated for POMW treatment. Biodegradation constituted the first biological stage, while ultrafiltration (UF) and reverse osmosis (RO) membrane units were combined as the second membrane separation stage [56,61].

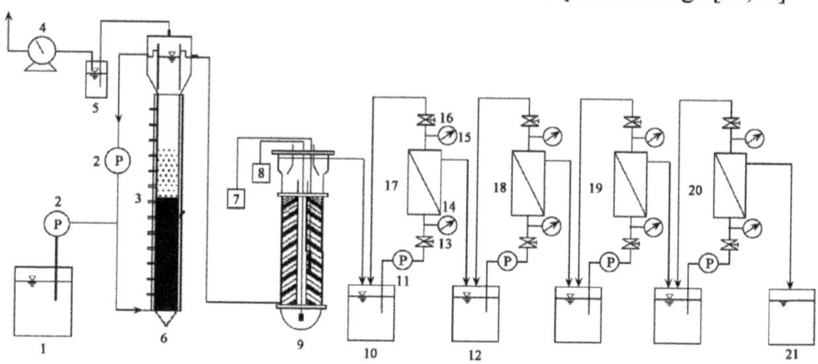

Figure 12 Schematic diagram of a pilot plant for palm oil mill effluent treatment [56].

The investigation on the feasibility and suitability of the membrane separation technology in POME treatment was carried out extensively in a pilot plant (Fig. 12) with the capacity of 450 L/h [64,65]. Three designs of industrial scale membrane based POMW treatment plant were investigated and optimized for evaluation of performance and cost, in order to develop and design an industrial scale membrane plant suitable for a typical palm oil mill in Malaysia [61] (Fig. 13).

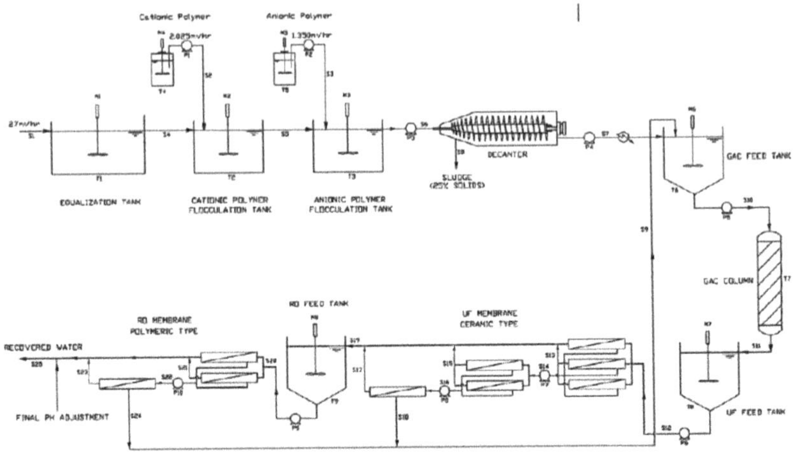

Figure 1 Process flow diagram of membrane based POME treatment system for design A [61].

Recently membrane bioreactors (MBRs) have been developed. However, high loading of MBR system was affected due to biofouling on the membrane surface. A treatment process is necessary to remove the high content of organics in POMW that would otherwise severely resulted in fouling of the membrane and to a shorter membrane life-span (Table 5). Powdered activated carbon, zeolite, and *Moringa oleifera* were used as biofouling reducers (BFRs) and added into aerobic MBR [66].

A membrane anaerobic system (MAS) as depicted in Fig. 14 was investigated in order to treat POMW [67].

Table 5 Membrane filtration systems used for pretreatment of POMW.

	Filtration System	Membranes (manufacturer)	Material	Cut-off	Feed	Ref.
UF	Dead-end	DSS-GR61PP (Alfa Laval)	PS	20000 g/mol	Pre-treated POMW	[58]
	Dead-end	DSS-GR70PP (Alfa Laval)	PS	20000 g/mol	Pre-treated POMW	[68]
	Dead-end	DSS- GR81PP (Alfa Laval)	PES	10000 g/mol	Pre-treated POMW	[68]
	Dead-end	DSS- GR95PP (Alfa Laval)	PES	2000 g/mol	Pre-treated POMW	[68]
	Cross-flow	(PCI-Memtech)	Ceramic Polymeric	/	Pre-treated POMW	[61]
	Cross-flow	/	/	200000 g/mol	/	[67]
	/	/	/	100000 g/mol	Pre-treated POMW	[56]
	/	/	PVDF	200000 g/mol	Pre-treated POMW	[57]
RO	Cross-flow	(PCI-Memtech)	Polymeric	/	UF permeates	[61]
	Cross-flow	(PCI-Memtech)	PVDF		Pre-treated POMW	[69]
	/	ESPA-2 (Hydranautics)	/	/	UF permeates	[56]
	/	/	TFC	/	UF permeates	[57]

Note: PES: polyethersulfone; PS: polysulfone; PVDF: polyvinylidene difluoride; TFC: thin-film composite.

3.3 Dairy industry

The dairy industry involves processing raw milk into products such as milk, butter, cheese, yogurt, condensed milk, dried milk (milk powder), and ice cream, using methods such as chilling, pasteurization, and homogenization. Water is used in all of the activities of the dairy industry, including cleaning, sanitization, heating, cooling, and floor washing. Consequently, nowadays, the dairy industry is considered the major source of food processing wastewater in regard to its large water consumption [70]. The effluent is characterized by variable volumes, flow rates and organic matter content. Dairy wastewater (DW) is distinguished by the higher BOD and COD values along with high levels of dissolved or suspended organic components (whey proteins, lactose, fats, oils, grease and minerals), nutrients (ammonia or minerals and phosphates), and cleaning chemicals (acids, alkalis and detergents) [71,72]. DWs, therefore, require specialized treatments to meet effluent discharge standards and to reduce the risk of environmental problems such as eutrophication in rivers, lakes and coastal waters.

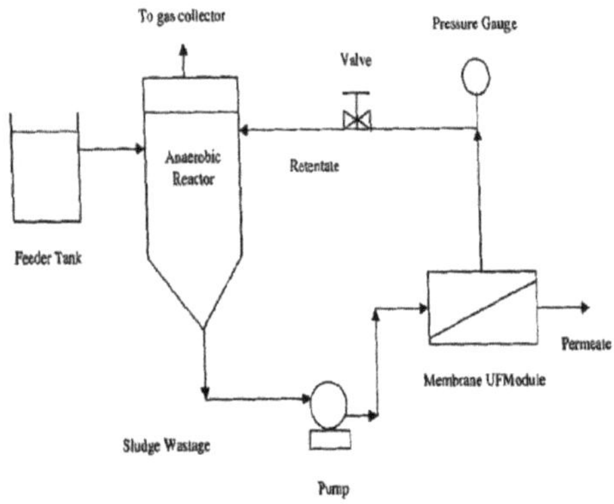

Figure 14 Membrane anaerobic system [67].

The highly variable nature of dairy wastewaters in terms of volume and flow rates as well as in terms of the pH and suspended solids content makes it difficult to choose an effective wastewater treatment regime. Conventional dairy wastewater treatment plants mainly based on activated sludge processes that involve the aerobic microbial metabolism of fats, lactose and proteins [73], anaerobic treatments [74–77], coagulation [78,79], and ecological treatment system [80]. Recently, there is an emerging interest focused on the energy recovery from wastes; in this field, microbial fuel cells (MFC) are gaining promising interest and a bio-electrochemical treatment system along with bioelectricity generation was used with real field dairy based wastewater as substrate [81].

Membrane technologies applied in the dairy industry since the early 1970s have been considered promising to treat DWs in order to produce reusable water [82]. Ultrafiltration (UF) of dairy wastewater yields a high permeate flux at low transmembrane pressure, but its permeate water is not reusable as it contains too much lactose [83]. Instead, nanofiltration (NF) and reverse osmosis (RO) permit the recovery of lactose and milk proteins for non-human consumption. In fact, the concentrate retentate could be precipitated by coagulation and reutilized as feed supplement for animals, or could be treated by anaerobic digestion to collect renewable energy sources (H_2 and CH_4). Thus, it is regarded as an economical and environment-friendly process for treatment of dairy wastewater. Permeate water obtained from NF/RO treatment of dairy wastewater can be discharged into river or reused, but with the increase of organic solutes and inorganic salts in retentate during a concentration process; concentration polarization and osmotic pressure increase rapidly, leading to a large flux decline. Consequently, the advantages of membrane filtration in wastewater treatment are adversely affected by concentration polarization and subsequent membrane fouling as these factors cause flux decline and permeate quality deterioration. One of the principal limitations on the optimal performance of membrane processes is membrane fouling, which leads to a decrease in membrane flux with time.

Milk proteins, lactose and mineral salts present in DW are the possible ingredients of fouling layer on/in NF membranes [84]. In particular, available data on the electrophoretic mobility of α-lactalbumin and β-lactoglobulin allowed to suggest that serum proteins would also participate in the flux variations with a possible specific impact of α-lactalbumin as an internal foulant in UF [85].

Membrane selection and operating conditions have been important issues in minimizing membrane fouling. To control the fouling and to improve the productivity and life of membranes, use of coagulant and adsorbent before membrane application were done in primary and secondary effluent treatments [72]. Frappart et al. [86] confirmed the high potential of high shear dynamic filtration in reverse osmosis due to a minimization of concentration polarization, resulting in higher permeate fluxes and better solute rejection than with cross-flow filtration modules equipped with the same membrane.

Another problem related to NF/RO treatment of dairy wastewater is the difficulty of nutrient recovery. That is, because cleaning chemicals contained in dairy wastewater, are all retained by NF/RO membranes, contaminate the nutrients (lipids, proteins and lactose), and lipids have a negative impact on anaerobic digestion because these are difficult to degrade and thus plug the sludge bed [87]. A two-stage UF/NF process was proposed for utilization of whey protein and lactose, as proteins were retained by UF membrane and lactose in UF permeate was concentrated by NF [88]. For dairy wastewater treatment, a similar approach was also applied [89,90]. This approach could simultaneously eliminate pollution, produce reusable water, and recycle waste.

In the dairy industry, the use of acid, alkaline cleaners and sanitizers affects wastewater characteristics and typically results in a highly variable pH, which will bring an impact on the wastewater on-line treatment (Fig. 15). The effect of pH on the treatment of dairy wastewater by nanofiltration was investigated by Luo & Ding [91] using a rotating disk membrane module. The role of calcium and inorganic phosphate over the wide pH range was discussed by taking calculated salt equilibrium of milk as a function of acidic pH [85].

There is a growing interest in combining membranes with biological wastewater treatment. The membrane bioreactors (MBR) offer distinct advantages over traditional biological processes: higher biodegradation efficiency, smaller footprint, better quality of treated water, the absolute control of solids and hydraulic retention time, retention of all microorganisms and viruses, and easy control of operating conditions [92]. In Table 6, the membrane treatments used for dairy wastewater are summarized.

Table 6 Membrane treatment used for dairy wastewater.

	Filtration System	**Membrane (manufacturer)**	**Material**	**Cut-off**	**Feed**	**Ref.**
MF	Cross-flow	(Orelis)	Ceramic	0.45 μm	Pre-treated DW	[72]
UF	Dead-end	(Millipore Corporation)	Cellulose acetate	1 kDa 10 kDa	Pre-treated DW	[72]
	Dead-end	UP005P (Microdyn-Nadir)	PES	5 kDa	Model DW	[89]
	Dead-end	UH030P (Microdyn-Nadir)	PES	30 kDa	Model DW	[89]
	Dead-end	Ultracel PLGC (Millipore)	Regenerated cellulose	10 kDa	Model DW	[89]
	Cross-flow	HFK-131 (Koch)	PES	150-300 $g \cdot mol^{-1}$	Skim milk	[85]
	Cross-flow (integrated into a Jet Loop MBR)	(Jiuwu Hitech, Chinese)	Ceramic (Al_2O_3/ZrO_2)	50 nm	DW	[92]
	Rotating disk membrane	UP005P (Microdyn-Nadir)	PES	5 kDa	Model DW	[93]

	Rotating disk membrane	P010F (Microdyn-Nadir)	PES	10 kDa	Model DW	[93]
	Rotating disk membrane	P020F (Microdyn-Nadir)	PES	20 kDa	Model DW	[93]
	Rotating disk membrane	UH030P (Microdyn-Nadir)	Permanently hydrophilic PES	30 kDa	Model DW	[93]
	Rotating disk membrane	UH050P (Microdyn-Nadir)	Permanently hydrophilic PES	50 kDa	Model DW	[93]
	Rotating disk membrane	PES50 (Microdyn-Nadir)	PES	50 kDa	Model DW	[93]
	Rotating disk membrane	US100P (Microdyn-Nadir)	Permanently hydrophilic PS	100 kDa	Model DW	[93]
NF	Dead-end	(Permionics Pvt.)	PA/PE	300 Da	Pre-treated DW	[72]
	Dead-end	NF 270 (DOW-Filmtec)	PA	150 Da	UF permeates	[89]
	Dead-end	NF 90 (DOW-Filmtec)	PA	90 Da	UF permeates	[89]
	Dead-end	Nanomax 50 (Millipore)	PA	400 Da	UF permeates	[89]
	Dead-end	Desal-5 DL (GE Osmonics)	PA	327 Da	UF permeates	[89]
	Dead-end	Desal-5 DK (GE Osmonics)	PA	225 Da	UF permeates	[89]
	Cross-flow	Desal-5 DL (Osmonics)	PA	5-10 $kg \cdot mol^{-1}$	Skim milk	[85]
	Cross-flow	TFC-SR3 (Koch)	PA	200 Da	Skim milk UF permeates	[94]
	Cross-flow	NF 90 (DOW-Filmtec)	PA	100 Da	Biologically pre-treated DW	[95]
	Cross-flow	Desal-5 DL (GE Osmonics)	PA	150-300 Da	Biologically pre-treated	[95]

					DW	
	Cross-flow	FM NP010 (Mycrodin-Nadir GmbH)	PES	1000 Da	Biologically pre-treated DW	[95]
	Cross-flow	NF 270 (DOW-Filmtec)	Semiaromatic piperazine-based PA	200-300 Da	Biologically pre-treated DW	[95]
	Rotating disk membrane	NF270 (DOW-Filmtec)	PA/PS	~ 270 Da	Model DW	[84,91,96,97]
RO	/	Duratherm HF (GE Water & Process Technologies)	Composite	/	Pre-filtered low-pollutant DW	[98]
	Dead-end	(Permionics PVt.)	PA/PE	/	Pre-treated DW	[72]
	Dead-end	(Osmonics)	Spiral-wound cellulose acetate	/	MF permeates	[72]
	Cross-flow	TFC HR SW 2540	Spiral-wound composite (TFC)/PA	/	DW	[99]
	Cross-flow	TFC HR (Koch)	Spiral-wound composite (TFC)/PA	/	Skim milk	[85]
	Cross-flow	TFC (GE Osmonics)	Aromatic PA-urea	/	NF permeates	[95]
	Shear-enhanced	Desal AG (Osmonics)	PA/PS	/	Model DW	[86]

Note: PA: polyamide; PE: polyester; PES: polyethersulfone; PS: polysulfone; TFC: thin-film composite.

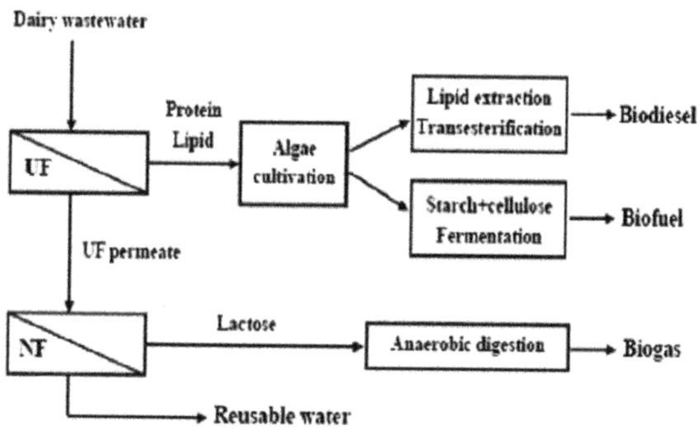

Figure 15 Schematic diagram of a two-step UF/NF process for dairy wastewater treatment and utilization for biorefinery processes [89].

3.4 Tomato-processing wastewater

The tomato industry produces huge amounts of wastewater effluents during tomato manufacturing. Tomato-processing wastewaters generally contain high organic content along with sufficient particulate and colloidal fractions that are not only slowly biodegradable but also exhibit very poor settling characteristics [100]. The wastewater stream produced during tomato manufacturing is characterized by dark color, bad smell, and high content of organics, as well as suspended solids particles. The concentration of pollutants in the effluent can vary considerably with time and space due to the changes in the harvested fruit composition and season. A discharge in the municipal sewage system of these streams is not directly possible because of the high organic contents exceeding the legally tolerated limits [101].

Recently, a process consisting of a biological pretreatment step and a batch nanofiltration process step was developed [101]. The spiral-wound module used for the separation step was a thin-film composite Desal-5 membrane (model DK2540) produced and supplied by Osmonics. Results showed that membrane processes can be successfully used to purify wastewater streams when operated at pressures far away from fouling conditions.

3.5 Fish-processing wastewater

Fish processing requires large amounts of water, primarily for washing and cleaning purposes, but also as media for storage and refrigeration of fish products before and during processing. In addition, water is an important lubricant and transport medium in the various handling and processing steps of bulk fish processing. Consequently, fish processing operations produce wastewater, which contains organic contaminants in soluble, colloidal and particulate forms. Depending on the particular operation, the degree of contamination may be small (e.g., washing operations), mild (e.g., fish filleting), or heavy (e.g., bloodwater drained from fish storage tanks). In fishery wastewater the contaminants present are undefined mixtures of mostly organic substances. Moreover, it is difficult to generalize the extent of the problem created by this wastewater as it depends on the effluent strength, wastewater discharge rate and the absorbing capacity of the receiving water body [102]. Generally fish-processing and seafood-processing units release food waste with high content of organic nitrogen which is easily biodegradable [103].

Fish processing wastewaters have been treated using physical-chemical methods and biological treatments where microorganisms are involved in degradation of organic matter or a combination of both biological and chemical techniques [102,104–107].

A limiting factor for reuse of treated fish canning wastewater in industrial plants and other purposes is the high salt content, which persists even after conventional treatment. So, for the reuse of fish canning industrial wastewater, it was treated by combining conventional treatments, such as sedimentation, chemical coagulation-flocculation and aerobic biological degradation (activated sludge process) followed by a polishing step by reverse osmosis (RO) and ultraviolet (UV) disinfection. The final clarified effluent was found to have the quality requirements to be recycled or reused in the industrial plant, allowing the reduction of the effluent to be discharged, the water use and the costs of tap water for industrial use [104].

It is clear that the modern trend is to remove the polluting species from processing waters by various pressure-driven membrane techniques either integrated or not with other processes. Cross-flow MF is used in pre-treatment step for removal of suspended matter and during UF step, the proteins are concentrated. However, for complete recovery of proteins, NF step is required [108,109].

Also, the MBR technology is an attractive option for the treatment of such industrial wastewaters [110,111]. The experimental results have strongly confirmed the efficiency of the MBR, with good removal efficiencies in terms of BOD_5 and COD [103]. A new pilot-scale hybrid biofilm-suspended biomass membrane bioreactor was used to treat two wastewater streams generated in a fish canning factory [111].

3.6 Distillery wastewater

Distillery spent wash is the unwanted residual liquid waste generated during alcohol production in distilleries and the pollution caused by it is one of the most critical environmental issue. There are a number of large-scale distilleries integrated with sugar mills. The waste products from sugar mill comprise bagasse (residue from the sugarcane crushing), pressmud (mud and dirt residue from juice clarification) and molasses (final residue from sugar crystallization section). Bagasse is used in paper manufacturing and as fuel in boilers; molasses as raw material in distillery for alcohol production while pressmud has no direct industrial application. The major sugar producing countries are listed in Table 7.

Table 7 Major sugar producing countries.

Period	Production (MT)	Export in (MT)	Population (millions)	Per capita consumption
Brazil	34	21.95	190	58
Australia	6.83	4.750	20	47
China	16.79	-	1 314	11
EU	21.567	2.400	490	34
India	30.8	3.298	1 117	20
Mexico	5.978	1	107	52
Pakistan	4.891	-	165	25
SADC	5.834	2.8	157	22
Thailand	8.033	6.25	65	36
United States	8.701	-	301	29

Note: MT: metric tonn; SADC: Southern African Development Community

3.7 Molasses spent wash

The effluents from molasses based distilleries contain large amounts of molasses spent wash (MSW) leading to extensive soil and water pollution. MSW is one of the most difficult waste products to dispose of because of low pH, elevated temperature, high concentrations of organic and inorganic substances, colorants (melanoidin, phenolics, caramel and melanin), polymers with relatively low biodegradability [112], and high ash content. The BOD and COD values typically ranged between 35000–50000 and 100000–150000 mg/L, respectively [113]. The production and characteristics of spentwash are highly variable and dependent on feedstocks and various aspects of the ethanol production process. Together with its pollution and toxic profile, MSW has been widely reviewed [113–115]. The MSW is a potential water pollutant in two ways. First, the highly coloured nature of MSW can block out sunlight from rivers and streams, thus reducing oxygenation of the water by photosynthesis and hence becomes detrimental to aquatic life. Secondly, it has a high pollution load which would result in eutrophication of contaminated water bodies. Moreover, undiluted effluent has shown toxic effect on fishes and other aquatic organisms. MSW also leads to significant levels of soil pollution and acidification in the cases of inappropriate land discharge.

Elimination of pollutants and color from distillery effluent is becoming increasingly important from environmental and aesthetic point of view. Stillage, fermenter and condenser cooling water and fermenter wastewater are the primary polluting streams of a typical distillery. Due to the large volumes of effluent and presence of certain recalcitrant compounds, the treatment of this stream is rather challenging by conventional methods. Therefore, to supplement the existing treatments, a number of studies encompassing physicochemical and biological advanced treatment processes have been conducted [113–116].

Membrane processes can result in significant color removal thereby permitting reuse of the treated effluent [117,118]. MSW can be pre-treated with ceramic membranes prior to anaerobic digestion to reduce COD and improve the efficiency of the anaerobic process possibly due to the removal of inhibiting substances (Fig. 16). Reverse osmosis (RO) has also been employed for distillery wastewater treatment. Since spent wash is a complex, multicomponent stream that is known to cause considerable fouling, an understanding of the components those are primarily responsible for this phenomenon would assist in appropriate feed pre-treatment, for the efficient operation of the membrane system. For this reason, some of the membrane processes developed in last decade are summarized in Table 8.

Table 8 Membrane processes used in hard drinks industries.

	Filtration System	Membranes (manufacturer)	Material	Cut-off	Feed	Ref.
UF	/	TFC-PA (Permionics)	/	/	Distillery effluent	[119]
NF	/	/	Composite polyamide	/	Pre-treated distillery effluent	[120]
RO	/	(Permionics)	/	/	UF permeates	[119]
	/	ESPA2 (Hydranautics) LFC3 (Hydranautics) CPA2 (Hydranautics) BW30 (Filmtec) BW30LE (Filmtec) SG (Osmonics) SE (Osmonics) CE (Osmonics)	/	/	Pre-treated distillery effluent	[121]
	MBR	/	Polymeric	/	Synthetic wastewater	[112]
	MBR equipped with filters prepared from waste fly ash	/	/	/	Pre-treated distillery effluent	[122]
	MBR equipped with a mesh filter	/	/	/	Distillery effluent	[123]

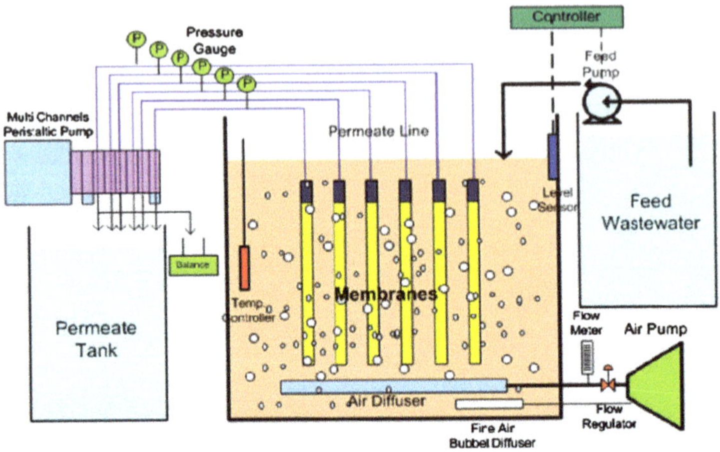

Figure 16 Schematic diagram of the HT-MBR system [112].

3.8 Winery wastewater

Wineries and other grape-processing industries are important distillery industries. These industries generate large volumes of wastewater annually [124]. Wine distillery wastewater, or vinasse, is produced by the distillation of wine, wine lees, or fermented grape juice to extract ethanol or flavor compounds. The winery wastewaters mainly originate from various washing operations during the crushing and pressing of grapes, as well as rinsing of fermentation tanks, barrels, and other equipment as illustrated in Fig. 17 [125]. Winery wastewater contains various pollutants such as sugars, ethanol, esters, glycerol, organic acids, polyphenolic compounds, and numerous population of bacteria and yeasts [126]. The chemical analysis of winery wastewater has indicated that sugars constituted large portion of the COD, whereas organic acids played a more prominent role in the acidity of the wastewater. Due to the acidity, color, and high COD, this wastewater is treated prior to discharge, in order to avoid eutrophication of the receiving environment. Cleaning agents (NaOH, KOH) could be present in winery wastewaters. Numerous phenolic compounds are present in wines as a result of their extraction from the skin, flesh, and seeds of grapes. Phenolic compounds, though form a relatively small portion of the wastewater, offer great negative effects upon treatment systems as well as environmental damage if released untreated. The inhibitory effects of wine distillery wastewater on plant growth can be attributed to the high percentage of organic compounds and salts, and thus high electrical conductivity of winery wastewater makes water uptake by seeds difficult and causes retardation of germination. Consequently, winery effluents are environmentally undesirable and require appropriate treatment. On the other hand, such effluents are valuable source for the recovery of polyphenols. As a result, numerous methods have been reported to treat or dispose of the phenolic-rich wastes that originate from wineries and distilleries. Among treatment methods, physical and chemical processes (such as filtration, oxidation by ozone, chlorine dioxide, hydrogen peroxide, and radiation as well as adsorption) and biological treatment technoques (including aerobic or anaerobic digestion) have been most popular [8,124–129].

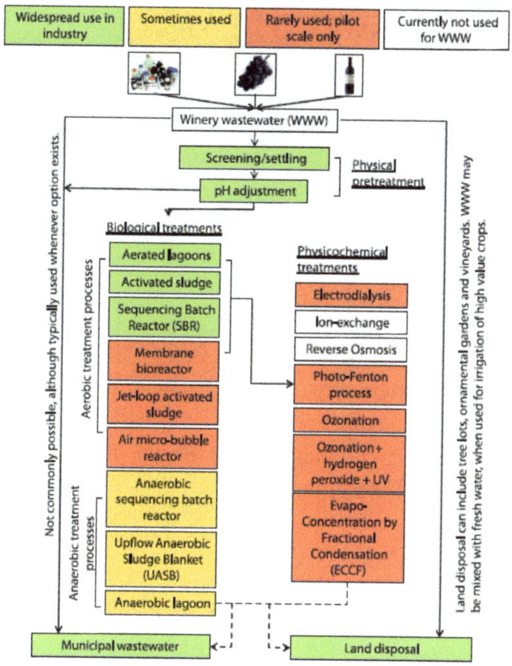

Figure 17 Schematic of process used in winery wastewater treatment [125].

Table 9 Membrane processes applied in winery wastewater treatment.

	Filtration System	Membranes (manufacturer)	Material	Cut-off	Feed	Ref.
MF	Total recirculation mode	V0.2 (Synder Filtration)	PVDF	/	Winery effluents from second racking	[132]
	Total recirculation mode	MFP5 (Alfa Laval)	Fluoro polymer	/	Winery effluents from second racking	[132]
	Concentration mode	(PAM Membranas Seletivas)	PI	/	Winery effluents from second racking	[132]
UF	Total recirculation mode	GR95PP (Alfa Laval)	/	7600 Da	Winery effluents from second racking	[133]
NF	Total recirculation mode	CA 400-22, CA 400-26, CA 400-28 (Laboratory made)	Cellulose acetate	/	UF permeates	[134]
	Total recirculation mode	NF270 (Filmtec Corp.)	/	/	UF permeates	[134]
	Total recirculation mode	ETNA01PP (Alpha Laval)	/	/	UF permeates	[134]
RO	/	AG2521TF (Desal)	PA	/	Pre-treated winery wastewater	[130]

Note: PA: polyamide; PI: polyimide; PVDF: polyvinylidene difluoride.

Biological treatment is particularly well suited to the treatment of winery wastewater, because the majority of the organic components present in the waste stream are readily biodegradable. Nevertheless, pollutants in winery wastewater like various recalcitrant high molecular weight compounds (e.g. polyphenols, tannins and lignins) are not mineralized by the biological treatment [130].

Consequently, other technologies for more exhaustive winery wastewater treatment must be applied to reach effluent quality standards for discharge in the environment. Membrane separation processes are becoming quite popular in wastewater treatment and reclamation since these combine process stability and excellent effluent quality.

Since winery effluents are rich in phenolic compounds, membrane treatments were applied with the aim of both reducing polluting compounds and recovering high-added values phenols. Beyond the application of UF for the fractionation of phenolic compounds and their separation from other co-extracted components in winery sludge [131], membrane processes as shown in Table 9 have been applied to winery wastewaters. Among these, reverse osmosis (Fig. 18) followed by solar photo-Fenton oxidation of the RO concentrate proved to be most successful combined process for the integrated treatment of winery effluents [130].

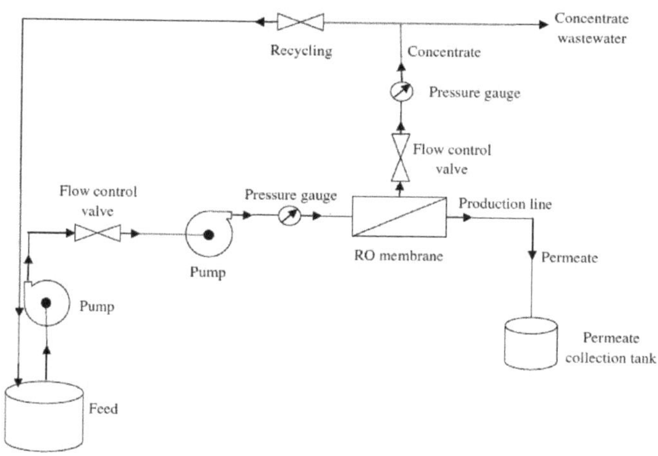

Figure 18. Flow diagram of RO unit [130].

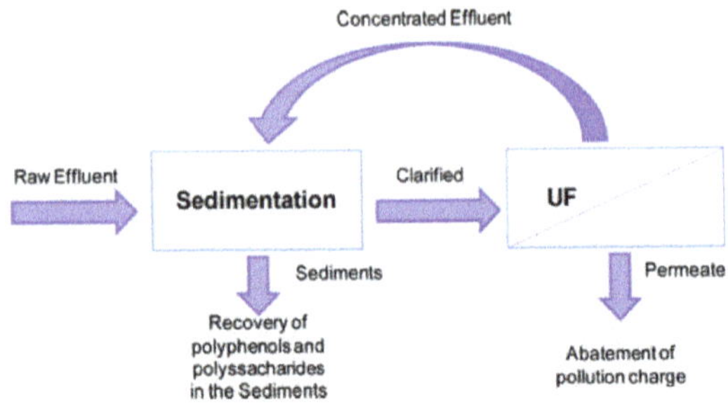

(a)

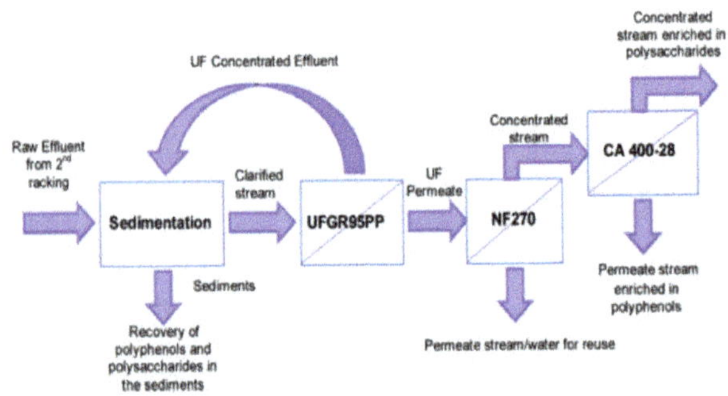

(b)

Figure 19 (a,b) Development of a process to treat winery effluent and recover polyphenols and polysaccharides [133,134].

Another important enzymatic activity in wine making and especially in wine clarification process is pectinases. Free enzyme membrane reactor (FEMR) is the main membrane bioreactor configuration used in wine clarification. The soluble enzyme is confined to the retentate side of the membranes where it is in contact with the substrate. In the following Table 10 some examples and membrane materials applied for the hydrolysis using pectinases are reported. These applications are referred both to the wine and fruit juice treatments as described in Fig. 19(a,b).

Table 10 List of biocatalysts and membrane materials used in wine production.

Biocatalyst	**Membrane**	**Bioreactor configuration**	**Application**	**Reference**
pectin lyase from *Penicillum italicum*		STR with an ultrafiltration unit	--	[144]
polygalacturonase from *A. niger*	30 kDa flat regenerated cellulose membrane	STR with an ultrafiltration unit	--	[145]
Polygalacturonase and pectin lyase from *A. niger*	Spiral wound polysulfone membrane (10 kDa)	STR with an ultrafiltration unit	--	[146]
Endo-polygalacturonase from *Aspergillus pulverulentus*	Amicon 10 kDa	STR with an ultrafiltration unit	Production of pectic oligosaccharides	[147]
Polygalacturonase from *A. niger*	Titania microfiltration	Biocatalytic membrane reactor	Pre-treatment of fruit juice	([148]
Amylase and pectinase	Polysulfone single-hollow fiber	Biocatalytic membrane reactor	Fruit juice processing	[149]
Commerical pectinase	Hollow fiber ultrafiltration	Biocatalytic membrane reactor	Fruit juice processing	[150]
Endopeptidase from *A. niger*	10 kDa Spiral wound polysulfone	STR with an ultrafiltration unit	Apple pectin hydrolysis	[146]

Membrane treatment methods have been investigated to remove heavy metals. Indeed, in most cases of winery effluent, the heavy metals, especially zinc and copper, did not meet the limits for the discharge as imposed by the most restrictive regulations at international level [135].

Winery wastewaters have different compositions and hence are difficult to treat by means of the conventional activated sludge processes because of the high organic loading associated with their production. To face this situation, membrane bioreactors (MBR) have been widely applied to treat winery wastewaters since 2007 [129,136–143]. MBR offers several benefits, such as rapid start-up, good effluent quality, low footprint area, and absence of voluminous secondary settler. Additionally, its operation is not affected by the settling properties of the sludge.

Cell immobilization is rapidly expanding research area. The main purpose to use this technique is to improve wine production. Many supports (inorganic, organic and natural materials) have been used for cell immobilization in this field. Some of the supports and their main applications have been listed in Table 11.

Table 11 Porous supports used in wine production processes.

Support		Immobilized cell	Application	System configuration	Reference
Inorganic support	Mineral kissis	*Saccaromices*	Aroma improvement	Batch and continuous wine-making	[151]
	γ- aluminium	*Saccaromices*		Continuous wine-making	[152]
Organic support	Cellulose covered with Ca-alginate	*Saccaromices and Candida*	Enhance glycerol formation in wine		[153]
	Ca-alginate beds	*Saccaromices*	Must fermentation	Immobilized cell-recycle batch process	[154]
Natural support	Delignified cellulose Gluten pellets	*Saccaromices*	Fermentation Production of wine with less alcohol content	-	[155,156] [153,157]

Compared with organic supports, inorganic supports have been more advantageous, because of providing improved fermentation productivity and in most cases better wine aroma [158]

Conclusions

Membrane applications for the treatment of agro-industrial wastewater focuse on separating water from contaminants, using semi-permeable membranes under applied driving forces. Pressure is applied to reverse the natural equilibrium between the clean water and wastewater. The basic principle of natural equilibrium is that the clean water tends to migrate to the wastewater side to equalize the concentrations across the membrane. Mechanical pressure is used to force water molecules from the wastewater side to the clean water side and, thus, a "high-tech" filtration of the wastewater occurs. In this way, the pollutants can be separated in terms of process intensification strategies. In the past, the energy needed to apply the pressure and the fragility of the membrane surface made use of these alternatives economically unjustifiable. But due to stringent regulation before waste discharge forcing the industry to opt for the advanced treatment mechanism. Various industries have opted as discussed but still there are gaps between fundamental research and application in real life of membrane application in pollution remediation. Also problems with membrane applications include fouling and fragility of the membrane surface. Toxic synthetic compounds can oxidize the surface of the membrane and thus are injurious to membranes. Innovations in membrane technology have advanced the "cleanability" and research on reuse of membranes. The suitability of stainless steel and ceramic materials for membranes has greatly improved the use of these materials in developing advanced wastewater treatment techniques.

References

[1] R. Rajagopal, N.M.C. Saady, M. Torrijos, J.V. Thanikal, Y.-T. Hung, Sustainable Agro-Food Industrial Wastewater Treatment Using High Rate Anaerobic Process, Water. 5 (2013) 292–311. https://doi.org/10.3390/w5010292

[2] J.P. Kushwaha, A review on sugar industry wastewater: sources, treatment technologies, and reuse, Desalin. Water Treat. 53 (2017) 309–318. https://doi.org/10.1080/19443994.2013.838526

[3] A. Chaudhary, A. Sharma, B. Singh, Study of physio-chemical characteristics and biological treatment of molasses-based distillery effluent, Int. J. Bioassays. 2 (2013) 612–615.

[4] P.M. Ayyasamy, R. Yasodha, S. Rajakumar, P. Lakshmanaperumalsamy, P.K.S.M. Rahman, S. Lee, Impact of Sugar Factory Effluent on the Growth and Biochemical Characteristics of Terrestrial and Aquatic Plants, Bull. Environ. Contam. Toxicol. 81 (2008) 449–454. https://doi.org/10.1007/s00128-008-9523-5

[5] C.F. Bennani, B. Ousji, D.J. Ennigrou, Desalination and Water Treatment Reclamation of dairy wastewater using ultrafiltration process, Desalin. Water Treat. Publ. 55 (2014) 37–41.

[6] F. Battista, D. Fino, F. Erriquens, G. Mancini, B. Ruggeri, Scaled-up experimental biogas production from two agro-food waste mixtures having high inhibitory compound concentrations, Renew. Energy. 81 (2015)

[7] N. Adhoum, L. Monser, Decolourization and removal of phenolic compounds from olive mill wastewater by electrocoagulation, Chem. Eng. Process. Process Intensif. 43 (2004) 1281–1287. https://doi.org/10.1016/j.cep.2003.12.001

[8] R. Ganesh, R. Rajinikanth, J.V. Thanikal, R.A. Ramanujam, M. Torrijos, Anaerobic treatment of winery wastewater in fixed bed reactors, Bioprocess Biosyst. Eng. 33 (2010) 619–628. https://doi.org/10.1007/s00449-009-0387-9

[9] C. Wei, T. Zhang, C. Feng, H. Wu, Z. Deng, C. Wu, B. Lu, Treatment of food processing wastewater in a full-scale jet biogas internal loop anaerobic fluidized bed reactor, Biodegradation. 22 (2011) 347–357. https://doi.org/10.1007/s10532-010-9405-5

[10] R.P. Singh, M.H. Ibrahim, N. Esa, M.S. Iliyana, Composting of waste from palm oil mill: a sustainable waste management practice, Rev. Environ. Sci. Bio/Technology. 9 (2010) 331–344. https://doi.org/10.1007/s11157-010-9199-2

[11] Y. Jiang, H. Wang, Y. Shang, K. Yang, Simultaneous removal of aniline, nitrogen and phosphorus in aniline-containing wastewater treatment by using sequencing batch reactor, Bioresour. Technol. 207 (2016) 422–429. https://doi.org/10.1016/j.biortech.2016.02.014

[12] A. Solovchenko, A.M. Verschoor, N.D. Jablonowski, L. Nedbal, Phosphorus from wastewater to crops: An alternative path involving microalgae, Biotechnol. Adv. 34 (2016) 550–564. https://doi.org/10.1016/j.biotechadv.2016.01.002

[13] P.O. Sahu, P.K. Chaudhari, The Characteristics , Effects , and Treatment of Wastewater in Sugarcane Industry, Water Qual. Expo. Heal. 7 (2015) 435–444. https://doi.org/10.1007/s12403-015-0158-6

[14] C. Becerra-Castro, A.R. López, I. Vaz-Moreira, E.F. Silva, C.M. Manaia, O.C. Nunes, Wastewater reuse in irrigation: a microbiological perspective on implications in soil fertility and human and environmental health, Environ. Int. 75 (2015) 117–135. https://doi.org/10.1016/j.envint.2014.11.001

[15] A. Hinkova, Z. Bubnık, P. Kadlec, J. Pridal, Potentials of separation membranes in the sugar industry, Sep. Purif. Technol. 26 (2002) 101–110. https://doi.org/10.1016/S1383-5866(01)00121-6

[16] C.H. Neoh, Z.Z. Noor, N.S.A. Mutamim, K.C. Lim, Green technology in wastewater treatment technologies: Integration of membrane bioreactor with various wastewater treatment systems, Chem. Eng. J. 283 (2016) 582–594. https://doi.org/10.1016/j.cej.2015.07.060

[17] J. Akunna, J. Bartie, Wastewater Treatment Infrastructure and Design, in: Colin A. Booth and Susanne M. Charlesworth (Ed.), Water Resour. Built Environ. Manag. Issues Solut., First Edit, John Wiley & Sons, Ltd., 2014: pp. 350–370. https://doi.org/10.1002/9781118809167.ch26

[18] M. Khan, U. Kalsoom, T. Mahmood, M. Riaz, R. Khan, Characterization and treatment of industrial effluents from sugar industry, J. Chem. Soc. Pak. 25 (2003).

[19] A. Cassano, C. Conidi, E. Drioli, Comparison of the performance of UF membranes in olive mill wastewaters treatment, Water Res. 45 (2011) 3197–3204. https://doi.org/10.1016/j.watres.2011.03.041

[20] J.M.O. Pulido, A review on the use of membrane technology and fouling control for olive mill wastewater treatment, Sci. Total Environ. 563–564 (2016) 664–675. https://doi.org/10.1016/j.scitotenv.2015.09.151

[21] A.Y. Gebreyohannes, R. Mazzei, L. Giorno, Trends and current practices of olive mill wastewater treatment: Application of integrated membrane process and its future perspective, Sep. Purif. Technol. 162 (2016) 45–60. https://doi.org/10.1016/j.seppur.2016.02.001

[22] A.C. Caputo, F. Scacchia, P.M. Pelagagge, Disposal of by-products in olive oil industry: waste-to-energy solutions, Appl. Therm. Eng. 23.2 (2003) 197–214. https://doi.org/10.1016/S1359-4311(02)00173-4

[23] G. Taralas, M.G. Kontominas, Thermochemical treatment of solid and wastewater effluents originating from the olive oil food industry, Energy and Fuels. 19 (2005) 1179–1185. https://doi.org/10.1021/ef040078r

[24] N. Adhoum, M. Lotfi, Decolourization and removal of phenolic compounds from olive mill wastewater by electrocoagulation, Chem. Eng. Process. Process Intensif. 43.10 (2004) 1281–1287. https://doi.org/10.1016/j.cep.2003.12.001

[25] L. Bertin, F. Ferri, A. Scoma, L. Marchetti, F. Fava, Recovery of high added value natural polyphenols from actual olive mill wastewater through solid phase extraction, Chem. Eng. J. 171 (2011) 1287–1293. https://doi.org/10.1016/j.cej.2011.05.056

[26] J. Beltran de Heredia, J. Garcia, Process integration: Continuous Anaerobic digestion−ozonation treatment of olive mill wastewater, Ind. Eng. Chem. Res. 44 (2005) 8750–8755. https://doi.org/10.1021/ie040232x

[27] A. Fiorentino, A. Gentili, M. Isidori, M. Lavorgna, A. Parrella, F. Temussi, Olive oil mill wastewater treatment using a chemical and biological approach, J. Agric. Food Chem. 52 (2004) 5151–5154. https://doi.org/10.1021/jf049799e

[28] I.P. Marques, Anaerobic digestion treatment of olive mill wastewater for effluent re-use in irrigation, Desalination 137 (2001) 233–239. https://doi.org/10.1016/S0011-9164(01)00224-7

[29] A. Rozzi, F. Malpei, Treatment and disposal of olive mill effluents, Int. Biodeterior. Biodegradation. 38 (1996) 135–144. https://doi.org/10.1016/S0964-8305(96)00042-X

[30] A. Ranalli, L. Lucera, S. Contento, antioxidizing potency of phenol compounds in olive oil mill wastewater, J. Agric. Food Chem. 51 (2003) 7636–7641. https://doi.org/10.1021/jf034879o

[31] F. Visioli, A. Romani, N. Mulinacci, S. Zarini, D. Conte, F.F. Vincieri, C. Galli, Antioxidant and other biological activities of olive mill waste waters, J. Agric. Food Chem. 47 (1999) 3397–3401. https://doi.org/10.1021/jf9900534

[32] G. Bisignano, A. Tomaino, R. Lo Cascio, G. Crisafi, N. Uccella, A. Saija, On the in-vitro antimicrobial activity of oleuropein and hydroxytyrosol, J. Pharm. Pharmacol. 51 (1999) 971–974. https://doi.org/10.1211/0022357991773258

[33] S. Takaç, A. Karakaya, Recovery of Phenolic Antioxidants from Olive Mill Wastewater, Recent Patents Chem. Eng. 2 (2009) 230–237. https://doi.org/10.2174/2211334710902030230

[34] F. Federici, F. Fava, N. Kalogerakis, D. Mantzavinos, Valorisation of agro-industrial by-products, effluents and waste: Concept, opportunities and the case of

olive mill waste waters, J. Chem. Technol. Biotechnol. 84 (2009) 895–900. https://doi.org/10.1002/jctb.2165

[35] O.A. Mudimu, M. Peters, F. Brauner, G. Braun, Overview of membrane processes for the recovery of polyphenols from olive mill wastewater olive mill wastewater, Am. J. Environ. Sci. 8 (2012) 195–201. https://doi.org/10.3844/ajessp.2012.195.201

[36] H. Dhaouadi, B. Marrot, Olive mill wastewater treatment in a membrane bioreactor: Process feasibility and performances, Chem. Eng. J. 145 (2008) 225–231. https://doi.org/10.1016/j.cej.2008.04.017

[37] T. Coskun, E. Debik, N.M. Demir, Treatment of olive mill wastewaters by nanofiltration and reverse osmosis membranes, Desalination. 259 (2010) 65–70. https://doi.org/10.1016/j.desal.2010.04.034

[38] C. Russo, A new membrane process for the selective fractionation and total recovery of polyphenols, water and organic substances from vegetation waters (VW), J. Memb. Sci. 288 (2007) 239–246. https://doi.org/10.1016/j.memsci.2006.11.020

[39] M. Pizzichini, C. Russo, Process for recovering the components of olive mill wastewater with membrane technologies, WO2005/123603, 2005.

[40] A. El-Abbassi, A. Hafidi, M.C. García-Payo, M. Khayet, Concentration of olive mill wastewater by membrane distillation for polyphenols recovery, Desalination. 245 (2009) 670–674. https://doi.org/10.1016/j.desal.2009.02.035

[41] S. Khoufi, F. Aloui, S. Sayadi, Pilot scale hybrid process for olive mill wastewater treatment and reuse, Chem. Eng. Process. Process Intensif. 48 (2009) 643–650. https://doi.org/10.1016/j.cep.2008.07.007

[42] O. Yahiaoui, H. Lounici, N. Abdi, N. Drouiche, N. Ghaffour, A. Pauss, N. Mameri, Treatment of olive mill wastewater by the combination of ultrafiltration and bipolar electrochemical reactor processes, Chem. Eng. Process. Process Intensif. 50 (2011) 37–41. https://doi.org/10.1016/j.cep.2010.11.003

[43] E. Garcia-Castello, A. Cassano, A. Criscuoli, C. Conidi, E. Drioli, Recovery and concentration of polyphenols from olive mill wastewaters by integrated membrane system, Water Res. 44 (2010) 3883–3892. https://doi.org/10.1016/j.watres.2010.05.005

[44] M. Stoller, M. Bravi, Critical flux analyses on differently pretreated olive vegetation waste water streams: Some case studies, Desalination. 250 (2010) 578–582. https://doi.org/10.1016/j.desal.2009.09.027

[45] E.O. Akdemir, A. Ozer, Investigation of two ultrafiltration membranes for treatment of olive oil mill wastewater, Desalination. 249 (2009) 660–666. https://doi.org/10.1016/j.desal.2008.06.035

[46] A. Cassano, C. Conidi, L. Giorno, E. Drioli, Fractionation of olive mill wastewaters by membrane separation techniques, J. Hazard. Mater. 248–249 (2013) 185–193. https://doi.org/10.1016/j.jhazmat.2013.01.006

[47] C.A. Paraskeva, V.G. Papadakis, E. Tsarouchi, D.G. Kanellopoulou, P.G. Koutsoukos, Membrane processing for olive mill wastewater fractionation, Desalination. 213 (2007) 218–229. https://doi.org/10.1016/j.desal.2006.04.087

[48] M. Stoller, Technical optimization of a dual ultrafiltration and nanofiltration pilot plant in batch operation by means of the critical flux theory: A case study, Chem. Eng. Process. Process Intensif. 47 (2008) 1165–1170. https://doi.org/10.1016/j.cep.2007.07.012

[49] A. El-Abbassi, M. Khayet, A. Hafidi, Micellar enhanced ultrafiltration process for the treatment of olive mill wastewater, Water Res. 45 (2011) 4522–4530. https://doi.org/10.1016/j.watres.2011.05.044

[50] M. Stoller, Effective fouling inhibition by critical flux based optimization methods on a NF membrane module for olive mill wastewater treatment, Chem. Eng. J. 168 (2011) 1140–1148. https://doi.org/10.1016/j.cej.2011.01.098

[51] A. El-Abbassi, A. Hafidi, M. Khayet, M.C. García-Payo, Integrated direct contact membrane distillation for olive mill wastewater treatment, Desalination. 323 (2013) 31–38. https://doi.org/10.1016/j.desal.2012.06.014

[52] A. El-Abbassi, H. Kiai, A. Hafidi, M.C. García-Payo, M. Khayet, Treatment of olive mill wastewater by membrane distillation using polytetrafluoroethylene membranes, Sep. Purif. Technol. 98 (2012) 55–61. https://doi.org/10.1016/j.seppur.2012.06.026

[53] A. El-Abbassi, M. Khayet, H. Kiai, A. Hafidi, M.C. García-Payo, Treatment of crude olive mill wastewaters by osmotic distillation and osmotic membrane distillation, Sep. Purif. Technol. 104 (2013) 327–332. https://doi.org/10.1016/j.seppur.2012.12.006

[54] C.A. Paraskeva, V.G. Papadakis, D.G. Kanellopoulou, P.G. Koutsoukos, K.C. Angelopoulos, Membrane filtration of olive mill wastewater and exploitation of its fractions, Water Environ. Res. 79 (2007). https://doi.org/10.2175/106143006X115345

[55] T.Y. Wu, A.W. Mohammad, J.M. Jahim, N. Anuar, Pollution control technologies for the treatment of palm oil mill effluent (POME) through end-of-pipe processes, J. Environ. Manage. 91 (2010) 1467–1490. https://doi.org/10.1016/j.jenvman.2010.02.008

[56] Y. Zhang, L. Yan, X. Qiao, L. Chi, X. Niu, Z. Mei, Z. Zhang, Integration of biological method and membrane technology in treating palm oil mill effluent, J. Environ. Sci. 20 (2008) 558–564. https://doi.org/10.1016/S1001-0742(08)62094-X

[57] A.L. Ahmad, C.Y. Chan, Sustainability of palm oil industries: an innovative treatment via membrane technology, J. Appl. Sci. 9 (2009) 3074–3079. https://doi.org/10.3923/jas.2009.3074.3079

[58] T.Y. Wu, A.W. Mohammad, J.M. Jahim, N. Anuar, Palm oil mill effluent (POME) treatment and bioresources recovery using ultrafiltration membrane: Effect of pressure on membrane fouling, Biochem. Eng. J. 35 (2007) 309–317. https://doi.org/10.1016/j.bej.2007.01.029

[59] P.E. Poh, M.F. Chong, Development of anaerobic digestion methods for palm oil mill effluent (POME) treatment, Bioresour. Technol. 100 (2009) 1–9. https://doi.org/10.1016/j.biortech.2008.06.022

[60] K. Vijayaraghavan, D. Ahmad, M. Ezani Bin Abdul Aziz, Aerobic treatment of palm oil mill effluent, J. Environ. Manage. 82 (2007) 24–31. https://doi.org/10.1016/j.jenvman.2005.11.016

[61] A.L. Ahmad, M.F. Chong, S. Bhatia, A comparative study on the membrane based palm oil mill effluent (POME) treatment plant, J. Hazard. Mater. 171 (2009) 166–174. https://doi.org/10.1016/j.jhazmat.2009.05.114

[62] P.F. Rupani, P.S. Rajeev, M.H. Irahim, N. Esa, Review of current palm oil mill effluent (POME) Treatment methods: Vermicomposting as a sustainable practice, World Appl. Sci. J. 11 (2010) 70–81.

[63] J.C. Igwe, C.C. Onyegbado, A review of palm oil mill effluent (Pome) water treatment, Glob. J. Environ. Res. 1 (2007) 54–62.

[64] A.L. Ahmad, S. Ismail, S. Bhatia, Water recycling from palm oil mill effluent (POME) using membrane technology, Desalination. 157 (2003) 87–95. https://doi.org/10.1016/S0011-9164(03)00387-4

[65] A.L. Ahmad, M.F. Chong, S. Bhatia, S. Ismail, Drinking water reclamation from palm oil mill effluent (POME) using membrane technology, Desalination. 191 (2006) 35–44. https://doi.org/10.1016/j.desal.2005.06.033

[66] A. Damayanti, Z. Ujang, M.R. Salim, The influenced of PAC, zeolite, and Moringa oleifera as biofouling reducer (BFR) on hybrid membrane bioreactor of palm oil mill effluent (POME), Bioresour. Technol. 102 (2011) 4341–4346. https://doi.org/10.1016/j.biortech.2010.12.061

[67] T.Y. Wu, A.W. Mohammad, J.M. Jahim, N. Anuar, A holistic approach to managing palm oil mill effluent (POME): Biotechnological advances in the sustainable reuse of POME, Biotechnol. Adv. 27 (2009) 40–52. https://doi.org/10.1016/j.biotechadv.2008.08.005

[68] A.W. Mohammad, P.T. Yap, T.Y. Wu, Performance of hydrophobic ultrafiltration membranes in the treatment and protein recovery from palm oil mill effluent (POME), Desalin. Water Treat. 10 (2009) 332–338. https://doi.org/10.5004/dwt.2009.932

[69] A.L. Ahmad, M.F. Chong, S. Bhatia, Mathematical modeling of multiple solutes system for reverse osmosis process in palm oil mill effluent (POME) treatment, Chem. Eng. J. 132 (2007) 183–193. https://doi.org/10.1016/j.cej.2006.12.022

[70] F.X. Milani, D. Nutter, G. Thoma, Environmental impacts of dairy processing and products: A review, J. Dairy Sci. 94 (2011) 4243–4254. https://doi.org/10.3168/jds.2010-3955

[71] T.J. Britz, C. van Schalkwyk, Y.-T. Hung, Treatment of dairy processing wastewater, in: Waste Treat. Food Process. Ind., CRC Press, Boca Raton, FL, 2006: pp. 1–28.

[72] B. Sarkar, P.P. Chakrabarti, A. Vijaykumar, V. Kale, Wastewater treatment in dairy industries - possibility of reuse, Desalination. 195 (2006) 141–152. https://doi.org/10.1016/j.desal.2005.11.015

[73] C. Tocchi, E. Federici, L. Fidati, R. Manzi, V. Vincigurerra, M. Petruccioli, Aerobic treatment of dairy wastewater in an industrial three-reactor plant: Effect of aeration regime on performances and on protozoan and bacterial communities, Water Res. 46 (2012) 3334–3344. https://doi.org/10.1016/j.watres.2012.03.032

[74] M. Jedrzejewska-Cicinska, K. Kozak, M. Krzemieniewski, A comparison of the technological effectiveness of dairy wastewater treatment in anaerobic UASB reactor and anaerobic reactor with an innovative design, Environ. Technol. 28 (2007) 1127–1133. https://doi.org/10.1080/09593332808618868

[75] S. Venkata Mohan, V. Lalit Babu, P.N. Sarma, Anaerobic biohydrogen production from dairy wastewater treatment in sequencing batch reactor (AnSBR): Effect of organic loading rate, Enzyme Microb. Technol. 41 (2007) 506–515. https://doi.org/10.1016/j.enzmictec.2007.04.007

[76] B. Demirel, O. Yenigun, T.T. Onay, Anaerobic treatment of dairy wastewaters: A review, Process Biochem. 40 (2005) 2583–2595. https://doi.org/10.1016/j.procbio.2004.12.015

[77] M. Passeggi, I. López, L. Borzacconi, Modified UASB reactor for dairy industry wastewater: Performance indicators and comparison with the traditional approach, J. Clean. Prod. 26 (2012) 90–94. https://doi.org/10.1016/j.jclepro.2011.12.022

[78] P. Seesuriyachan, A. Kuntiya, K. Sasaki, C. Techapun, Biocoagulation of dairy wastewater by Lactobacillus casei TISTR 1500 for protein recovery using micro-aerobic sequencing batch reactor (micro-aerobic SBR), Process Biochem. 44 (2009) 406–411. https://doi.org/10.1016/j.procbio.2008.12.006

[79] S. Tchamango, C.P. Nanseu-Njiki, E. Ngameni, D. Hadjiev, A. Darchen, Treatment of dairy effluents by electrocoagulation using aluminium electrodes, Sci. Total Environ. 408 (2010) 947–952. https://doi.org/10.1016/j.scitotenv.2009.10.026

[80] S.L. Lansing, J.F. Martin, Use of an ecological treatment system (ETS) for removal of nutrients from dairy wastewater, Ecol. Eng. 28 (2006) 235–245. https://doi.org/10.1016/j.ecoleng.2006.04.006

[81] S. Venkata Mohan, G. Mohanakrishna, G. Velvizhi, V.L. Babu, P.N. Sarma, Bio-catalyzed electrochemical treatment of real field dairy wastewater with simultaneous power generation, Biochem. Eng. J. 51 (2010) 32–39. https://doi.org/10.1016/j.bej.2010.04.012

[82] Y. Pouliot, Membrane processes in dairy technology-From a simple idea to worldwide panacea, Int. Dairy J. 18 (2008) 735–740. https://doi.org/10.1016/j.idairyj.2008.03.005

[83] A. Chollangi, M.M. Hossain, Separation of proteins and lactose from dairy wastewater, Chem. Eng. Process. Process Intensif. 46 (2007) 398–404. https://doi.org/10.1016/j.cep.2006.05.022

[84] J. Luo, L. Ding, Y. Wan, P. Paullier, M.Y. Jaffrin, Fouling behavior of dairy wastewater treatment by nanofiltration under shear-enhanced extreme hydraulic conditions, Sep. Purif. Technol. 88 (2012) 79–86. https://doi.org/10.1016/j.seppur.2011.12.008

[85] M. Rabiller-Baudry, H. Bouzid, B. Chaufer, L. Paugam, D. Delaunay, O. Mekmene, S. Ahmad, F. Gaucheron, On the origin of flux dependence in pH-modified skim milk filtration, Dairy Sci. Technol. 89 (2009) 363–385. https://doi.org/10.1051/dst/2009018

[86] M. Frappart, M. Jaffrin, L.H. Ding, Reverse osmosis of diluted skim milk: Comparison of results obtained from vibratory and rotating disk modules, Sep. Purif. Technol. 60 (2008) 321–329. https://doi.org/10.1016/j.seppur.2007.09.007

[87] M. Passeggi, I. López, L. Borzacconi, Integrated anaerobic treatment of dairy industrial wastewater and sludge, Water Sci. Technol. 59 (2009) 501–506. https://doi.org/10.2166/wst.2009.010

[88] M.S. Yorgun, I. Akmehmet Balcioglu, O. Saygin, Performance comparison of ultrafiltration, nanofiltration and reverse osmosis on whey treatment, Desalination. 229 (2008) 204–216. https://doi.org/10.1016/j.desal.2007.09.008

[89] J. Luo, L. Ding, B. Qi, M.Y. Jaffrin, Y. Wan, A two-stage ultrafiltration and nanofiltration process for recycling dairy wastewater, Bioresour. Technol. 102 (2011) 7437–7442. https://doi.org/10.1016/j.biortech.2011.05.012

[90] Y.-W. Gong, H.-X. Zhang, X.-N. Cheng, Treatment of dairy wastewater by two-stage membrane operation with ultrafiltration and nanofiltration, Water Sci. Technol. 65 (2012) 915–919. https://doi.org/10.2166/wst.2012.937

[91] J. Luo, L. Ding, Influence of pH on treatment of dairy wastewater by nanofiltration using shear-enhanced filtration system, Desalination. 278 (2011) 150–156. https://doi.org/10.1016/j.desal.2011.05.025

[92] B. Farizoglu, S. Uzuner, The investigation of dairy industry wastewater treatment in a biological high performance membrane system, Biochem. Eng. J. 57 (2011) 46–54. https://doi.org/10.1016/j.bej.2011.08.007

[93] J. Luo, W. Cao, L. Ding, Z. Zhu, Y. Wan, M.Y. Jaffrin, Treatment of dairy effluent by shear-enhanced membrane filtration: The role of foulants, Sep. Purif. Technol. 96 (2012) 194–203. https://doi.org/10.1016/j.seppur.2012.06.009

[94] G. Rice, A. Barber, A. O'Connor, G. Stevens, S. Kentish, Fouling of NF membranes by dairy ultrafiltration permeates, J. Memb. Sci. 330 (2009) 117–126. https://doi.org/10.1016/j.memsci.2008.12.048

[95] R. Hepsen, Y. Kaya, Optimization of membrane fouling using experimental design: An example from dairy wastewater treatment, Ind. Eng. Chem. Res. 51 (2012) 16074–16084. https://doi.org/10.1021/ie302450s

[96] J. Luo, L. Ding, Y. Wan, M.Y. Jaffrin, Threshold flux for shear-enhanced nanofiltration: Experimental observation in dairy wastewater treatment, J. Memb. Sci. 409–410 (2012) 276–284. https://doi.org/10.1016/j.memsci.2012.03.065

[97] J. Luo, L. Ding, Y. Wan, P. Paullier, M.Y. Jaffrin, Application of NF-RDM (nanofiltration rotating disk membrane) module under extreme hydraulic conditions for the treatment of dairy wastewater, Chem. Eng. J. 163 (2010) 307–316. https://doi.org/10.1016/j.cej.2010.08.007

[98] A. Suárez, T. Fidalgo, F.A. Riera, Recovery of dairy industry wastewaters by reverse osmosis. Production of boiler water, Sep. Purif. Technol. 133 (2014) 204–211. https://doi.org/10.1016/j.seppur.2014.06.041

[99] M. Vourch, B. Balannec, B. Chaufer, G. Dorange, Treatment of dairy industry wastewater by reverse osmosis for water reuse, Desalination 219 (2008) 190–202. https://doi.org/10.1016/j.desal.2007.05.013

[100] Z. Xu, G. Nakhla, J. Patel, Characterization and modeling of nutrient-deficient tomato-processing wastewater treatment using an anaerobic/aerobic system, Chemosphere 65 (2006) 1171–1181. https://doi.org/10.1016/j.chemosphere.2006.03.063

[101] M. Iaquinta, M. Stoller, C. Merli, Optimization of a nanofiltration membrane process for tomato industry wastewater effluent treatment, Desalination. 245 (2009) 314–320. https://doi.org/10.1016/j.desal.2008.05.028

[102] P. Chowdhury, T. Viraraghavan, A. Srinivasan, Biological treatment processes for fish processing wastewater - A review, Bioresour. Technol. 101 (2010) 439–449. https://doi.org/10.1016/j.biortech.2009.08.065

[103] P.C. Sridang, J. Kaiman, A. Pottier, C. Wisniewski, Benefits of MBR in seafood wastewater treatment and water reuse: study case in Southern part of Thailand, Desalination. 200 (2006) 712–714. https://doi.org/10.1016/j.desal.2006.03.509

[104] R.O. Cristóvão, C.M. Botelho, R.J.E. Martins, J.M. Loureiro, R.A.R. Boaventura, Fish canning industry wastewater treatment for water reuse – a case study, J. Clean. Prod. 87 (2015) 603–612. https://doi.org/10.1016/j.jclepro.2014.10.076

[105] R. Cristóvão, C. Botelho, R. Martins, R. Boaventura, Chemical and Biological Treatment of Fish Canning Wastewaters, Int. J. Biosci. Biochem. Bioinforma. 2 (2012) 7763. https://doi.org/10.7763/IJBBB.2012.V2.108

[106] R.O. Cristóvão, C. Gonçalves, C.M. Botelho, R.J.E. Martins, J.M. Loureiro, R.A.R. Boaventura, Fish canning wastewater treatment by activated sludge: Application of factorial design optimization. Biological treatment by activated sludge of fish canning wastewater., Water Resour. Ind. 10 (2015) 29–38. https://doi.org/10.1016/j.wri.2015.03.001

[107] M.D. Afonso, R. Borquez, Review of the treatment of seafood processing wastewaters and recovery of proteins therein by membrane separation processes - Prospects of the ultrafiltration of wastewaters from the fish meal industry, Desalination 142 (2002) 29–45. https://doi.org/10.1016/j.desal.2008.01.068

[108] M. Kuca, D. Szaniawska, Application of microfiltration and ceramic membranes for treatment of salted aqueous effluents from fish processing, Desalination 241 (2009) 227–235.

[109] R. Pérez-Gálvez, E.M. Guadix, J.P. Bergé, A. Guadix, Operation and cleaning of ceramic membranes for the filtration of fish press liquor, J. Memb. Sci. 384 (2011) 142–148. https://doi.org/10.1016/j.memsci.2011.09.019

[110] P.C. Sridang, A. Pottier, C. Wisniewski, A. Grasmick, Performance and microbial surveying in submerged membrane bioreactor for seafood processing wastewater treatment, J. Memb. Sci. 317 (2008) 43–49. https://doi.org/10.1016/j.memsci.2007.11.011

[111] P. Artiga, G. García-Toriello, R. Méndez, J.M. Garrido, Use of a hybrid membrane bioreactor for the treatment of saline wastewater from a fish canning factory, Desalination. 221 (2008) 518–525. https://doi.org/10.1016/j.desal.2007.01.112

[112] M.R. Bilad, P. Declerck, A. Piasecka, L. Vanysacker, X. Yan, I.F.J. Vankelecom, Treatment of molasses wastewater in a membrane bioreactor: Influence of membrane pore size, Sep. Purif. Technol. 78 (2011) 105–112.

[113] D. Pant, A. Adholeya, Biological approaches for treatment of distillery wastewater: A review, Bioresour. Technol. 98 (2007) 2321–2334. https://doi.org/10.1016/j.biortech.2006.09.027

[114] Y. Satyawali, M. Balakrishnan, Wastewater treatment in molasses-based alcohol distilleries for COD and color removal: A review, J. Environ. Manage. 86 (2008) 481–497. https://doi.org/10.1016/j.jenvman.2006.12.024

[115] S. Mohana, B.K. Acharya, D. Madamwar, Distillery spent wash: Treatment technologies and potential applications, J. Hazard. Mater. 163 (2009) 12–25. https://doi.org/10.1016/j.jhazmat.2008.06.079

[116] L.A. Ioannou, G.L. Puma, D. Fatta-Kassinos, Treatment of winery wastewater by physicochemical, biological and advanced processes: A review, J. Hazard. Mater. 286 (2015) 343–368. https://doi.org/10.1016/j.jhazmat.2014.12.043

[117] S.K. Nataraj, K.M. Hosamani, T.M. Aminabhavi, Distillery wastewater treatment by the membrane-based nanofiltration and reverse osmosis processes, Water Res. 40 (2006) 2349–2356. https://doi.org/10.1016/j.watres.2006.04.022

[118] P.A. Shivajirao, Treatment of distillery wastewater using membrane technologies, Int. J. Adv. Eng. Res. Stud. I (2012) 275–283.

[119] Z.V.P. Murthy, L.B. Chaudhari, Treatment of Distillery Spent Wash By Combined UF and RO Processes, Glob. NEST J. 11 (2009) 235–240.

[120] U.K. Rai, M. Muthukrishnan, B.K. Guha, Tertiary treatment of distillery wastewater by nanofiltration, Desalination. 230 (2008) 70–78. https://doi.org/10.1016/j.desal.2007.11.017

[121] C. Sagne, C. Fargues, R. Lewandowski, M.-L. Lameloise, M. Decloux, Screening of reverse osmosis membranes for the treatment and reuse of distillery condensates into alcoholic fermentation, Desalination. 219 (2008) 335–347. https://doi.org/10.1016/j.desal.2007.05.020

[122] R. Gupta, Y. Satyawali, V.S. Batra, M. Balakrishnan, Submerged membrane bioreactor using fly ash filters: trials with distillery wastewater, Water Sci. Technol. 58 (2008) 1281–1284. https://doi.org/10.2166/wst.2008.352

[123] Y. Satyawali, M. Balakrishnan, Treatment of distillery effluent in a membrane bioreactor (MBR) equipped with mesh filter, Sep. Purif. Technol. 63 (2008) 278–286. https://doi.org/10.1016/j.seppur.2008.05.008

[124] X.L. Melamane, P.J. Strong, J.E. Burgess, Treatent of wine distillery wastewater: a review with ephasis on anaerobic membrane reactors, S. Afr. J. Enol. Vitic. 28 (2007) 25–36.

[125] K.P.M. Mosse, A.F. Patti, E.W. Christen, T.R. Cavagnaro, Review: Winery wastewater quality and treatment options in Australia, Aust. J. Grape Wine Res. 17 (2011) 111–122. https://doi.org/10.1111/j.1755-0238.2011.00132.x

[126] G. Lofrano, S. Meric, A comprehensive approach to winery wastewater treatment: a review of the state-of the-art, Desalin. Water Treat. 57 (2016) 3011–3028. https://doi.org/10.1080/19443994.2014.982196

[127] P.J. Strong, J.E. Burgess, Treatment Methods for Wine-Related and Distillery Wastewaters: A Review, Bioremediat. J. 12 (2008) 70–8770. https://doi.org/10.1080/10889860802060063

[128] G. Andreottola, P. Foladori, G. Ziglio, Biological treatment of winery wastewater: An overview, Water Sci. Technol. 60 (2009) 1117–1125. https://doi.org/10.2166/wst.2009.551

[129] N. Basset, S. López-Palau, J. Dosta, J. Mata-Álvarez, Comparison of aerobic granulation and anaerobic membrane bioreactor technologies for winery wastewater treatment, Water Sci. Technol. 69 (2014) 320–327. https://doi.org/10.2166/wst.2013.713

[130] L.A. Ioannou, C. Michael, N. Vakondios, K. Drosou, N.P. Xekoukoulotakis, E. Diamadopoulos, D. Fatta-Kassinos, Winery wastewater purification by reverse osmosis and oxidation of the concentrate by solar photo-Fenton, Sep. Purif. Technol. 118 (2013) 659–669. https://doi.org/10.1016/j.seppur.2013.07.049

[131] C.M. Galanakis, E. Markouli, V. Gekas, Recovery and fractionation of different phenolic classes from winery sludge using ultrafiltration, Sep. Purif. Technol. 107 (2013) 245–251. https://doi.org/10.1016/j.seppur.2013.01.034

[132] A. Giacobbo, J.M. Do Prado, A. Meneguzzi, A.M. Bernardes, M.N. De Pinho, Microfiltration for the recovery of polyphenols from winery effluents, Sep. Purif. Technol. 143 (2015) 12–18. https://doi.org/10.1016/j.seppur.2015.01.019

[133] A. Giacobbo, M. Oliveira, E.C.N.F. Duarte, H.M.C. Mira, A.M. Bernardes, M.N. De Pinho, Ultrafiltration based process for the recovery of polysaccharides and polyphenols from winery effluents, Sep. Sci. Technol. 48 (2013) 438–444. https://doi.org/10.1080/01496395.2012.725793

[134] A. Giacobbo, A.M. Bernardes, M.N. de Pinho, Nanofiltration for the Recovery of Low Molecular Weight Polysaccharides and Polyphenols from Winery Effluents, Sep. Sci. Technol. 48 (2013) 2524–2530. https://doi.org/10.1080/01496395.2013.809762

[135] G. Andreottola, M. Cadonna, P. Foladori, G. Gatti, F. Lorenzi, P. Nardelli, Heavy metal removal from winery wastewater in the case of restrictive discharge regulation, Water Sci. Technol. 56 (2007) 111–120. https://doi.org/10.2166/wst.2007.479

[136] P. Artiga, M. Carballa, J.M. Garrido, R. Méndez, Treatment of winery wastewaters in a membrane submerged bioreactor, Water Sci. Technol. 56 (2007) 63–69. https://doi.org/10.2166/wst.2007.473

[137] A. Shah, J. Bulleri, R. Ross, J. Carter, M. Long, Successful plant scale winery wastewater treatment using membrane bioreactor in Northern California, in: Proc. Water Environ. Fed. (2008) 3408–3425. https://doi.org/10.2175/193864708788733107

[138] V. Ferre, A. Trepin, T. Giménez, S. Lluch, Design and performance of full-scale iMBR plants treating winery wastewater effluents in Italy and Spain, in: 5th Int. Spec. Conf. Sustain. Vitic. Winer. Waste Ecol. Impacts Manag., Trento, Italy, 2009: pp.

[139] G. Guglielmi, G. Andreottola, P. Foladori, G. Ziglio, Membrane bioreactors for winery wastewater treatment: Case-studies at full scale, Water Sci. Technol. 60 (2009) 1201–1207. https://doi.org/10.2166/wst.2009.560

[140] D. Bolzonella, F. Fatone, P. Pavan, F. Cecchi, Application of a membrane bioreactor for winery wastewater treatment, Water Sci. Technol. 62 (2010) 2754–2759. https://doi.org/10.2166/wst.2010.645

[141] N. Basset, C. Vidal, A. Coll, I. Fernández, J. Dosta, AnMBR technologies (CSTR and UASB type) for winery wastewater treatment at low temperatures, (2012).

[142] C. Valderrama, G. Ribera, N. Bah, M. Rovira, T. Giménez, R. Nomen, S. Lluch, M. Yuste, X. Martinez-Lladó, Winery wastewater treatment for water reuse purpose: Conventional activated sludge versus membrane bioreactor (MBR). A comparative case study, Desalination 306 (2012) 1–7. https://doi.org/10.1016/j.desal.2012.08.016

[143] N. Basset, E. Santos, J. Dosta, J. Mata-Álvarez, Start-up and operation of an AnMBR for winery wastewater treatment, Ecol. Eng. 86 (2016) 279–289. https://doi.org/10.1016/j.ecoleng.2015.11.003

[144] I. Alkorta , C. Garbisu, M.J. Llama, J.L. Serra, β-Transelimination of citrus pectin catalyzed by penicillium italicum pectin lyase in a membrane reactor, Applied Biochemistry and Biotechnology 55 (1995) 249- 259. https://doi.org/10.1007/BF02786864

[145] K. Bélafi-Bakò, M. Eszterle, K. Kiss, N. Nemestòthy, L. Gubicza, Hydrolysis of pectin by Aspergillus niger polygalacturonase in a membrane bioreactor, J. Food Eng. 78 (2007) 438-442. https://doi.org/10.1016/j.jfoodeng.2005.10.012

[146] J.M Rodriguez-Nogales, N. Ortega , M. Perez-Mateos, M. Busto, Pectin hydrolysis in a free enzyme membrane reactor: An approach to the wine and juice clarification, Food Chem. 107 (2008) 112-119. https://doi.org/10.1016/j.foodchem.2007.07.057

[147] E. Olano-Martin, K.C. Mountzouris , G.R.Gibson, R.A. Rastall, Continuous production of pectic oligosaccharides in an enzyme membrane reactor, J. Food Sci. 66 (2001) 966-971. https://doi.org/10.1111/j.1365-2621.2001.tb08220.x

[148] A.R. Szaniawski, H.G. Spencer, Effects of pectin concentration and crossflow velocity on permeability in the microfiltration of dilute pectin solutions by macroporous titania membranes containing immobilized pectinase, Biotechnol. Prog. 12 (1996) 403-405. https://doi.org/10.1021/bp9500829

[149] M.E. Carrín., L.N. Ceci, J.E.Lozano, Ultrafiltration fibers like bioreactors, Latin American Applied Research 31 (2001) 241-245.

[150] M.E Carrín, L.N Ceci, J.E. Lozano, Effects of pectinase immobilization during hollow fiber ultrafiltration of apple juice, J. Food Proc. Eng. 23 (2000) 281-298. https://doi.org/10.1111/j.1745-4530.2000.tb00516.x

[151] V. Bakoyianis, A.A Koutinas, A catalytic multistage fixed-bed tower bioreactor in an industrial-scale pilot plant for alcohol production, Biotechnol. Bioeng. 49 (1996) 197–203. https://doi.org/10.1002/(SICI)1097-0290(19960120)49:2<197::AID-BIT8>3.0.CO;2-L

[152] P. Loukatos, M. Kiaris, I. Ligas, G .Bourgos, M. Kanellaki, M. Komaitis, A.A. Koutinas, Continuous wine making by γ-alumina-supported biocatalyst, Appl. Biochem. Biotechnol. 89 (2000) 1–13. https://doi.org/10.1385/ABAB:89:1:1

[153] M. Ciani, L. Ferraro, Enhanced glycerol content in wines made with immobilized *Candida stellata* Cells, Appl. Env. Microbiol. 62 (1996) 128–132.

[154] G. Suzzi, P. Romano, L. Vannini, L. Turbanti, P. Domizio, Cell-recycle batch fermentation using immobilized cells of flocculent Saccharomyces cerevisiae wine strains, World J. Microbiol. Biotechnol. 12 (1996) 25–27. https://doi.org/10.1007/BF00327794

[155] E.P. Bardi, A.A. Koutinas, Immobilization of yeast on delignified cellulosic material for room temperature and low-temperature wine making, J. Agric. Food Chem. 42 (1994) 221–226. https://doi.org/10.1007/BF00327794

[156] E.P. Bardi, A.A. Koutinas, M. Kanellaki, Room and low temperature brewing with yeast immobilized on gluten pellets, Process Biochem. 32 (1997) 691–696. https://doi.org/10.1016/S0032-9592(97)00030-7

[157] E.P. Bardi, A.A. Koutinas, M. Soupioni, M. Kanellaki, Immobilization of yeast on delignified cellulosic material for low temperature brewing, J. Agric. Food Chem. 44 (1996), 463–467. https://doi.org/10.1021/jf9501406

[158] Y. Kourkoutas, A. Ekatorou, I.M. Banat, R. Marchant, A.A. Koutinas, Immobilization technologies and support materials suitable in alcohol beverages production: a review, Food Microb. 21 (2004) 377-397. https://doi.org/10.1016/j.fm.2003.10.005

Keyword Index

About the Editors

Dr. Inamuddin is currently working as Assistant Professor in the Chemistry Department, Faculty of Science, King Abdulaziz University, Jeddah, Saudi Arabia. He is a permanent faculty member (Assistant Professor) at the Department of Applied Chemistry, Aligarh Muslim University, Aligarh, India. He obtained Master of Science degree in Organic Chemistry from Chaudhary Charan Singh (CCS) University, Meerut, India, in 2002. He received his Master of Philosophy and Doctor of Philosophy degrees in Applied Chemistry from Aligarh Muslim University (AMU) in 2004 and 2007, respectively. He has extensive research experience in multidisciplinary fields of Analytical Chemistry, Materials Chemistry, and Electrochemistry and, more specifically, Renewable Energy and Environment. He has worked on different research projects as project fellow and senior research fellow funded by University Grants Commission (UGC), Government of India, and Council of Scientific and Industrial Research (CSIR), Government of India. He has received Fast Track Young Scientist Award from the Department of Science and Technology, India, to work in the area of bending actuators and artificial muscles. He has completed four major research projects sanctioned by University Grant Commission, Department of Science and Technology, Council of Scientific and Industrial Research, and Council of Science and Technology, India. He has published 85 research articles in international journals of repute and seventeen book chapters in knowledge-based book editions published by renowned international publishers. He has published nineteen edited books with Springer, United Kingdom, Nova Science Publishers, Inc. U.S.A., CRC Press Taylor & Francis Asia Pacific, Trans Tech Publications Ltd., Switzerland and Materials Science Forum, U.S.A. He is the member of various editorial boards of the journals and also serving as associate editor for a journal Environmental Chemistry Letter, Springer Nature. He has attended as well as chaired sessions in various international and national conferences. He has worked as a Postdoctoral Fellow, leading a research team at the Creative Research Initiative Center for Bio-Artificial Muscle, Hanyang University, South Korea, in the field of renewable energy, especially biofuel cells. He has also worked as a Postdoctoral Fellow at the Center of Research Excellence in Renewable Energy, King Fahd University of Petroleum and Minerals, Saudi Arabia, in the field of polymer electrolyte membrane fuel cells and computational fluid dynamics of polymer electrolyte membrane fuel cells. He is a life member of the Journal of the Indian Chemical Society. His research interest includes ion exchange materials, a sensor for heavy metal ions, biofuel cells, supercapacitors and bending actuators.

Prof. Abdullah M. Asiri is the Head of the Chemistry Department at King Abdulaziz University since October 2009 and he is the founder and the Director of the Center of Excellence for Advanced Materials Research (CEAMR) since 2010 till date. He is the Professor of Organic Photochemistry. He graduated from King Abdulaziz University (KAU) with B.Sc. in Chemistry in 1990 and a Ph.D. from University of Wales, College of Cardiff, U.K. in 1995. His research interest covers color chemistry, synthesis of novel photochromic and thermochromic systems, synthesis of novel coloring matters and dyeing of textiles, materials chemistry, nanochemistry and nanotechnology, polymers and plastics. Prof. Asiri is the principal supervisors of more than 20 M.Sc. and six Ph.D. theses. He is the main author of ten books of different chemistry disciplines. Prof. Asiri is the Editor-in-Chief of King Abdulaziz University Journal of Science. A major achievement of Prof. Asiri is the discovery of tribochromic compounds, a new class of compounds which change from slightly or colorless to deep colored when subjected to small pressure or when grind. This discovery was introduced to the scientific community as a new terminology published by IUPAC in 2000. This discovery was awarded a patent from European Patent office and from UK patent. Prof. Asiri involved in many committees at the KAU level and on the national level. He took a major role in the advanced materials committee working for KACST to identify the national plan for science and technology in 2007. Prof. Asiri played a major role in advancing the chemistry education and research in KAU. He has been awarded the best researchers from KAU for the past five years. He also awarded the Young Scientist Award from the Saudi Chemical Society in 2009 and also the first prize for the distinction in science from the Saudi Chemical Society in 2012. He also received a recognition certificate from the American Chemical Society (Gulf region Chapter) for the advancement of chemical science in the Kingdome. He received a Scopus certificate for the most publishing scientist in Saudi Arabia in chemistry in 2008. He is also a member of the editorial board of various journals of international repute. He is the Vice- President of Saudi Chemical Society (Western Province Branch). He holds four USA patents, more than one thousand publications in international journals, several book chapters and edited books.

Prof. Ali Mohammad is retired as the professor of Applied Chemistry from the Department of Applied Chemistry, Faculty of Engineering and Technology, Aligarh Muslim University, Aligarh, India. He served the department for about 30 years. He also served the department as the chair person for 6 years. He also worked as UGC-Emeritus Fellow after his retirement at the same department for two years. His scientific interests include physico-analytical aspects of solid-state reactions, micellar thin layer chromatography, surfactants analysis, and green chromatography. He is the author or coauthor of 250 scientific publications including research articles, reviews, and book chapters. He has supervised 53 students for Ph.D./M.Phil. and M.Tech. degrees. He has also served as Editor of the Scientific Journal, "Chemical and Environmental Research" published in India since 1992 to 2012 and as the Associate Editor for Analytical Chemistry section of the Journal of Indian Chemical Society. He has published four edited books with Springer, the United Kingdom and three by Materials Research Forum LLC, U.S.A. He has been the member of the editorial boards of Acta Chromatographica, Acta Universitatis Cibiniensis Seria F. Chemia, Air Pollution, and Annals of Agrarian Science. He has attended as well as chaired sessions in various international and national conferences. He is the life member of several Indian Scientific and Chemical Societies. He has also served as Visiting Professor at King Saudi University, Riyadh, Kingdom of Saudi Arabia. Dr. Mohammad obtained his M.Sc. (1972), M.Phil. (1975), Ph.D. (1978), and D.Sc. (1996) degrees from Aligarh Muslim University, Aligarh, India.

www.ingramcontent.com/pod-product-compliance
Ingram Content Group UK Ltd.
Pitfield, Milton Keynes, MK11 3LW, UK
UKHW020658041026
15590UKWH00001B/350

9 781945 291623